Modelling Land-Use Change

The GeoJournal Library

Volume 90

Managing Editor: Max Barlow, Toronto, Canada

Founding Series Editor:
　　　　　　　Wolf Tietze, Helmstedt, Germany

Editorial Board: Paul Claval, France
　　　　　　　Yehuda Gradus, Israel
　　　　　　　Sam Ock Park, South Korea
　　　　　　　Herman van der Wusten, The Netherlands

The titles published in this series are listed at the end of this volume.

Modelling Land-Use Change

Progress and applications

Edited by

ERIC KOOMEN

Department of Spatial Economics/SPINlab,
Vrije Universiteit Amsterdam,
The Netherlands

JOHN STILLWELL

School of Geography,
University of Leeds,
United Kingdom

ALDRIK BAKEMA

Netherlands Environmental Assessment Agency,
Bilthoven, The Netherlands

and

HENK J. SCHOLTEN

Department of Spatial Economics/SPINlab,
Vrije Universiteit Amsterdam,
The Netherlands

 Springer

A C.I.P. Catalogue record for this book is available from the Library of Congress

ISBN 978-1-4020-5647-5 (HB)
ISBN 978-1-4020-5648-2 (e-book)

Published by Springer,
P.O. Box 17, 3300 AA Dordrecht, The Netherlands.

www.springer.com

Printed on acid-free paper

Contents

Contents vii

Contributing authors

Antonio Aledo, Departamento de Sociología y Teoría de la Educación, Universidad de Alicante, Ap. Correos 99, 03080 Alicante, Spain. E-mail: antonio.aledo@ua.es

Eyal Ashbel, Department of Geography, Hebrew University of Jerusalem, Mount Scopus, 91905 Jerusalem, Israel. E-mail: eashbel@netvision.net.il

Aldrik Bakema, Netherlands Environmental Assessment Agency (MNP), PO Box 303, 3720 AH Bilthoven, The Netherlands. E-mail: aldrik.bakema@mnp.nl

José I. Barredo, Land Management and Natural Hazards Unit, Joint Research Centre of the European Commission (JRC), TP 261, I-21020 Ispra, Italy. E-mail: jose.barredo@jrc.it

Michael Batty, Centre for Advanced Spatial Analysis (CASA), University College London, 1-19 Torrington Place, London WC1E 6BT, United Kingdom. E-mail: mbatty@geog.ucl.ac.uk

Juan Bellot, Departamento de Ecología, Universidad de Alicante, Ap. Correos 99, 03080 Alicante, Spain. E-mail: juan.bellot@ua.es

Adi Ben-Nun, Department of Geography, Hebrew University of Jerusalem, Mount Scopus, 91905 Jerusalem, Israel. E-mail: bennun@cc.huji.ac.il

Patrick Bogaert, Département d'Agronomie, Université Catholique de Louvain, Place Croix du Sud 2, B-1348 Louvain-la-Neuve, Belgium. E-mail: bogaert@enge.ucl.ac.be

Andreu Bonet, Departamento de Ecología, Universidad de Alicante, Ap. Correos 99, 03080 Alicante, Spain. E-mail: andreu@ua.es

Judith Borsboom-van Beurden, Netherlands Environmental Assessment Agency (MNP), PO Box 303, 3720 AH Bilthoven, The Netherlands. E-mail: judith.borsboom@mnp.nl

Matthias Bürgi, Research Unit of Land Use Dynamics, Swiss Federal Research Institute (WSL), Zürcherstrasse 111, 8903 Birmensdorf, Switzerland. E-mail: matthias.buergi@wsl.ch

Giancarlo Carrai, SVALTEC S.r.l., Via del Campofiore 106, 50136 Florence, Italy. E-mail: gc.carrai@svaltec.it

João Corte-Real, Centro de Geofísica de Évora e Departamento de Física da Universidade de Évora, Rua Romão Ramalho, 59, 7000 Évora, Portugal. E-mail: jmcr@uevora.pt

Jasper Dekkers, Department of Spatial Economics/SPINlab, Vrije Universiteit, De Boelelaan 1105, 1081 HV, Amsterdam, The Netherlands. E-mail: jdekkers@feweb.vu.nl

Nicolas Dendoncker, Département de Géographie, Université Catholique de Louvain, Place L. Pasteur 3, B-1348 Louvain-la-Neuve, Belgium. E-mail: dendoncker@geog.ucl.ac.be

Tomaz Dentinho, Gabinete de Gestão e Conservação da Natureza, Departamento de Ciências Agrárias, Universidade dos Açores, 9701-851 Angra do Heroísmo, Portugal. E-mail: tomaz.dentinho@mail.angra.uac.pt

Denise Eisenhuth, Departamento de Ecología, Universidad de Alicante, Ap. Correos 99, 03080 Alicante, Spain. E-mail: d.m.eisenhuth@ua.es

Guy Engelen, Centre for Integrated Environmental Studies, Flemish Institute for Technological Research (VITO), Boeretang 200, BE-2400 Mol, Belgium. E-mail: guy.engelen@vito.be

Dick Ettema, Faculty of Geosciences, Utrecht University, PO Box 80115, 3508 TC Utrecht, The Netherlands. E-mail: d.ettema@geo.uu.nl

Daniel Felsenstein, Department of Geography, Hebrew University of Jerusalem, Mount Scopus, 91905 Jerusalem, Israel. E-mail: msdfels@mscc.huji.ac.il

Chang-Chun Feng, Center of Real Estate Research and Appraisal, Peking University, Rm. 3273, Bld. Yifuerlou, 100871 Beijing, China. E-mail: fcc@urban.pku.edu.cn

Naftali Goldshlager, Soil Erosion Station, Israeli Ministry of Agriculture, Ruppin Institute, Post. Emek-Hefer 40250, Israel. E-mail: gold_n@macam.ac.il

Joana Gonçalves, Gabinete de Gestão e Conservação da Natureza, Departamento de Ciências Agrárias, Universidade dos Açores, 9701-851 Angra do Heroísmo, Portugal. E-mail: joanagoncalves78@gmail.com

Dagmar Haase, Department of Applied Landscape Ecology, Centre for Environmental Research (UFZ), Permoserstrasse 15, D-04318 Leipzig, Germany. E-mail: dagmar.haase@ufz.de

Stephen Hallett, National Soil Resources Institute, Cranfield University at Silsoe, MK45 4DT Bedfordshire, United Kingdom. E-mail: s.hallett@Cranfield.ac.uk

Anna M. Hersperger, Research Unit of Land Use Dynamics, Swiss Federal Research Institute (WSL), Zürcherstrasse 111, 8903 Birmensdorf, Switzerland. E-mail: anna.hersperger@wsl.ch

Peter Heuberger, Netherlands Environmental Assessment Agency (MNP), PO Box 303, 3720 AH Bilthoven, The Netherlands. E-mail: peter.heuberger@mnp.nl

Annelie Holzkämper, Department of Applied Landscape Ecology, Centre for Environmental Research (UFZ), Permoserstrasse 15, D-04318 Leipzig, Germany. E-mail: annelie.holzkaemper@ufz.de

Louisa J.M. Jansen, Land/natural resources consultant, Via Girolamo Dandini 21, 00154 Rome, Italy. E-mail: Louisa.Jansen@tin.it

Kor de Jong, Faculty of Geosciences, Utrecht University, PO Box 80115, 3508 TC Utrecht, The Netherlands. E-mail: k.dejong@geo.uu.nl

Eric Koomen, Department of Spatial Economics/SPINlab, Vrije Universiteit, De Boelelaan 1105, 1081 HV, Amsterdam, The Netherlands. E-mail: ekoomen@feweb.vu.nl

Mario Köstl, Austrian Research Centers - ARC Systems Research GmbH, Donau-City-Strasse 1, 1200 Vienna, Austria. E-mail: mario.koestl@arcs.ac.at

Marianne Kuijpers-Linde, Netherlands Environmental Assessment Agency (MNP), PO Box 303, 3720 AH Bilthoven, The Netherlands. E-mail: marianne.kuijpers@mnp.nl

Carlo Lavalle, Land Management and Natural Hazards Unit, Joint Research Centre of the European Commission (JRC), TP 261, I-21020 Ispra, Italy. E-mail: carlo.lavalle@jrc.it

Wolfgang Loibl, Austrian Research Centers - ARC Systems Research GmbH, Donau-City-Strasse 1, 1200 Vienna, Austria. E-mail: wolfgang.loibl@arcs.ac.at

Willem Loonen, Netherlands Environmental Assessment Agency (MNP), PO Box 303, 3720 AH Bilthoven, The Netherlands. E-mail: willem.loonen@mnp.nl

Pedro Lourenço, Departmento de Ciências e Engenharia do Ambiente, Faculdade de Ciências e Tecnologia, Universidade Nova de Lisboa, Campus do Monte de Caparica, 2829-516 Caparica, Portugal. E-mail: pmbl@fct.unl.pt

Subrata K. Mandal, National Institute of Public Finance and Policy, 18/2 Satsang Vihar Rd. Special Institutional Area, 110067 New Delhi, India. E-mail: Subrata@nipfp.org.in

Maarten van der Meulen, Research Institute for Knowledge Systems (RIKS), PO Box 463, 6200 AL Maastricht, The Netherlands. E-mail: maarten@riks.nl

João Pedro Nunes, Departamento Ciências e Engenharia do Ambiente, Faculdade Ciências e Tecnologia, Universidade Nova de Lisboa, Campus do Monte de Caparica, 2829-516 Caparica, Portugal. E-mail: jpcn@fct.unl.pt

Koen P. Overmars, Department of Environmental Sciences, Wageningen University, PO Box 47, 6700 AA Wageningen, the Netherlands. E-mail: koen.overmars@wur.nl

Juan Peña, Departamento de Ecología, Universidad de Alicante, Ap. Correos 99, 03080 Alicante, Spain. E-mail: jpl@ua.es

Massimiliano Petri, Department of Civil Engineering, University of Pisa, Via Diotisalvi 2, 56126 Pisa, Italy. E-mail: m.petri@ing.unipi.it

Kampanart Piyathamrongchai, Centre for Advanced Spatial Analysis (CASA), University College London, 1-19 Torrington Place, London WC1E 6BT, United Kingdom. E-mail: ucfakpi@ucl.ac.uk

Eike Rommelfanger, Institute of Biometry and Population Genetics, Justus Liebig University, Heinrich-Buff-Ring 26-32, D-35390 Gießen, Germany. E-mail: eike.f.rommelfanger@agrar.uni-giessen.de

Mark Rounsevell, Départment de Géographie, Université Catholique de Louvain, Place L. Pasteur 3, B-1348 Louvain-la-Neuve, Belgium. E-mail: rounsevell@geog.ucl.ac.be

Juan Rafael Sánchez, Departamento de Ecología, Universidad de Alicante, Ap. Correos 99, 03080 Alicante, Spain. E-mail: jr.sanchez@ua.es

Henk J. Scholten, Department of Spatial Economics/SPINlab, Vrije Universiteit, De Boelelaan 1105, 1081 HV, Amsterdam, The Netherlands. E-mail: hscholten@feweb.vu.nl

Jan Ole Schroers, Institute of Agricultural and Food Systems Management, Justus Liebig University, Senckenbergstrasse 3, D-35390 Gießen, Germany. E-mail: jan.o.schroers@agrar.uni-giessen.de

Júlia Seixas, Departmento de Ciências e Engenharia do Ambiente, Faculdade de Ciências e Tecnologia, Universidade Nova de Lisboa, Campus do Monte de Caparica, 2829-516 Caparica, Portugal. E-mail: mjs@fct.unl.pt

Ralf Seppelt, Department of Applied Landscape Ecology, Centre for Environmental Research (UFZ), Permoserstrasse 15, D-04318 Leipzig, Germany. E-mail: ralf.seppelt@ufz.de

Patrick Sheridan, Institute of Agricultural and Food Systems Management, Justus Liebig University, Senckenbergstrasse 3, D-35390 Gießen, Germany. E-mail: patrick.sheridan@agrar.uni-giessen.de

Maxim Shoshany, Department of Transportation and Geoinformation Engineering, Faculty of Civil and Environmental Engineering, Technion Israel Institute of Technology, 32000 Haifa, Israel. E-mail: maximsh@techunix.technion.ac.il

Michael Sonis, Department of Geography, Bar-Ilan University, 52900 Ramat-Gan, Israel. E-mail: sonism@mail.biu.ac.il

Klaus Steinnocher, Austrian Research Centers - ARC Systems Research GmbH, Donau-City-Strasse 1, 1200 Vienna, Austria. E-mail: klaus.steinnocher@arcs.ac.at

John Stillwell, School of Geography, University of Leeds, Leeds LS2 9JT, United Kingdom. E-mail: j.c.h.stillwell@leeds.ac.uk

Hanneke Tijbosch, Netherlands Environmental Assessment Agency (MNP), PO Box 303, 3720 AH Bilthoven, The Netherlands. E-mail: hanneke.tijbosch@falw.vu.nl

Harry Timmermans, Urban Planning Group/EIRASS, Eindhoven University of Technology, PO Box 513, 5600 MB Eindhoven, The Netherlands. E-mail: h.j.p.timmermans@bwk.tue.nl

Tanja Tötzer, Austrian Research Centers - ARC Systems Research GmbH, Donau-City-Strasse 1, 1200 Vienna, Austria. E-mail: tanja.toetzer@arcs.ac.at

Peter H. Verburg, Department of Environmental Sciences, Wageningen University, PO Box 47, 6700 AA Wageningen, The Netherlands. E-mail: peter.verburg@wur.nl

Roger White, Department of Geography, Memorial University of Newfoundland, St. John's, Newfoundland A1B 3X9, Canada. E-mail: roger@mun.ca

Zhi-Gang Wu, Center of Real Estate Research and Appraisal, Peking University, Rm. 3273, Bld. Yifuerlou, 100871 Beijing, China. E-mail: wuzhigang@pku.edu.cn

Su-Hong Zhou, Center for Urban and Regional Studies, School of Geography and Planning, Zhongshan University, 510275 Guangzhou (Canton), China. E-mail: eeszsh@zsu.edu.cn

Preface

The transformation of land use and land cover is driven by a range of different factors and mechanisms. Climate, technology and economics are key determinants of land-use change at different spatial and temporal scales. Whilst the implications of climatic warming at a global level are hugely worrying for low lying parts of the world, the processes of urbanisation continue in a seemingly uninterrupted manner. As time goes by, the use of land in both natural and man-made environments is influenced by the pressures associated with development. The demand for land for new residential housing in northwest European countries has been a huge challenge for governments striving to protect greenfield sites in recent years, whilst brownfield regeneration has been a common response to the decline of staple manufacturing in older industrial heartlands. The variety of forces that drive change in the use of land is extensive and complex, including spatial planning policies designed at local, regional, national and supra-national levels.

Given this complexity and in order to understand the mechanisms of change and the impact of policies, researchers and practitioners have turned their attention to formulating, calibrating and testing models that simulate land-use dynamics. These land-use change models help us to understand the characteristics and interdependencies of the components that constitute spatial systems. Moreover, when utilized in a predictive capacity, they provide valuable insights into possible land-use configurations in the future. Models of land-use change incorporate concepts and knowledge from a wide range of disciplines. Geography, as a spatial science, contributes significantly to the understanding of land-use change whilst demography and economics help explain underlying trends. Model building relies heavily on mathematics and (geographical) information science, but also includes many elements from the softer sciences, such as management studies and environmental science.

This book offers a cross-sectional overview of current research progress in the field of land-use modelling. The contributions that are included in the chapters of the book range from methodology and model calibration to the actual application of systems and studies of recent policy implementation and evaluation. The contributors originate from academic and applied research institutes around the world and thus offer an international mix of theoretical and practical perspectives in different case study contexts. The book is an indispensable guide for researchers and practitioners interested in state-of-the-art land-use modelling, its background and its application. A special website (www.lumos.info/ModellingLand-UseChange/Exercises.htm) provides demonstration versions of well-known land-use models that give detailed insights into the way these models work. Additional exercises and assignments help students to critically assess the potential of these instruments.

The Editors
January, 2007

Acknowledgements

This book is the result of the joint efforts of many individuals and organisations. We are particularly grateful to the authors and researchers that contributed the text and educational materials for the book. Special thanks go to the graphics team in the School of Geography at the University of Leeds for improving the original maps, graphs and diagrams and Rosan van Wilgenburg of the SPINlab at the Vrije Universiteit Amsterdam for compiling the educational material available from the special website. In addition, we thank the Dutch National research programme 'Climate Changes Spatial Planning' for sponsoring part of the work involved in editing the book. Furthermore, we would like to thank the Organising Committee of the European Regional Science Association for allowing a special 'Modelling Land-Use Change' session to be held at the ERSA2005 conference in Amsterdam. It was the success of this occasion that provided the inspiration for the book. Finally, the first editor is grateful to the SIGTE-group at the Universitat de Girona, Catalunya, for hosting him during the last months of 2005 when the foundations of this book were established.

Chapter 1

MODELLING LAND-USE CHANGE
Theories and methods

E. Koomen[1] and J. Stillwell[2]

[1]Department of Spatial Economics/SPINlab, Vrije Universiteit Amsterdam, The Netherlands;
[2]School of Geography, University of Leeds, UK

Abstract: This first chapter explains some of the basic theoretical ideas, concepts and methodologies that underpin the modelling of land-use change. It represents an overview of the types of approaches that have been adopted by researchers hitherto. It also provides a rationale for the structure of the book and a synopsis of the contents that follow.

Key words: Land-use change modelling; theory; methodology; book structure.

1.   INTRODUCTION

The existence of the well-known computer-game, *SIM-City^{TM}* (http://simcity.ea.com/), has taken the modelling of land-use change beyond its original domain of researchers and policymakers. Simulating the complex interaction of natural and social systems has now come within reach of computer games enthusiasts, both young and old. However, the popularity of the products generated by the games industry has not stretched as far as the land-use models that have been developed by researchers and planning practitioners. Some commentators might suggest that, during the last decade, these systems have tended to remain relatively under-used 'black boxes', producing little more than nicely coloured maps. Perhaps the lack of attention to the development of useful applications in the field of land use is related to the extensive array of existing models, the different approaches they take, and the relative complexity of their underpinning theories and methods of application.

E. Koomen et al. (eds.), Modelling Land-Use Change, 1–21.
© 2007 *Springer.*

This book aims to address this paradox by providing an overview of recent land-use modelling efforts and by clarifying their background and application possibilities. It does so by presenting a wide range of approaches (both geographically and thematically) that analyse and explain past land-use changes and simulate possible future changes. As an initial introduction to the simulation of land-use change, we begin with a discussion of some of the basic characteristics of land-use change models and the theories and methods on which they are based. Thereafter the structure of the book is explained and a synopsis of its contents is given.

2.   CHARACTERISING LAND-USE CHANGE MODELS

Land-use change is a complex, dynamic process that links together natural and human systems. It has direct impacts on soil, water and atmosphere (Meyer and Turner, 1994) and is thus directly related to many environmental issues of global importance. The large-scale deforestations and subsequent transformations of agricultural land in the tropics are examples of land-use change with strong likely impacts on biodiversity, soil degradation and the earth's ability to support human needs (Lambin *et al.*, 2003). Land-use change is also one of the important factors in the climate change cycle and the relationship between the two is interdependent; changes in land use may affect the climate whilst climatic change will also influence future land-use (Dale, 1997; Watson *et al.*, 2000). On a smaller scale, in the densely populated parts of the urbanised western world, land-use change is the expression of continuing urbanisation pressure on ever scarcer open spaces (e.g. Bell and Irwin, 2002; Rietveld and Wagtendonk, 2004), many of which have been designated by planning authorities as greenfield areas for conservation reasons. This issue is often referred to as urban sprawl, a topic of debate in the United States especially (e.g. Brueckner, 2000; Glaeser and Kahn, 2004). Modelling land-use change helps understand the processes of continuing urbanisation and can also be of value in informing policymakers of possible future conditions under different scenarios. Land-use change models can therefore be defined as tools to support the analysis of the causes and consequences of land-use change (Verburg *et al.*, 2004a). Many authors (e.g. Lambin *et al.*, 2001) make a distinction between the *land cover* that can be observed (e.g. grass, building) and the *land use,* the actual use to which the land is put (e.g. grassland for livestock grazing, residential area). For convenience, we use the term land use predominantly in this book, referring to both land cover and actual land use.

Recent inventories of operational models for land-use change are numerous. Briassoulis (2000) offers a very extensive discussion of the most common land-use change models and their theoretical backgrounds. Waddell and Ulfarsson (2003) and Verburg *et al.* (2004a) present more concise overviews that focus on the future directions of research in this field, whilst more detailed, technical information on the actual models is provided by Agarwal *et al.* (2002) and the U.S. Environmental Protection Agency (U.S. EPA, 2000). All inventories show a very heterogeneous group of instruments with considerable differences regarding their background, starting points, range of applications *et cetera*. We will refrain here from classifying existing models, but rather discuss a number of characteristics that can be used to differentiate the most common modelling approaches.

One of the most important distinctions refers to *static* as opposed to *dynamic* models. Static (or cross-sectional) models directly calculate the situation at a given point in time, whereas dynamic models work with intermediate time-steps, each of which might become the starting-point for calculating the subsequent situation. Dynamic modelling, therefore, takes possible developments during the simulation period into account, providing a richer behaviour and the possibility to better mimic actual spatial developments.

Land-use change models can also be characterised as dealing with either *transformation* or *allocation*. Transformation models start from the current land use and simulate the possible conversion into another land-use type, e.g. based on a transformation probability or the status of surrounding locations. Allocation models, on the other hand, allocate a certain type of land use to a location based on its characteristics. Current land use may thus be one of the factors influencing locational characteristics, but it is not necessarily preserved in future land use. This approach to simulation basically starts with an empty map.

From a theoretical perspective, there is a clear difference between models starting from a direct emphasis on *land use* and those whose initial consideration is the *land user*. Many models focus purely on land use, merely simulating its state at a certain location. Other approaches take land users as the starting point and try to understand their behaviour. The description of the spatial decisions of (groups of) individuals is then used to deduce the land-use changes.

Approaches to simulating land-use change may be either *deterministic* or *probabilistic*. The former applies strict cause-effect relations, whereas the latter considers the probability of land-use changes taking place. The essence of this second approach is the introduction of an element of uncertainty. A type of use is attached to a location based on an estimated probability, rather than following a straightforward deductive approach. In some cases, a random error-term is added to express the uncertainty in the explanatory factors.

Another common distinction is the one made between *sector-specific* and *integrated* models. Sector-specific models focus on one part of the land-use system (e.g. housing, employment, agriculture) and describe that part as precisely as possible. Integrated models consider the mutual relationships between these sectors, thus approaching the land-use system in a very comprehensive and inter-dependent (or systems-oriented) manner. Truly integrated models also incorporate the feedbacks of the land-use system with other related systems such as climate, hydrology or transport.

In relation to the spatial level of detail, both *zones* and *grids* are used. Zones are relatively homogeneous, often irregularly shaped areas or vector polygons, e.g. socio-economic or administrative regions that more often than not have little functional coherence. Grids, on the other hand, are collections of (mostly square) cells defined in a regular raster pattern that are often used in geographical information systems (GIS). Models that use grids often make use of geographical information from other sources, thus having access to valuable base data.

As land-use change models can differ from each other on all of the above mentioned characteristics, classifying them into homogenous groups is difficult, if not impossible. They do rely, however, on a limited number of basic principles to allocate land use and these theories are discussed in the next section.

3.   THEORIES AND METHODS OF LAND-USE MODELLING

Models for simulating future land use exist in many different types and forms, but they all rely on a limited number of theories and methods. Economic theories, for example, are often used to explain land-use patterns and their dynamics (e.g. Bockstael and Irwin, 2000; Irwin and Geoghegan, 2001). The underlying idea is that those who can afford the most money for the land are the ones using it. But disciplines such as geography and mathematics have also contributed to the understanding and simulation of changes in land use. In order to provide some background for the models that will be presented in later chapters of the book, we introduce some of the basic principles of land-use modelling. For an introduction to a number of additional aspects that are relevant for the simulation of land-use change, such as policy perspectives, driving forces, data considerations, evaluation and visualisation methods, the reader is referred to a previous text, *Land Use Simulation for Europe* (Stillwell and Scholten, 2001).

3.1 Economic principles

For a number of reasons, land is a special economic asset. Firstly, the supply of land is fixed, creating specific demand-supply relations. Secondly, every parcel of land has a fixed location with its associated unique features in terms of soil quality, gradient, altitude, accessibility *et cetera*. The marketable asset is therefore far from homogeneous, severely hampering the price-making analysis. Thirdly, the land use at a certain location influences its surroundings. The impact may be either negative or positive; basic infrastructure or industry causes visual disturbance for many, but small-scale agriculture may increase the aesthetic or natural value of the landscape. This impact, that economists call an externality of land use, often gives rise to government intervention. Examples of this intervention include prohibition of dwellings in the proximity of big industrial estates or airports, economic activities that are relocated to the outskirts of cities, or farmers that receive subsidies to provide 'nature' as an additional product under sub-optimal agrarian conditions. Combined with the limited supply and the heterogeneity of land, the externalities and the resulting government interventions are expressed in a segmented land market, where different prices are used for 'green' (agriculture, nature) and 'red' (housing, employment, infrastructure) functions and where considerable spatial price differences exist within sector-specific markets (e.g. Buurman *et al.*, 2001).

The focus on land in economic theories has changed over time. The early and well-known theories of Ricardo and, in a more spatial context, Von Thünen, have laid the foundation of land-price and land-use theories. These are to a certain extent still valid and used in current research. Ricardo (1817, in Kruijt *et al.*, 1990) explained land prices in terms of differences in soil fertility levels or, more generally speaking, in terms of land quality. Better quality land is more profitable than lower quality land, and this difference leads to payment of a higher price for the land. Von Thünen (1826) focused on the impact of distance and hence transportation costs, to explain land-use patterns and land prices. Current economic analysis of land use often takes bid rent theory (Alonso, 1964) as a starting point, focusing on the relationship between urban land use and the value of urban land. Individual households and companies weigh up the land price, transportation costs and the amount of land they need. This leads to a simple model with decreasing land prices as you move away from the city centre. The land use resulting from these assumptions is that of a typical monocentric city. Commercial activities are concentrated in the city centre (central business district); industrial and housing functions will have less money available for a central location and will select a location at a greater distance from the centre; the edge of the city is identified where the offer of the urban bidders is equal to that of the agrarian bidders.

Another important concept related to economic science and used to explain land-use patterns is discrete choice theory. Nobel prize winner McFadden has made important contributions to this approach of modelling choices between mutually exclusive alternatives (e.g. McFadden, 1978). In this theory, the probability that an individual selects a certain alternative is dependent on the utility of that specific alternative in relation to the total utility of all alternatives. This probability is, given its definition, expressed as a value between 0 and 1, but it will never reach these extremes. When translated into land use, this approach explains the probability of a certain type of land use at a certain location based on the utility of that location for that specific type of use in relation to the total utility of all possible uses. The utility of a location can be interpreted as the *suitability* for a certain use. This can be formulated as follows:

$$X_{ci} = e^{\beta * S_{ci}} / \sum_k e^{\beta * S_{ck}} \qquad (1)$$

where:

X_{ci} is the probability of cell c being used for land-use type i;

e is the base of the natural logarithm ($= 2.71828$);

S_{ci} is the suitability of cell c for land-use type i; dependent on different factors;

S_{ck} is the suitability of cell c for all (k) land-use types; and

β is a parameter to adjust the sensitivity of the model.

The suitability of a location for a certain use can be explained by a range of different factors. This may refer, for example, to physical suitability, as is the case with the soil type that largely determines the most profitable type of agricultural use. Other important aspects that influence suitability include accessibility of relevant facilities or spatial policies that will restrict or encourage certain land-use types. Suitability is assessed by potential users and can also be interpreted as a bid price. After all, the user deriving the highest benefit from a location will offer the highest price.

The renewed interest for geography in economics (e.g. Krugman, 1999) offers interesting concepts to analyse the spatial interaction between actors (represented by, for example, residences or industries) in terms of centripetal forces leading to concentration, and centrifugal forces leading to a spatial spread of functions.

3.2 Spatial interaction

A classical group of land-use models is based on spatial interaction modelling theory. Spatial interaction in a social, geographical context refers to every movement in space as a consequence of a human process (Haynes and Fotheringham, 1984). By analogy with Newton's first law, these models assume that the interaction between two entities depends on their own mass (or size) and is inversely proportionate to the distance between them. Early applications of this principle can be found in studies of migration (Ravenstein, 1885; Young, 1924) and trade (Reilly, 1931). Their main assumption was that the volume of interaction, being migration or commercial transactions between two cities, for example, depended on the size of the two cities and the distance between them. Thus, bigger cities were expected to attract more migrants or trade than smaller ones and this flow of migrants or trade was expected to be strong when distances were small.

This way, the concepts of scale and distance are introduced in the description of spatial relations, indicating that their influence is relative; size matters, especially when distances are small. This simple gravity principle has been adjusted and extended in several different ways. An important extension is the inclusion of more than two objects. The total interaction in a system is supposed to be equal to the sum of all interactions between all pairs of objects or, in other words, the interaction potential of an object is equal to the sum of all potential interactions with other objects.

Lowry (1964) was the first to develop an urban land-use model based on two dependent gravity models. The first model relates the population distribution to residential areas on the basis of fixed employment locations. The demand for trade can then be deducted from the population distribution. The second gravity model allocates retail businesses based on the newly determined demand. The changed distribution of services results in an adapted demand for labour force that can be introduced in turn in the population model. This dynamic interaction will continue until a previously defined small amount of allocation difference occurs. The Lowry model is spatially explicit on the level of homogeneous urban zones. Current land-use models display a higher level of detail in both their spatial resolution and allocation principle, but often fall back on this type of model for the sector-specific demand for land.

The primary architect of contemporary spatial interaction modelling is Alan Wilson, whose seminal work in the late 1960s (Wilson, 1967) on entropy maximisation led to the inclusion of balancing factors in the gravity model equations that served to ensure constraints were satisfied. A family of models was developed (Wilson, 1970), variants of which could be applied in situations of differing known information. In the context of migration, spatial interaction models based on these principles have been extended by Stillwell (1991) and Fotheringham (1991) and used recently in an applied context to model flows within the UK for the Office of the Deputy Prime Minister (Champion *et al.*, 2003).

A related type of research focuses on the interaction between land use and transport. Central to this approach is the assumption that land use is influenced by the available infrastructure network and *vice versa*: the transportation demand depends on the spatial configuration of the different, mostly urban, land-use types. One of the first researchers to model the interdependence of these systems was Putman (1983), but many others have created similar structures, often referred to as LUTI models, more detailed overviews of which are provided in Wegener (1998) and Kanaroglou and Scott (2002), for example. Most of the original LUTI models were based on a classic spatial interaction framework and adopted a relatively coarse zonal scale. In the newest wave of these models, however, research attention has shifted towards activity-based microsimulation (Timmermans, 2003). This is a trend that is also visible in general (non-transport related) land-use change models as will be discussed later.

3.3 Cellular automata

The cellular automata (CA) methods deriving from mathematics are very well suited for imitating complex spatial processes on the basis of simple decision rules (Wolfram, 1984). Every cell has a certain state (or function) that is influenced by its surrounding cells as well as the characteristics of the cell itself. The degree and direction of interaction between the functions is determined through so-called transition rules. The application of CA in geographical modelling was originally proposed by Tobler (1979) and the concept has subsequently been applied to model urban form (Batty, 1997; Yeh and Li, 2001), urban growth (Clarke *et al.*, 1997; Couclelis, 1997; Clarke and Gaydos, 1998), land-use planning (Wu, 1998; Li and Yeh, 2000) and urban and regional development and planning (Samat, 2002; Engelen *et al.*, 1999; White and Engelen, 2000).

A strong dimension of this approach is the simulation of the interaction of a location with its direct surroundings that has empirically proven to be an important driver of land-use change (O´Sullivan and Torrens, 2000; Verburg *et al.*, 2004b). A crucial component of this local interaction approach is 'emergence', discussed by Holland (1998) amongst others. In CA models this phenomenon refers to global patterns that appear spontaneously from the collective behaviour of individual cells influencing each other. This rich behaviour leads to simulation results that are very hard, if not impossible, to predict from the behaviour of the individual cells.

Additional, location-based information is often used in creating the transition rules in CA models, for example relating to the physical suitability or policy restrictions within a cell. The model thus moves beyond the classical focus on spatial interaction to achieve more realistic simulations. Classical CA models have a limited theoretical relationship with the decision-making process that leads to changes in land use. Hence, modern CA applications also incorporate components from other disciplines to obtain a more realistic simulation of land-use changes, an example of which is the Markov model that uses transition probabilities to describe the possible spatial developments of a location (Balzter *et al.*, 1998; Li and Reynolds, 1997). The probability of a cell changing its function is determined here by the initial state of the cell, the surrounding cells and a transition matrix with its transition probabilities. The interesting aspect of this approach is that consecutive changes in land use known from the literature or from experience (e.g. a succession in vegetation types or the changeover from agricultural to residential use) can be included explicitly as being probable whereas other transitions (e.g. industry to agriculture) can be described as being improbable or, in some cases, impossible. The cell changes its status according to these estimated probabilities rather than from the deterministic transition rules of the classical CA models.

Another option to control the spatial interaction behaviour of individual cells in CA models is the inclusion of higher level constraints on, for example, the magnitude of land-use changes. This can be implemented through a regional level spatial interaction model as is the case in the *Environment Explorer* model (White and Engelen, 2000) and related *MOLAND* framework (Engelen *et al.*, this book).

3.4 Statistical analysis

Statistical analysis is an essential tool for almost all models of land-use change. Regression analysis, for example, helps to quantify the contribution of the individual forces that drive land-use change, as demonstrated by Rietveld and Wagtendonk (2004) and Verburg *et al.* (2004c) and thus provides the information needed to properly calibrate models of land-use change. An important aspect of analysing land-use patterns is addressed in spatial econometrics and relates to issues such as spatial dependency and spatial heterogeneity (Irwin and Geoghegan, 2001). The analysis of spatial dependence may point to structural interdependencies between, for example, land-use types and can be useful in formulating the transition rules in CA-models. See Anselin and Florax (1995) for an extensive discussion of this topic.

Many examples exist of models that rely solely on a statistical description of observed past land-use changes to simulate future patterns (e.g. Schneider and Pontius, 2001; Serneels and Lambin, 2001). These empirical-statistical models have the advantage of being relatively easy to construct, but they miss a theoretical foundation as no attempt is made to understand and simulate the processes that actually drive land-use change. The applicability of these purely statistical models is therefore limited. They can be used to simulate possible spatial developments within a relatively short time-span under 'business as usual' conditions, but they are not suited to simulate possible changes according to diverging socio-economic future scenarios, for example. A combination with theoretical insights in land-use change processes is therefore welcomed to add a notion of causality to statistical models (Veldkamp and Lambin, 2001; Parker *et al.*, 2003). Examples of this combination are provided by Chomitz and Gray (1996) and Geoghegan *et al.* (2004).

3.5 Optimisation techniques

Another modelling approach is optimisation. By applying mathematical optimisation techniques such as linear integer programming or neural networks, the optimal land-use configuration is calculated here given a set of prior conditions, criteria and decision variables (e.g. Aerts, 2002; Pijanowski *et al.*, 2002). The simplest applications aim to optimise a single objective (for example, profit maximisation) for a specific group of decisionmakers (e.g. project developers). But there are also mathematic programming techniques that can determine the optimal solution for different, divergent objectives. This is especially interesting for policymakers who are interested in the optimal configuration of an area based on different, often conflicting, policy goals. This approach is further discussed in Part III of this book.

3.6 Rule-based simulation

The central element in rule-based simulation is the imitation of a known process. This approach is generally used in the field of physical sciences and is often applied in combination with a GIS. Rule-based simulation models can be used to imitate processes that can be described by strict, quantitative, location-based rules. These are normally natural processes such as soil erosion or landscape dynamics. The latter is modelled, for example, in the Landscape Modelling Shell (*LAMOS*, see Lavorel *et al.*, 2000) that integrates a quantitative description of different landscape processes, such as vegetation succession, disturbance and dispersal, to simulate possible landscape-ecological patterns.

The rule-based approach has also been applied, however, in studies with a more socio-scientific orientation such as land-use change. Examples of the application of rule-based simulation models include the original California Urban Futures (*CUF*) model and the *What If?* system. A typical feature of these models is that they allow users to include explicit decision rules that steer their behaviour (Klosterman and Petit, 2005). This flexible characteristic allows the models to simulate the consequences of spatial decisions and thus makes them useful as planning support tools. *CUF* (Landis, 1994) simulates alternative residential-development scenarios for cities based on specified policy changes at various levels of government. Projected population growth based on past trends is allocated and the profitability of each land parcel if developed is ascertained based on this demand, but also on user-specified development regulations and incentives. *What If?* (Klosterman, 1999) is a self-contained visualisation tool that accepts user-defined spatial data, growth rules and parameters to map land-use allocation alternatives.

Rule-based simulation is also an important element in many integrated models of global change as described by Alcamo *et al.* (1998) and Cramer *et al.* (2001), for example. These models typically apply relatively simple descriptions of the relations between various subsystems to simulate their interaction and assess the resulting state of, for example, land use, vegetation cover or greenhouse gas concentrations. The subsystems (e.g. economy, emissions, vegetation, agriculture and atmosphere) are often modelled in more elaborate, individual models.

3.7 Multi-agent models

Human decision making and interaction are the central elements in multi-agent (MA) models. The key concept here is that of the agents or decisionmakers. Parker *et al.* (2003) define agents as being autonomous, yet sharing an environment through communication and interaction, and they take decisions linking their behaviour to their environment. Autonomy means that the actors control their own actions and internal status in order to achieve their goals.

In MA models, as a minimum, actors have a strategy that makes them react to their environment and the actions of other actors. More advanced models of human decision making apply the rational choice theory. These models assume agents being fully informed, taking long-term decisions and having infinite analytical capacities. It is very difficult, however, to combine these models with the decision-making processes related to land-use change. It remains to be seen whether those complex decision-making models can be used to simulate land-use changes. Because of different spatial dependencies and feedback mechanisms, it is virtually impossible for an individual actor to consider all possible consequences of his own acts and those of all other actors. Hence, many MA models apply a type of limited rationality for the choice behaviour of their actors (Parker *et al.*, 2003). A recent overview by Berger and Parker (2002) on MA models for land-use changes shows different applications from the whole world on topics such as crop choices, deforestation and urbanisation. The choice behaviour therein is modelled with the assistance of relatively simple rules of thumb (heuristics), limited rationality or (economic) utility functions. MA models appear mainly effective in combination with CA models. The CA part then describes the natural system (the interaction between ecological processes and the physical subsoil), while the MA part describes the human part (choice behaviour of actors). The potential of CA and MA models currently under development in academic institutions to act as planning support systems with practical application in the real world has been reviewed by Torrens (2003). Several examples of this approach are described in Part IV of this book.

3.8 Microsimulation

Microsimulation is related to the simulation of processes at the level of individuals. Within land-use models, the idea is to include all individual actors who influence changes in land use. In this sense, this approach deviates from the multi-actor approach that uses a cross-sectional (average) description of the relevant decision-making groups. An important advantage of this method is that land-use changes are modelled on the scale level on which the actual choices are also made. Microsimulation demands enormous amounts of data and therefore computing power to simulate the actions of all relevant individuals. But as more detailed spatial data and faster computers become available, this approach is gaining popularity. For a description of the choice behaviour of individuals, one is often referred to (microeconomic) discrete choice theory, such as is used in the *UrbanSim* model (Waddell, 2002; Felsenstein *et al.* this book). The big challenge continues to be the reconciliation of microsimulation with the macro-scale socio-economic processes, such as structural economic and demographic developments (Alberti and Waddell, 2000).

3.9 Application of the theories and methods

All of the described theories and methods have their own advantages and disadvantages. The economic approach is useful to model choice behaviour in sector-specific submarkets, like the agricultural or urban land market. CA models on the other hand, are apt to model land-use changes when the interaction with the surroundings is important. This is the case, for example, when physical or ecological aspects are dominant as with deforestation. Optimisation models can be used to determine the optimal land-use configuration according to certain (policy) goals and are mostly applied to inform decision makers of possible solutions for land-use management issues.

Table 1-1 links this book's land-use change models with the theories and methods introduced in this section. Due to the complexity of most of the models, these links are not always straightforward and may even depend on the application that is described. The table merely indicates the prevalent theoretical and methodological background. The models that basically aim at explaining current and past land-use changes (Chapters 6, 7 and 8) offer a relatively straightforward approach, as do the land-use optimisation efforts described in the subsequent three chapters. None of the individual approaches can, however, provide a basis for a comprehensive, integrated and spatially explicit model to simulate future land-use change in a complex modern society. Such models therefore often combine different approaches into one hybrid model as was also advocated by Torrens (2001). The land-use simulation models presented in this book indeed rely on a combination of different theories and methods, as is discussed below.

The recently developed models that focus on the behaviour of agents in particular incorporate many theories and methods in their frameworks. The renowned *UrbanSim* model, discussed in Chapter 12, is an example of an interesting hybrid model that combines a microsimulation approach with discrete choice theory for the location choice of individual households and general economic theory for macro-economic evolution. Statistical analysis is used to actually quantify the behaviour of these agents. The new *PUMA* model (Chapter 14) also combines an agent-based approach with economic theory (utility-maximising functions) and statistical analysis to describe the choice behaviour of households.

Most of the other much applied integrated land-use simulation models also rely on a combination of theories and methods. The *LUMOS* toolbox (Chapter 16), for example, provides a framework for land-use simulation that, amongst others, contains the *Environment Explorer* and *Land Use Scanner* models. The former model is comparable to the *MOLAND* modelling framework (Chapter 17) and is essentially a CA model, but it is combined with a spatial interaction model to constrain regional land-use demand. Statistical analyses and expert judgement are furthermore used to estimate local transition potentials. The *Land Use Scanner* (Chapters 16 and 20) applies an allocation algorithm that is based on economic, discrete choice theory. The additional application of constraints on regional demand and the supply of land, however, enforce a bidding process that is in line with other economic (bid-rent) theory. This model also relies on expert judgement to define local suitability and prospected claims for the different land-use types following the specified scenario conditions, adding a rule-based element to the simulations. The *CLUE-s* model (Chapter 18) provides a framework for land-use simulation that, depending on the constructed configuration, can contain elements of statistics, cellular automata and a rule-based approach.

Table 1-1. Theoretical and methodological background and case study area of the land-use simulation models presented in this book

Model name or method (chapter number)	Economic principles	Spatial interaction	Cellular Automata	Statistical analysis	Optimisation	Rule-based	Multi-agent models	Microsimulation	Described case study area
Markov model (6)				X					Marina Baixa, Spain
Statistical analysis (7)				X					Southern Belgium
Spatial interaction (8)		X							Corvo island, Azores, Portugal
Genetic algorithm (9)					X				Netherlands
Linear programming (10)					X				Hawalbag, India
GeneticLand (11)					X				Southern Portugal
UrbanSim (12)	X			X			X	X	Tel Aviv region, Israel
Multi-agent simulation (13)				X			X	X	Rhine valley, Austria
PUMA (14)	X			X			X	X	Randstad, Netherlands
DSSM (15)			X	X	X				Chiang Mai, Thailand
LUMOS (16)	X	X	X		X				Netherlands
MOLAND (17)			X	X					Urban areas across Europe
CLUE-s (18)				X	X	X			Netherlands and Malaysia
SELES environment (19)				X		X			Leipzig-Grünau, Germany
Land Use Scanner (20)	X					X			Netherlands and Elbe area
ProLand and *UPAL* (21)	X					X			Hesse, Germany

Note that the first chapters of the book are not included in this table because they focus on the analysis of land-use change rather than its simulation.
LUMOS and *CLUE-s* provide frameworks for land-use simulation consisting of various models and configuration possibilities that each rely on different theoretical and methodological backgrounds.

4.   STRUCTURE OF THE BOOK

This book presents a cross-sectional overview of recent research progress related to the modelling of land-use change. The contributions range from analysing past land-use changes to simulating future changes to help policymakers take their decisions. The case studies that are presented in the book originate from academic and applied research institutes around the world and are grouped in coherent parts that roughly correspond to different phases of model development. These phases consist of the *analysis* of the land-use change process, the *exploration* of new methods and theoretical insights and the actual *application* of land-use simulation models. Each of the phases is subdivided in two parts, providing a total of six parts for the remaining 20 chapters as is explained below and presented in Figure 1-1.

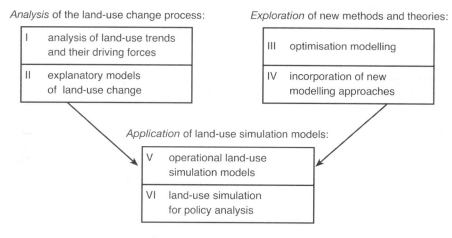

Figure 1-1. Basic layout of the book consisting of six coherent parts.

It is clear that any model development should start with a thorough *analysis of land-use trends and their driving forces* and this is the topic of Part I. Louisa Jansen, Giancarlo Carrai and Massimiliano Petri perform an analysis of the land-use change dynamics at cadastral parcel level in Albania. They apply an object-oriented geo-database approach to describe and understand spatial developments in the turbulent time of transition from a socialist state to a decentralised, market-oriented economy. Anna Hersperger and Matthias Bürgi analyse the driving forces of landscape change in the urbanizing Limmat Valley, Switzerland based on 70-year time series of historic maps. Michael Sonis, Maxim Shoshany and Naftali Goldshlager offer a new method of matrix land-use analysis for analysing landscape change in the Carmel area of Israel. Zhi-Gang Wu, Su-Hong Zhou and Chang-Chun Feng describe land-use development in one the fastest developing regions in the world: the Pearl River Delta metropolis, China. They point at the friction between the legal framework and the rapid economic changes.

In Part II, various *explanatory models of land-use change* are presented. These models typically analyse, explain and simulate land-use changes from a thorough study of past developments. Juan Peña and colleagues analyse the trends and driving factors in land-use change in Marina Baixa, Spain and go on to develop a simple Markov chain model to simulate the possible future state of the area. Nicolas Dendoncker, Patrick Bogaert and Mark Rounsevell employ a number of statistical techniques to empirically derive suitability maps that are then used to downscale aggregated land-use data. Joana Gonçalves and Tomaz Dentinho present a spatial interaction model that explains the historic development of a small island of the Azores group by means of a simulation of land-use changes in the past 400 years.

The book's attention then shifts towards the exploration of new methodological and theoretical insights in land-use modelling. Specific attention is paid in Part III to recent research progress in *optimisation modelling*. Willem Loonen, Peter Heuberger and Marianne Kuijpers-Linde demonstrate the benefits of genetic algorithms for optimising land-use patterns in two different environmental problems. Subrata Mandal uses linear programming in a case study of a watershed in the Indian Himalaya to examine sustainable land-use and water management in mountain ecosystems. Júlia Seixas, João Pedro Nunes, Pedro Lourenço and João Corte-Real report their experience of using evolutionary algorithms for optimising land use from the perspectives of soil erosion and carbon sequestration.

Part IV concentrates on the *incorporation of new modelling approaches* in land-use simulations. Approaches such as microsimulation, agent-based modelling and dynamic simulation have received a lot of attention in recent land-use modelling research and suggest much promise for further model development. Daniel Felsenstein, Eyal Ashbel and Adi Ben-Nun describe the incorporation of a microsimulation approach in the well-known *UrbanSim* model to simulate employment deconcentration in the Tel Aviv region. Wolfgang Loibl, Tanja Tötzer, Mario Köstl and Klaus Steinnocher present an urban growth model that simulates location decisions of households and company start-ups, based on a multi-agent system. Dick Ettema, Kor de Jong, Harry Timmermans and Aldrik Bakema outline the modelling framework that underpins *PUMA*, a multi-agent modelling system for urban systems. Kampanart Piyathamrongchai and Michael Batty discuss a new model that integrates the local dynamics of CA with a regional-level dynamic simulation model.

In the last two parts of the book the emphasis shifts from research progress to the actual application of land-use simulation models. Part V introduces several well-established *operational land-use simulation models* that can be considered the current state of the art. Judith Borsboom-van Beurden, Aldrik Bakema and Hanneke Tijbosch describe the *LUMOS* land-use modelling system that they apply in land-use related environmental impact assessments in the Netherlands. Guy Engelen and others introduce the *MOLAND* modelling framework for urban and regional land-use dynamics that was applied in many urban regions in Europe. Peter Verburg and Koen Overmars explain the *CLUE-s* model for dynamic simulation of land-use change trajectories by means of two scenario-based case studies.

In the final section of the book (Part VI), a number of recent case study applications of *land-use simulation for policy analysis* are outlined. Dagmar Haase, Annelie Holzkämper and Ralf Seppelt look at residential vacancies and demolition in urban land-use planning policy in Eastern Germany. Jasper Dekkers and Eric Koomen provide an assessment of the suitability of scenario-based modelling by simulating future land use for water management. Finally, Patrick Sheridan, Eike Rommelfanger and Jan Ole Schroers review EU Common Agricultural Policy reform and evaluate its impact on agricultural land use and plant species richness.

Together with this book a website (www.lumos.info/ModellingLand-UseChange/Exercises.htm) has been developed that provides a number of exercises covering the analysis of land-use change, general modelling approaches and demonstration versions of some of the well-known land-use models discussed in this book. The provided land use modelling software and included base-data give detailed insights into the way these models work. The related exercises and assignments help students to critically assess the potential of these instruments. Further installation instructions and explanatory texts are available on the website.

REFERENCES

Aerts, J. (2002) *Spatial Decision Support for Resource Allocation*, PhD Dissertation, Universiteit van Amsterdam.

Agarwal, C., Green, G.M., Grove, J.M., Evans, T.P. and Schweik, C.M. (2002) *A review and assessment of land-use change models: dynamics of space, time, and human choice*. Gen. Tech. Rep. NE-297, Newtown Square, PA: U.S. Department of Agriculture, Forest Service, Northeastern Research Station. P.61.

Alberti, M. and Wadell, P. (2000) An integrated urban development and ecological simulation model, *Integrated Assesment*, 1: 215–227.

Alcamo, J., Leemans, R. and Kreileman, E. (1998) *Global Change Scenarios of the 21st Century. Results from the IMAGE 2.1 Model.* Elsevier, Amsterdam, P. 296.

Alonso, W.A. (1964) *Location and Land Use: Toward a General Theory of Land Rent*, Harvard University Press, Cambridge.

Anselin, L., Florax, R. (eds) (1995) *New Directions in Spatial Econometrics*, Springer, Berlin.

Balzter, H., Braun, P.W. and Kohler, W. (1998) Cellular automata models for vegetation dynamics, *Ecological Modelling*, 107(2/3): 113–125.

Batty, M. (1997) Cellular automata and urban form: a primer, *Journal of the American Planning Association*, 63(2): 266–274.

Bell, K.P. and Irwin, E.G. (2002) Spatially explicit micro-level modelling of land use change at the rural–urban interface, *Agricultural Economics*, 27(3): 217–232.

Berger, T. and Parker, D.C. (2002) Examples of specific research, in Parker, D.C., Berger, T. and Manson, S.M. (eds) *Agent-Based Models of Land-Use and Land-Cover Change: Report and Review of an International Workshop*, Irvine, California October 4–7, LUCC Report Series No. 6. Bloomington, Indiana, LUCC International Project Office.

Bockstael, N.E. and Irwin, E.G. (2000) Economics and the land use-environment link, in Folmer, H. and Tietenberg, T. (eds) *The International Yearbook of Environmental and Resource Economics 1999/2000*, Edward Elgar Publishing, Northampton, MA, pp. 1–54.

Briassoulis, H. (2000) Analysis of land use change: theoretical and modeling approaches, in Jackson, W.R. (ed) *The web-book of regional science* Regional Research Institute, West Virginia University, USA.

Brueckner, J.K. (2000) Urban sprawl: diagnosis and remedies, *International Regional Science Review*, 23: 160–171.

Buurman, J.J.G., Rietveld, P. and Scholten, H.J. (2001) The land market: spatial economic perspective, Chapter 6 in Stillwell, J.C.H. and Scholten, H.J. (eds) *Land Use Simulation for Europe*, Kluwer Academic Publishers, Dordrecht, pp. 65–82.

Clarke, K. and L. Gaydos (1998) Loose-coupling a cellular automation model and GIS: long-term urban growth prediction for San Fransisco and Washington/Baltimore, *International Journal of Geographical Information Science*, 12(7): 699–714.

Clarke, K.C., Hoppen, C. and Gaydos, L. (1997) A self-modifying cellular automaton model model of historical urbanisation in the San Francisco Bay Area, *Environment and Planning B*, 24: 247–261.

Champion, A., Bramley, G., Fotheringham, A.S., Macgill, J. and Rees, P. (2003) A migration modelling system to support government decision-making, Chapter 15 in Geertman, S. and Stillwell, J.C.H. (eds) *Planning Support Systems in Practice*, Springer, Berlin, pp. 269–290.

Chomitz, K.M. and Gray, D.A. (1996) Roads, land use and deforestation: a spatial model applied to Belize, *The World Bank Economic Review*, 10: 487 512.

Couclelis, H. (1997) From cellular automata to urban models: new principles for model development and implementation, *Environment and Planning B*, 24: 165–174.

Cramer, W., Bondeau, A., Woodward, F.I., Prentice, I.C., Betts, R.A., Brovkin, V., Cox, P.M., Fisher, V., Foley, J., Friend, A.D., Kucharik, C., Lomas, M.R., Ramankutty, N., Sitch, S., Smith, B., White, A. and Young-Molling, C. (2001) Global response of terrestrial ecosystems structure and function to CO_2 and climate change: results from six dynamic global vegetation models, *Global Change Biology*, 7: 357–374

Dale, V.H. (1997) The relationship between land-use change and climate change, *Ecological Applications*, 7(3): 753–769.

Engelen, G., Geertman, S., Smits, P. and Wessels, C. (1999) Dynamic GIS and strategic physical planning support: a practical application, Chapter 5 in Stillwell, J.C.H., Geertman, S. and Openshaw, S. (eds) *Geographical Information and Planning*, Springer, Berlin, pp. 87–111.

Fotheringham, A.S. (1991) Migration and spatial structure; the development of the competing destinations model, Chapter 4 in Stillwell, J.C.H. and Congdon, P. (eds) *Migration Models Macro and Micro Approaches*, Belhaven Press, London, pp. 57–72.

Geoghegan, J. Schneider L. and Vance, C. (2004) Temporal dynamics and spatial scales: Modeling deforestation in the southern Yucatán peninsular region, *GeoJournal*, 61(4): 353–363.

Glaeser, E. and Kahn, M. (2004) Sprawl and urban growth, in Henderson, V. and Thisse, J.F. (eds) *Handbook of Regional and Urban Economics: Cities and Geography, Handbooks in Economics, Volume 4*, Elsevier North-Holland, Amsterdam.

Haynes, K.E. and Fotheringham, A.S. (1984) *Gravity and Spatial Interaction Models*, Sage Publications, Beverly Hills.

Holland, J.H. (1998) *Emergence: From Chaos to Order*. Perseus Books, Reading, MA.

Irwin, E. and Geoghegan, J. (2001), Theory, data, methods: developing spatially-explicit economic models of land use change, *Agriculture, Ecosystems and Environment*, 85: 7–24.

Kanaroglou, P. and Scott, D. (2002) Integrated urban transportation and land-use models for policy analysis, in Dijst, M., Schenkel, W. and Thomas, I. (eds) Governing Cities on the Move: Functional and Management Perspectives on Transformations of European Urban Infrastructures, Ashgate, Aldershot, UK, pp. 42–72.

Klosterman, R.E. (1999) The what if? collaborative planning support system, *Environment and Planning B*, 26: 393–408.

Klosterman, R.E. and Pettit, C.J. (2005) Guest editorial: an update on planning support systems, *Environment and Planning B*, 32: 477–484.

Krugman P. (1999) The role of geography in development, *International Regional Science Review*, 22: 142–161.

Kruijt, B., Needham, B. and Spit, T. (1990) *Economische grondslagen van grondbeleid, Stichting voor beleggings- en vastgoedkunde*, Universiteit van Amsterdam.

Lambin, E.F., Geist, H.J. and Lepers, E. (2003) Dynamics of land-use and land-cover change in tropical regions, *Annual Review of Environment and Resources*, 28: 205–241.

Lambin, E.F., Turner, B.L., Geist, H.J., Agbola, S.B., Angelsen, A., Bruce, J.W., Coomes, O.T., Dirzo, R., Fischer, G., Folke, C., George, P.S., Homewood, K., Imbernon, J., Leemans, R., Li, X., Moran, E.F., Mortimore, M., Ramakrishnan, P.S., Richards, J.F., Skanes, H., Stone, G.D., Svedin, U., Veldkamp, T.A., Vogel, C. and Xu, J. (2001) The causes of land-use and land-cover change, moving beyond the myths, *Global Environmental Change*, 11: 261–269.

Landis, J.D. (1994) The California urban futures model: a new generation of metropolitan simulation models, *Environment and Planning B*, 21: 399–420.

Lavorel, S., Davies, I.D. and Noble, I.R. (2000) *LAMOS*: a Landscape Modelling Shell, in Hawkes, B.C. and Flannigan, M.D. (eds) *Landscape Fire Modeling – Challenges and Opportunities*, Natural Resources Canada, Canadian Forest Service, Victoria, British Columbia.

Li, H. and Reynolds, J.F. (1997) Modeling effects of spatial pattern, drought, and grazing on rates of rangeland degradation: a combined Markov and cellular automaton approach, in Quattrochi, D.A. and Goodchild, M.F. (eds) *Scale in Remote Sensing and GIS*, Lewis Publishers, New York, pp. 211–230.

Li, X. and Yeh, A.G.O. (2000) Modelling sustainable urban development by the integration of constrained cellular automata and GIS, *International Journal of Geographical Information Science*, 14(2): 131–152.

Lowry, I.S. (1964) *A Model of Metropolis*, Rm-4035-RC, Rand Corporation, Santa Monica, CA.

Meyer, W.B. and Turner, B.L. (1994) *Changes in Land Use and Land Cover*, Cambridge University Press, Cambridge .

McFadden, D.L. (1978) Modelling the choice of residential location, in Karlsqvist, A., Lundqvist, L., Snickars, F. and Weibull, J.W. (eds) *Spatial Interaction Theory and Planning Models*, North-Holland, Amsterdam, pp. 75–96.

O´Sullivan, D. and Torrens, P.M. (2000) Cellular models of urban systems, *CASA Working Paper, Number 22*, Centre for Advanced Spatial Analysis, University College London.

Pijanowski, B.C., Brown, D.G., Manik, G. and Shellito, B. (2002) Using neural nets and GIS to forecast land use changes: a land transformation model, *Computers, Environment and Urban Systems*, 26(6): 553–575.

Parker, D.C., Manson, S.M., Janssen, M., Hoffmann, M.J. and Deadman, P.J. (2003) Multi-agent systems for the simulation of land use and land cover change: a review, *Annals of the Association of American Geographers*, 93(2): 316–340.

Putman, S.H. (1983) *Integrated Urban Models*, Pion, London.

Ravenstein, E.G. (1885) The laws of migration, *Journal of the Royal Statistical Society*, 48: 167–227.

Reilly, W.J. (1931) *The Law of Retail Gravitation*, Pilsbury, New York.

Rietveld, P. and Wagtendonk, A.J. (2004) The location of new residential areas and the preservation of open space; experiences in the Netherlands, *Environment and Planning A*, 36: 2047–2063.

Samat, N. (2002) A geographic information system and cellular automata spatial model of urban development for Penang State, Malaysia, *Unpublished PhD Thesis*, School of Geography, University of Leeds, Leeds.

Schneider, L.C. and Pontius, R.G. (2001) Modeling land-use change in the Ipswich Watershed, Massachusetts, USA, *Agriculture, Ecosystems and Environment*, 85: 83–94.

Serneels, S. and Lambin, E.F. (2001) Proximate causes of land-use change in Narok District, Kenya: a spatial statistical model, *Agriculture, Ecosystems & Environment*, 85: 65–81.

Stillwell, J.C.H. (1991) Spatial interaction models and the propensity to migrate over distance, Chapter 3 in Stillwell, J.C.H. and Congdon, P. (eds) *Migration Models Macro and Micro Approaches*, Belhaven Press, London, pp. 34–56.

Stillwell, J.C.H. and Scholten, H.J. (2001) *Land Use Simulation for Europe*, Kluwer Academic Publishers, Dordrecht.

Timmermans, H.J.P. (2003) The saga of integrated land use-transport modeling: How many more dreams before we wake up? in Proceedings of the 10th International Conference on Travel Behavior Research, Lucerne, August.

Tobler, W. R. (1979) Cellular geography, in Gale, S. and Olsson, G. (eds) *Philosophy in Geography*, Reidel, Dordrectht, pp. 379–386.

Torrens, P.M. (2001) Can geocomputation save urban simulation? Throw some agents into the mixture, simmer, and wait…, *CASA Working Paper Number 32*, Centre for Advanced Spatial Analysis, University College London, London.

Torrens, P.M. (2003) Cellular automata and multi-agent systems as planning support tools, Chapter 12 in Geertman, S. and Stillwell, J.C.H. (eds) *Planning Support Systems in Practice*, Springer, Berlin, pp. 205–222.

U.S. EPA (2000) *Projecting Land-use Change: A Summary of Models of Assessing the Effects of Community Growth and Change on Land-Use Patterns*, EPA/600/R-00/098, U.S. Environmental Protection Agency, Office of Research and Development, Cincinnati, OH, P. 226.

Veldkamp, A. and Lambin, E.F. (2001) Editorial; Predicting land-use change, *Agriculture, Ecosystems and Environment*, 85: 1–6.

Verburg, P.H., Schot, P.P., Dijst, M.J. and Veldkamp, A. (2004a) Land use change modelling: current practice and research priorities. *GeoJournal*, 61: 309–324.

Verburg, P.H., Ritsema van Eck, J., de Nijs, T., Schot, P. and Dijst, M. (2004b) Determinants of land use change patterns in the Netherlands, *Environment and Planning B*, 31(1): 125–150.

Verburg, P.H., de Nijs, T.C.M., Ritsema van Eck, J., Visser, H., de Jong, K. (2004c) A method to analyse neighbourhood characteristics of land use patterns, *Computers, Environment and Urban Systems*, 28: 667–690.

Von Thünen, J.H. (1826) *Der isolierte Staat, in Beziehung auf Landwirtschaft und Nationalökonomie*, Neudruck nach der Ausgabe letzter Hand (1842) Gustav Fisher Verlag, Stuttgart, 1966.

Waddell, P. (2002) Urbansim: modeling urban development for land use, *Transportation and Environmental Planning, Journal of the American Planning Association*, 68(3): 297–314.

Waddell, P. and Ulfarsson, G.F. (2003) Introduction to urban simulation: design and development of operational models, in Stopher, Button, Kingsley and Hensher (eds) *Handbook in Transport, Volume 5: Transport Geography and Spatial Systems*, Pergamon Press, New York.

Watson, R.T., Noble, I.R., Bolin, B., Ravindranath, N.H., Verardo, D.J. and Dokken, D.J. (eds) (2000) *Land Use, Land-Use Change, and Forestry*. A Special Report of the Intergovernmental Panel on Climatic Change, Cambridge University Press, Cambridge.

Wegener, M. (1998) Applied models of urban land use, transport and environment: state of the art and future developments, in Lundqvist, L. Mattson, L.G. and Kim, T.J. (eds) *Network Infrastructure and the Urban Environment: Recent Advances in Land use/Transportation Modelling*, Springer, Heidelberg, pp. 245–267.

White, R. and Engelen, G. (2000) High-resolution integrated modelling of the spatial dynamics of urban and regional systems, *Computers, Environment and Urban Systems*, 24: 383–400.

Wilson, A.G. (1967) A statistical theory of spatial distribution models, *Transportation Research*, 1: 253–269.

Wilson, A.G. (1970) *Entropy in Urban and Regional Modelling*, Pion, London.

Wolfram, S. (1984) Cellular automata as models of complexity, *Nature*, 311: 419–424.

Wu, F. (1998) Simulating urban encroachment on rural land with fuzzy-logic-controlled cellular automata in a geographical information system, *Journal of Environmental Management*, 53: 293–308.

Yeh, A. G.O. and Li, X. (2001) A constrained CA model for the simulation and planning of sustainable urban forms by using GIS, *Environment and Planning B*, 28: 733–753.

Young, E.C. (1924) *The Movement of Farm Population*, Cornell Agricultural Experimental Station, Bulletin 426, Ithaca, New York.

PART I: ANALYSIS OF LAND-USE TRENDS AND THEIR DRIVING FORCES

Chapter 2

LAND-USE CHANGE AT CADASTRAL PARCEL LEVEL IN ALBANIA

An object-oriented geo-database approach to analyse spatial developments in a period of transition (1991-2003)

L.J.M. Jansen[1], G. Carrai[2] and M. Petri[3]
[1]*Land/natural resources consultant, Rome, Italy;* [2]*SVALTEC S.r.l., Florence, Italy;* [3]*Department of Civil Engineering, University of Pisa, Italy*

Abstract: A case study in Albania is presented based on the EU Phare Land Use Policy II project results where GIS-oriented instruments and innovative methodologies were implemented to support decision making for land-use policy and planning. The developed Land-Use Information System for Albania allows the logical and functional hierarchical arrangement of land uses and data harmonisation with other land-use description systems. It is linked to the object-oriented Land-Use Change Analyses methodology that groups changes into conversions and modifications. The preferred change patterns indicate that land users take rational decisions when changing land use, even in the absence of any regulating plan, as is the case in post-communist Albania.

Key words: Land-use change dynamics; knowledge discovery in databases; object-oriented database approach; agriculture; urbanisation.

1.   INTRODUCTION

In Albania, the Government has distributed land to rural households instead of restitution of most of the fertile land to a small number of families that would have restored the highly unequal, pre-reform land distribution (Swinnen, 1999; 2000). The transition from 550 large agricultural cooperatives to 467,000 smallholder farms was associated with the fragmentation of land into 1.5 million parcels that often have limited or no access to infrastructure and mechanisation. Most of the agricultural land lies

E. Koomen et al. (eds.), Modelling Land-Use Change, 25–44.
© 2007 *Springer*

in sloping areas with soils having high erosion risk potentials. Most of the farms are subsistence ones and about 75% of farm production is for home consumption. The lack of information, inadequate extension services, almost no access to bank credit, lack of marketing channels and difficult access to transport are the major constraints for the Albanian farmer. Since around half of the Albanian population is employed in the agricultural sector, a national development priority is a sound land-use policy, allocating land to uses that prevent degradation and yield high long-term returns. The land users should ensure the long-term quality of land for human use, minimise social conflicts and protect ecosystems. All user categories should have enough land with an infrastructure balanced against environmental threats, at reasonable cost and having a well-defined tenure.

The EU Phare Land Use Policy (LUP) II project provided GIS-oriented instruments and innovative methodologies to support decision making for land-use policy and planning to the Ministry of Agriculture and Food in Albania. These methodologies and tools have been applied in three representative pilot communes in the northwest, centre and southeast of the country. This chapter illustrates the concepts adopted and results obtained for the analysis of land-use change dynamics over the period 1991-2003. Land-use change is one of the main driving forces of (global) environmental change and therefore central to sustainable development (Meyer and Turner, 1994; Walker *et al.*, 1997; Walker, 1998). Thus, analysis of past land uses and understanding processes and preferred pathways of change will support informed decision making for improved, sustainable and environmentally sound land uses in future.

2.   METHODOLOGY

This section gives a short description of the information system and its basic unit that were used in this study and briefly introduces the methods that were used in the analysis of the land-use changes. The methodology is described more extensively in two LUP II project documents (Agrotec S.p.A. Consortium, 2003a; 2003b).

2.1   The cadastral land parcel as a basic unit

For each piece of land, individuals choose a type of use from which they expect to derive the most benefits in the context of their knowledge, the individual's household, the community, the bio-physical environment and the political structure to which the individual may be subject. These choices vary in space and time resulting in a spatial pattern of land uses. The analysis at the level of the spatially explicit legal parcel unit of the multi-purpose cadastre may show the variability at the level of each cadastral zone while the aggregated level of the commune may show patterns that remain invisible at the detailed scale, and vice versa (Veldkamp *et al.*, 2001). The aggregated level of the commune is important in the land-use policy and planning process while the cadastral parcel unit is a level that corresponds with the decisions made by the individual landowner or land user. It should be clear though, that such decisions may be related to the size of the group that the individual belongs to (Verburg *et al.*, 2003). Individuals interact to form groups and organise collective action (e.g. farmer associations).

In general, land registration and the cadastre should be seen as part of the process of natural resources planning and management. The multi-purpose cadastre should therefore be seen as an integral part of the land management system. It is therefore important to establish linkages with a wider range of land-related data, especially those relating to the environment. In this manner, managing land and land information come together (Dale, 1995).

2.2 The Land-Use Information System for Albania

There is significant diversity of opinion about what constitutes a land use (UNEP/FAO, 1994). In the context of the project, land use is defined as *"the type of human activity taking place at or near the surface"* (Cihlar and Jansen, 2001). The developed Land-Use Information System for Albania (LUISA) has adopted, as guiding principles, two criteria that are commonly applied in international systems (Anderson *et al.*, 1976; IGU, 1976; ECE-UN, 1989; UN, 1989; 1998; CEC, 1993; 1995; 1999; FAO, 1998; APA, 1999): (1) *function* that refers to the economic purpose of the land use and can group many different land-use types in a single category; and (2) *activity* that refers to a process resulting in a similar type of product and is used at the lower levels of the hierarchy (Jansen and Di Gregorio, 1998; 2002). The adopted concept builds upon and exceeds experiences gained in two case studies (Jansen and Di Gregorio, 2003; 2004). Furthermore, LUISA arranges in a logical and functional manner land uses at different levels of detail and allows data harmonisation with other land-use description systems in use in the country (e.g. statistical office, cadastre and communes).

Categories present in the current version of LUISA represent the key categories of the Albanian law on the land: 'agricultural', 'forests', 'pastures and meadows' and 'non-agricultural' land uses (Figure 2-1). The set of classes in this legend is only a proportion of what one may actually find in Albania. The cadastre in Albania contains information on 1.5 million parcel units with an average size of less than 1 ha. Because of the scale of observation selected, i.e. the cadastral parcel unit, and in order to create in a timely manner a pragmatic land-use database of manageable size (i.e. all records created will need to be maintained and updated at regular intervals), the decision was made that only one land-use class is attached to each parcel unit. At aggregated cadastral parcel levels, mixed classes can be introduced but they do not exist at the most detailed level of LUISA.

The LUISA data, together with other data sets, have been structured according to the European Environmental Agency's Infrastructure for Spatial Information in Europe initiative (INSPIRE Environmental Thematic Coordination Group, 2002).

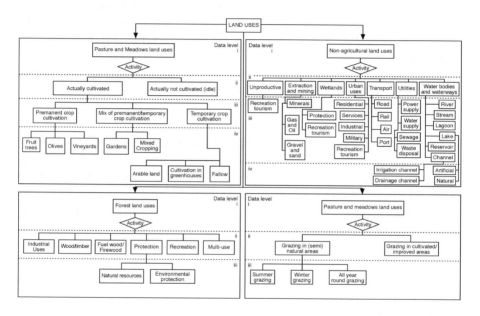

Figure 2-1. Overview of the LUISA legend with the four main categories of land use (Agrotec S.p.A. Consortium, 2003b).

2.3   The Land-Use Change Analyses methodology

LUISA contains many classes and thus will result in numerous possible land-use changes that do not facilitate a meaningful interpretation if not grouped in a functional and systematic manner. The developed object-oriented Land-Use Change Analyses (LUCA) methodology arranges the potential land-use changes in three main groups per land-use category in order to underline the change processes: (1) land-use *conversion*, i.e. where a certain land use has been changed into a land use that is very different and the change cannot easily be reversed; (2) land-use *modification*, i.e. changes that are related to one another and where the situation can be reversed; and (3) *no change*, i.e. areas that have remained under the same land use. The parent-child relationships created facilitate the analysis of the spatio-temporal dimensions, i.e. area and perimeter over time (Booch, 1994).

In principle, land-use modifications occur within a land-use category and land-use conversion occurs between land-use categories. The exception is the 'non-agricultural' land-use category that contains a larger variety of classes than the other categories; in this category modifications occur within one group (e.g. within 'urban uses') and conversions between groups (e.g. from 'unproductive' to 'urban uses'). Unlikely changes such as a 'residential area' having changed into 'arable land' have been excluded from the change analysis.

2.4 Knowledge Discovery in Databases

The Knowledge Discovery in Databases (KDD) process is an iterative procedure of selection, exploration and modelling of large amounts of data that was used to detect *a priori* unknown relationships in the data. The KDD process comprises many elements of which the two most important in the context of this chapter are (Bonchi and Pecori, 2003):

1. data-mining: the most important phase in which, through the use of specific algorithms, previously unknown patterns are extracted from the data that are channelled into a data model; and
2. pattern evaluation: an interpretation and evaluation of the identified patterns and data model in order to create new knowledge.

Some preliminary statistics on correlations between parameters were performed using the On-Line Analytical Process (OLAP) cube for multi-dimensional analysis in order to better understand which parameters to use in the KDD process. OLAP was performed with the following variables: (1) land-use change class, (2) land-use change period, (3) slope class and (4) land suitability.

The variables used as inputs into the decision tree that belongs to the data-mining phase of KDD have been used with the assumption that one of the variables, i.e. land use in 2003, is dependent on the other variables. The use of the variables to construct the decision tree is such that one starts at the initial node with all the available data; then at each step groups are created on the basis of an explanatory variable and in the successive step, each group created will be further subdivided by another explanatory variable and so on until the terminal node. Once a variable has been used, it cannot be used in successive steps (Lombardo *et al.*, 2002). From the initial node to the terminal node, a series of decision rules can be extracted of the type

IF-THEN. Each decision rule is characterised by a weight and a confidence level that measure the frequency and strength of the decision rule respectively. Decision rules that are valid for many cells have a major weight, whereas those that repeat themselves in the same manner have more significance. The method requires several runs in order to create groups that maximise the internal homogeneity and the external heterogeneity. To create the groups at each level of the procedure, a function is used as an efficiency index known as the 'function segmentation criteria' (Han and Kamber, 2000).

3.   RESULTS

3.1   Pilot area selection

The choice of pilot communes illustrates the diversity in landforms and (agro-)ecological conditions plus the variety in socioeconomic settings. The choice of Preza Commune was also governed by the fact that it already served as a pilot area in the LUP I project. The availability of suitable digital data sets was a prime criterion for selection.

3.2   The temporal changes in the communes

Each of the three land-use data sets available represents a critical moment in time: (1) the 1991 data represent the land uses under the former centralised government; (2) the 1996 data represent the time when distribution and registration of the land to the family households took place; and (3) the 2003 data represent the actual land uses in the market-oriented economy.

Table 2-1 shows the different types of land-use changes aggregated for the three communes, i.e. Preza, Ana-e-Malit and Pirg, in 1991-1996 and 1996-2003. The communes comprise 2552, 3357 and 2150 ha and are situated in the centre, northwest and southeast of the country respectively. In all three communes, the intensity of changes in 1991-1996, before the land distribution, is higher than in 1996-2003. The majority of parcels were not subject to any change in either period. In Ana-e-Malit and Pirg the area not subject to change increases in the second period, but in Preza it decreases. The main change in land use in both periods involves a land-use modification and in all three communes it is the 'medium-level-modification-in-agriculture', which means that classes in the 'agricultural' land-use category changed at level III, i.e. from permanent into temporary crop cultivation or vice versa. However, the extent of this modification is diminishing in 1996-2003 in Ana-e-Malit and Pirg, whereas Preza shows a clear increase. Land-use conversions are much less important in terms of their extent but their impact may be bigger than that of land-use modifications. The most common conversion is 'agriculture-to-nonagriculture', except in Preza in 1991-1996 where 'pasture-to-agriculture' conversion is dominant. The second most common conversion is 'agriculture-to-pasture' in Preza and Ana-e-Malit in both periods and in Pirg in 1996-2003. In Pirg, 'nonagriculture-to-agriculture' conversion is important in 1991-1996. It seems that in 1996-2003 in particular, agricultural lands were converted, whereas overall changes were affecting less parcels. In this period, land was privatized and apparently many new owners did not want or did not have the means to continue agricultural activities.

Table 2-1. Predominant types of land-use changes (claiming over 1% of the total area) in Preza, Ana-e-Malit and Pirg in 1991-1996 and 1996-2003

Type of land-use change	Preza		Ana-e-Malit		Pirg	
	91-96	96-03	91-96	96-03	91-96	96-03
No change	86.5	80.2	71.7	90.2	81.3	91.9
Medium level modification in agriculture	4.9	7.6	9.8	1.9	8.2	3.9
High level modification in nonagriculture		1.8			1.5	
Agriculture-to-forest			1.3			
Agriculture-to-pasture	1.6	1.1	5.6			1.8
Agriculture-to-nonagricultural		1.1	2.5	2.1	1.4	
Forest-to-pasture		1.1	2.9			
Forest-to-agriculture		3.2				
Pasture-to-agriculture	1.2		1.5			
Nonagricultural-to-agriculture					2.5	

Concerning the most important change, 'medium-level-modification-in-agriculture', more insight is gained when analysing what type of land-use classes result in this type of change. Selection of this change type in the three communes and grouping the class combinations of this change shows that in Preza and Ana-e-Malit in 1991-1996 the trend is to go from temporary to permanent crops, whereas in Pirg the trend in the same period is from permanent to temporary crops (Figure 2-2). In 1996-2003, the trend in Ana-e-Malit remains more or less the same. In Preza, however, the majority of changes still involve the change from temporary to permanent crops though the rate of change is at a lower level than in the previous period, while the change from permanent to temporary crops increases. In 1996-2003, the main trend in Pirg remains the change from permanent to temporary cropping but at a lower level than in the previous period and the change to permanent crops increases. In Pirg, many terraces with fruit trees, the main crop production system, were destroyed in the 1990s; in Preza and Ana-e-Malit projects are underway to plant useful trees (e.g. fruit trees, olives).

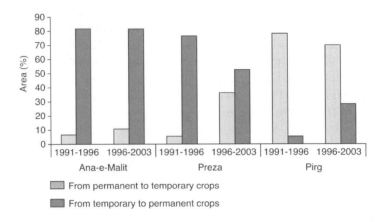

Figure 2-2. Detailed analysis of LUCA change type 'medium-level-modification-in-agriculture'.

The identified change dynamics have some important repercussions: the permanent cultivation land-use types are usually found on man-made terraces or in landscapes with slopes where the trees stabilise and protect the environment. Further analysis combining the land-use change data with a digital terrain model shows that one of the adverse affects of the change from permanent to temporary crops is increased erosion in hilly areas. Furthermore, there seems to be a shift in agricultural land uses because the area lost in one place and gained in another affects different parts of the commune territory. From the three-dimensional analysis of where such changes are found, it becomes clear that parts of the flat or almost flat areas favourable for agriculture are lost, whereas areas where less or even unfavourable terrain conditions (e.g. steep slopes) exist are gained. This consumption of prime agricultural land, in plains and river valleys of peri-urban areas, blurs the distinction between cities and countryside (Lambin *et al.*, 2003).

3.3   The spatial distribution of changes

As physical and social characteristics of communities vary in space and time, so do land-use choices, resulting in a spatial pattern of land-use types (Cihlar and Jansen, 2001). If one shows the land-use changes not in the format of statistics but as maps, one can easily identify in each commune areas that were more prone to land-use changes than others.

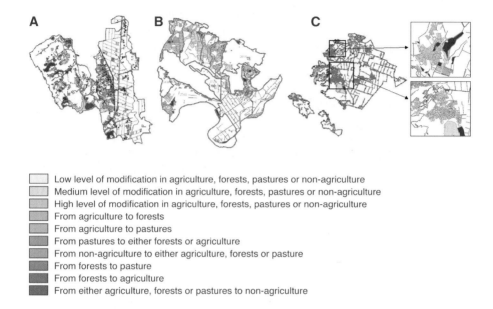

☐ Low level of modification in agriculture, forests, pastures or non-agriculture
☐ Medium level of modification in agriculture, forests, pastures or non-agriculture
☐ High level of modification in agriculture, forests, pastures or non-agriculture
☐ From agriculture to forests
☐ From agriculture to pastures
☐ From pastures to either forests or agriculture
☐ From non-agriculture to either agriculture, forests or pasture
☐ From forests to pasture
☐ From forests to agriculture
☐ From either agriculture, forests or pastures to non-agriculture

Figure 2-3. Distribution of land-use changes in the commune of (A) Preza, (B) Ana-e-Malit and (C) Pirg, in 1991-2003 (communes are not shown at same scale). (See also Plate 1 in the Colour Plate Section)

Figure 2-3 shows the distribution of changes over the territory of the communes ranked according to the environmental impact of the change and the fact that Albanian law protects agricultural land, forests and pastures from other uses. The changes with the strongest adverse environmental impact, occurring in protected lands are indicated at the bottom of the figure in the darkest colours. The changes in Preza seem to be divided clearly over the territory: most conversions are found in the western part that consists mainly of hills, whereas most modifications occur in the eastern part that consists of foothills and a plain (indicated by the channel system). In Ana-e-Malit, modifications occur mainly on the foothills and close to the main village of the commune where also the frequency of conversions is highest. In the flatter areas, indicated by the channel system, few changes occur. In Pirg, modifications occur in areas where the land parcels have been divided into many very small parcels close to the villages as shown in the two detailed windows. Also conversions occur in these areas but of a type that is considered to have a positive environmental impact. Large parcels are more often subject to conversions considered to have a negative impact than small land parcels. Also in this commune, the flat areas with channel systems are not subject to many changes.

The areas where land-use conversions occurred that cannot be easily reversed are mainly in the sloping and hilly parts of the communes. In the plains, land-use modifications were dominant, whereas the residential areas grew at the cost of neighbouring land uses.

3.4   Preferred pathways of change in Preza Commune

The change dynamics can be related to the landscape position of the cadastral parcel within the terrain and the land suitability for irrigated agriculture, as the communes are predominantly agricultural ones, as well as a set of variables related to what is found in or close to the land parcel. The area of Preza Commune that changed in 1991-1996 and/or 1996-2003 was examined more closely.

A preliminary statistical analysis using OLAP showed that:

- In 1991-1996, more stability concerning land uses existed with around 39% of the total area being classified as no land-use change or 'medium-level-modification-in-agriculture' changes homogeneously distributed within the area involving the various slope and land suitability classes.
- In the same period, transformations are uniformly distributed between the different land-use classes and slope categories. Moreover, there are no major conversions of land use but only some medium-level-modifications.
- In 1996-2003, contrary to the changes in the previous period, a portion of steep sloping lands has been abandoned (20%); this is probably related to abandonment of terraced areas.
- Moreover, in the same period, privatisation of agricultural lands led to encroachment of fields at the costs of forests. Conversion from forests into pastures and meadows is around 10%.
- It is interesting to note that there is a strong relation between slope class and land-use class, i.e. steep lands are always related to land uses like forestry and pastures and meadows.

The data for Preza Commune was used as input into the KDD process in order to identify which variables in the extracted decision rules are important and lead to specific pathways of change. The rules with major weights were chosen first, followed by those with high significance. The territory of Preza Commune was divided into cells of 50 by 50 metres to which a series of attributes are linked from the available data sets. The analysis aims at explaining which factors in or near the cells are important in a specific type of change in either period.

The analysis concerns in particular the 'medium-level-modification-in-agriculture' land-use change and focuses on areas that are either not cultivated or fallow, regrouped under uncultivated, as (temporary) abandonment of cultivated and especially terraced areas is a problem. For the two periods, a set of decision rules was extracted that describe the pathways of change. The complete set of rules for 1996-2003 is almost twice the number of the previous period (719 versus 366 rules), though there are less changes in that period. Two types of rules are extracted, i.e. *transformation* and *inertial rules*, with the description of their conditions (e.g. IF LU_1 and [conditions 1, 2, ...] THEN LU_2). Transformation rules describe a land-use change ($LU_1 \neq LU_2$), whereas inertial rules describe a land use not subject to change ($LU_1 = LU_2$). The extracted rules show that in 1996-2003, the vicinity of the examined cell does not influence the land-use change dynamics in particular. In 1991-1996, one finds the opposite, i.e. the vicinity of the cell is very important for change dynamics. One should also note that in 1991-1996, the extracted rules are essentially inertial rules and transformation rules are few and related to only a few cells, whereas in 1996-2003, there are more transformation rules than inertial rules. Furthermore, the transformation rules for 1991-1996 contain one principal condition that leads to a certain land-use change. In 1996-2003, a principal condition accompanied by more than one set of sub-conditions leads to the same land-use change. So, the preferred pathways of change are much more complex in the second period.

Tables 2-2 to 2-5 show those rules related to permanent cropping, temporary cropping and uncultivated areas. A change that becomes more evident is that remote areas with either permanent or temporary cropping, often on steeper terrain, and with a lack of infrastructure tend to become uncultivated. So, in these areas the agricultural intensity has decreased dramatically.

Application of the set of decision rules for 1996-2003 to the original data of 1996 resulted in a predicted land use for 2003 with a correlation coefficient of 0.75 with the observed 2003 data. The difference between the average square root of classification (0.15) and average absolute error (0.04) is low, which means the absence of classification outliers. In addition, the accuracy of prediction for each land-use class is above 0.70 with the exceptions of services and industrial areas because the first class is barely present in 1996 and the second absent at that date. Low values for these two classes, however, do not imply that the extracted decision rules involving these classes are erroneous, but they do indicate that these rules are not easily tested and evaluated.

Table 2-2. Transformation rules for land-use groups in 1991-1996

1991	Principal conditions	1996	Cells
Permanent cropping	High level of uncultivated in vicinity	Uncultivated areas	167
Temporary cropping	High level of olive trees in vicinity	Permanent cropping	28
	Medium level of uncultivated in vicinity AND Original parcel medium-to-low in size AND No water in vicinity	Uncultivated areas	15
	Low number of buildings in 500m AND High level of residence in vicinity	Urban areas	68

Table 2.3. Inertial rules for 1991-1996

Land use	Principal conditions	Secondary conditions, if applicable	Cells
Permanent Crops	Near to urban secondary road AND No pastures/meadows in vicinity AND No residence in vicinity AND No water in vicinity AND No uncultivated in vicinity AND No transport in vicinity		173
	Medium-to-high level of olive trees in vicinity		54
	Low level of uncultivated in vicinity AND Medium level of olive trees in vicinity AND No forests in vicinity		40
Temporary Crops	High level of crops in vicinity	AND No olive trees in vicinity AND No uncultivated in vicinity AND No water in vicinity AND No fruit trees in vicinity AND No pastures/meadows in vicinity	2245
		AND Low number of buildings in 500m	223
		AND Low level of fruit trees in vicinity	137
		AND No fruit trees in vicinity	125
	Medium-to-high level of crops in vicinity AND No buildings within 500m AND Poor road condition	AND No water in vicinity	283
		AND No distance from the edge of cell AND No olive trees in vicinity AND No uncultivated in vicinity	103
		No residence in vicinity AND Small distance to the edge of cell	93
	Medium-to-high level of crops in vicinity	AND No buildings within 500m AND Low erosion risk AND No olive trees in vicinity AND No forests in vicinity AND No fruit trees in vicinity AND Original parcel low in size AND Small distance to artificial watering canal	150
		AND No residence in vicinity AND Fair road condition	96
	Medium level of crops in vicinity AND No buildings within 500m AND No olive trees in vicinity AND No uncultivated in vicinity AND No water in vicinity		264

Table 2-4. Transformation rules for land-use groups in 1996-2003

1996	Principal conditions	Secondary conditions, if applicable	2003	Cells
Permanent cropping	Medium-to-high level of fruit trees in vicinity	AND Original parcel medium in size	Uncultivated	48
		AND No additional condition		23
Temporary cropping	100%>slope>75% AND Near to urban main road AND No pastures/meadows in vicinity		Permanent cropping	21
	100%>slope>75% AND Low number of buildings in 500m	AND Original parcel low in size AND Small distance to nearest road	Uncultivated	45
	AND No transport in vicinity AND Medium-to-low erosion risk	AND Original parcel high in size AND No pastures/meadows in vicinity		36
	100%>slope>75% AND Gravel loose road in vicinity AND Medium-to-low erosion risk	AND Low number of buildings in 500m AND No transport in vicinity AND Original parcel medium in size AND No residence in vicinity		71
		AND Very low drainage value AND Medium-to-small distance to natural watering canal		19
	Gravel loose road in vicinity AND No Residence in vicinity	AND Low number of buildings in 500m AND No transport in vicinity AND Original parcel low in size AND No pastures/meadows in vicinity AND Low level of olive trees in vicinity		48
		AND 75%>slope >50% AND Original parcel medium in size		26
	Slope < 25% AND Low number of buildings in 500m AND No forests in vicinity AND Near to urban secondary road AND No transport in vicinity AND No residence in vicinity AND Low drainage value AND No water in vicinity			53
	Medium-to-high distance from artificial watering canal AND Medium distance from nearest road			21
Uncultivated	Paved road in vicinity AND Original parcel medium in size	AND No crops in vicinity	Permanent cropping	49
		AND No pastures/meadows in vicinity AND No forests in vicinity		28

Table 2-5. Inertial rules for 1996-2003

Land use	Principal conditions	Secondary conditions, if applicable	Cells
Permanent cropping	No Fruit trees in vicinity AND Paved road in vicinity AND Medium-to-high distance from artificial watering canal AND Medium-to-low number of buildings in 500m		87
	No Fruit trees in vicinity AND No transport in vicinity AND No Water in vicinity AND No forests in vicinity AND No crops in vicinity AND Medium distance from artificial watering canal		54
Temporary cropping	Slope < 25% AND Original parcel low in size AND No buildings inside	AND Small distance to nearest road AND No water in vicinity AND Low number of buildings in 500m AND No residence in vicinity AND Unpaved road in vicinity	133
		AND Paved road in vicinity AND Medium distance from artificial watering canal	59
Unculti-vated	Unpaved road in vicinity AND Low number of buildings in 500m AND Medium distance from nearest road		64

3.5   Factors in the decision-making process that drive land management

The land-use change dynamics discussed previously, are related to changes in land management that, in turn, are driven by changes in decision making processes. This decision making is influenced by factors at different levels with direct or indirect causes (Lombardo *et al.*, 2002). A number of such factors, relevant for our case, are discussed below. This inventory is based on the findings of the LUP II project inventories and workshops.

The change in economic system in Albania has forced changes at all levels of organisation. Many land users have a sceptical approach to any form of collective action and at receiving advice from government related services. Farmers, for example, are reluctant to organise themselves on a voluntary basis in farmer associations and they hardly use the free agricultural extension services. The general lack of information hampers informed and strategic decision making by the rural households. Economic factors and policies, such as taxes, subsidies, credit access, technology, production and transportation costs, define a range of variables that have a direct impact on the decision making by land users. Market access is largely conditioned by government investments in transportation infrastructure and is identified as one of the major problems and constraints in the communes (Table 2-6). The lack of market access in certain areas has greatly influenced the agricultural production, identified as another major problem and constraint. With mainly semi-subsistence farming and no external demand (or the impossibility to respond to any external demand), the agricultural intensity has decreased dramatically. In the pilot areas, results from the socioeconomic study report that the production of most crops has declined drastically (e.g. wheat by 50%; tobacco, sunflower, sugar beet and soya by 25-33%), whereas the area of forage crops (e.g. alfalfa) increased by 17% and so did livestock production. The only crops experiencing an increase in area and production are vegetables, though mainly used for self-sufficiency purposes. Another result of the land distribution was the changed access to non-land assets such as agricultural equipment. If farmers have no or little access to machinery and labour needs to be executed manually, agricultural production will suffer. Thus, the tendency of rural households active in farming is to move towards a mixture of livestock and forage production. Crop types that are in competition with imports from EU countries in the internal market especially lose out in this competition and, as a result of their low quality and the lack of facilities, cannot be exported to an external market (e.g. CIS countries). It should therefore not come as a surprise that because of the many difficulties, 47% of the rural households in the pilot communes decided to be active in agriculture only part-time. The low agricultural productivity levels can be seen as an indicator of the non-ability of the land users to adapt to changed circumstances as described by Lambin *et al.* (2000).

Erosion and land degradation, flooding and sedimentation (especially in the floodplain of Ana-e-Malit) and pollution and solid waste problems mentioned in Table 2-6 can be seen as other indicators of the fact that, in the pilot communes, the ability to adapt to changed circumstances is very limited.

Table 2-6. Main problems and constraints in the pilot areas as identified by the communes and the LUP II project

Constraints and problems	Preza	Ana-e-Malit	Pirg
Agricultural production	xxx	xx	xxx
Marketing	xx	xxx	xxx
Land tenure (security and size)	xx	xx	xx
Settlement and peri-urban development	xxx	x	xx
Erosion and land degradation	xx	x	xxx
Flooding and sedimentation	x	xxx	x
Pollution and solid waste	xxx	xx	xx

xxx - very serious; xx - serious problem; x - moderate

Another factor influencing the decision making of the land users is land tenure. The farm sizes in the pilot communes are very small: 78% of households have a farm smaller than 1 ha distributed over 3 to 5 land parcels. Correcting land fragmentation is therefore considered important in Albania, as in many other parts of central Europe (van Dijk, 2003). Graefen (2002) confirms that land fragmentation is putting an additional burden on farm management. But the question is if land consolidation is meaningful considering the average farm size of a rural household, i.e. if four parcels of less than 1 ha farm are re-allocated one can still not make a decent living. In such cases, off-farm income can supplement the revenues from the farm, thus overcoming the farm size restriction. Small farms may make sense in some labour-abundant agricultural economies in the short run; in the longer run, the transition to a modern state means that farm size must be sufficiently large (Rozelle and Swinnen, 2005).

4.   CONCLUSION

For the first time in Albania, the temporal and spatial magnitude of change dynamics at cadastral level was studied in three pilot areas.

Modification is the predominant land-use change type and concerns agricultural lands where temporary crops are replaced by permanent crops or vice versa. In the understanding of the change processes of modification, the decision-making processes of the land users play a key role. Development of future trajectories that include intensification of agriculture should consequently include the decision-making processes of these farmers though policies usually address more aggregated levels (e.g. district or national levels). A study carried out at national and district levels may obscure the existing local variability of spatially explicit land-use changes, whereas it may show patterns that, at more detailed data levels, remain invisible (Jansen *et al.*, 2005). Understanding land-use change dynamics is foremost concerned with the quantities of change, i.e. the amount of area changed and the amounts of inputs used and/or production per unit area gained or lost as a function of management level.

In 1991-1996, the observed changes were still influenced by a central planning policy, most likely due to the persisting influence of former officials, technicians and experts still considered to be a reference in land use. With the collapse of central government, the absence of any planning authority and without any improvement in the land market, land uses were mainly preserved where environmental conditions were more favourable, and degradation occurred where environmental conditions were less favourable. With the beginning of a land market and corresponding lack of regulation and legislation in 1996-2003, land-use changes were more dynamic. The greater number of pathways for 1996-2003 seems to confirm that the new landowners of the cadastral parcels each went their own way without any level of governmental land-use planning involved.

The analysis of preferred pathways of change in Preza Commune indicates that the land users take rational decisions when they change land use because of, for example, low suitability or unsuitable soils for a particular use and they seem to abandon steep lands where erosion phenomena manifest themselves. The socioeconomic evolution confirms that before 1991 agricultural output was mainly increased by bringing more (terraced) land into production followed by the intensification of production through fertilizer use and/or irrigation. After 1996, the costs of maintenance of these terraced areas and, more important, the division of this area not according to contour lines but perpendicular to the terracing led to the prevalent use of these areas for pasture. Furthermore, the areas most suitable to agriculture, well served with infrastructure and close to urban centres, have in general maintained their production characteristics. In the case of urbanisation, green areas around buildings have been maintained for production of fruit and vegetables for self-sufficiency purposes of the family household. These developments are especially surprising in the absence of any regulating plan.

Trajectories of land-use change involve both positive and negative human-environment relationships. The extracted rules, i.e. the pathways of change, for Preza Commune could be particularly critical when both types of rules indicate negative developments at national level such as the trend confirming that individuals tend to exploit better environmental conditions for their own benefit while a planning policy should distribute resources and exploitations over the area in a well-balanced manner. Indirectly, these results should stimulate the Albanian Government to develop a land-use policy and strongly invest in land-use planning to prevent the permanent deterioration of the environment with non-reversible transformations.

Land-use change analyses assist the Government in defining those areas where certain land-use processes and patterns are undesired or cause negative environmental impacts that need to be mitigated. It will assist in prioritising areas for the definition of land-use planning interventions in the three pilot communes and development of sustainable future land-use trajectories. Spatial analysis can thus be instrumental in land-use planning and informed decision making. In addition, an analysis of change may not only help to identify vulnerable places but also vulnerable (groups of) people that on their own are incapable of responding in the face of environmental change.

ACKNOWLEDGEMENTS

Two authors worked as consultants in the EU Phare Land Use Policy II project (EU Phare AL98-0502): Giancarlo Carrai was the GIS/IT manager and Louisa J.M. Jansen the land use/land evaluation expert. They wish to gratefully acknowledge the Ministry of Agriculture and Food (MoAF) in

Tirana, Albania for organising the project, and AGROTEC S.p.A. (www.agrotec-spa.net), Rome, Italy, responsible for project execution. Permission to publish was granted by the EU Phare Programme Management Unit and MoAF.

The participants of the 45th ERSA Congress on Land Use and Water Management in a Sustainable Network Society – Land-use Change Modelling Session, 23-27 August 2005, Amsterdam, are thanked for their valuable feedback.

REFERENCES

Agrotec S.p.A. Consortium (2003a) LUP geo-database model - version 1.0, LUP II Technical report No. 1. EU Phare Project 98-0502 Land Use Policy II (author: G. Carrai).

Agrotec S.p.A. Consortium (2003b) A standard methodology for land-use description and harmonisation between data sets, Technical report No. 2. EU Phare Project 98-0502 Land Use Policy II (author: L.J.M. Jansen).

Anderson, J.R., Hardy, E.E., Roach J.T. and Witmer, R.E. (1976) A land use and land cover classification system for use with remote sensor data, U.S. Geological Survey Professional Paper 964. USGS, Washington, D.C.

APA (1999) *Land-based Classification Standard, Second Draft*, American Planning Association, Research Department, Chicago.

Bonchi, F. and Pecori, S. (2003) Data Mining ed estrazione della conoscenza da grandi database territoriali, in Santini, L.G. and Zotta, D. (eds) *Atti della Terza Conferenza Nazionale su Informatica e Pianificazione Territoriale - Input 2003*, 5–7 June 2003, Pisa. Alinea, Firenze (on CDROM).

Booch, G. (1994) *Object-oriented Analysis and Design with Applications*, Second edition, Benjamin/ Cummings Publishing Company, Redwood City, MA.

CEC (1993) *Nomenclature des activités de la communauté européenne (NACE). Première révision.* Règlement 761/93 du Conseil, Commission of the European Communities, Brussels.

CEC (1995) *CORINE – Guide Technique*, Commission of the European Communities, Brussels.

CEC (1999) *CORINE Land Cover*, Commission of the European Communities, Brussels.

Cihlar, J. and Jansen, L.J.M. (2001) From land cover to land use: a methodology for efficient land-use mapping over large areas, *The Professional Geographer*, 53(2): 275–289.

Dale, P.F. (1995) *Land reform, land registration and the cadastre – a comparative analysis between Bulgaria, Hungary and Rumania* (downloadable from OICRF website).

ECE-UN (1989) Proposed ECE Standard International Classification of Land Use, Economic Commission for Europe of the United Nations, Geneva.

FAO (1998) FAO Statistical Databases (FAOSTAT), FAO, Rome (www.fao.org).

Graefen, C. (2002) Land reform and land fragmentation and consequences for rural development in the CEE/CIS countries, Paper prepared for the International Symposium by FAO, GTZ, FIG, ARGE Landentwicklung und Technische Universität München, 25–28 February 2002, Munich.

Han, J. and Kamber, M. (2000) *Data Mining: Concepts and Techniques*, Morgan Kaufmann Publishers, San Mateo, CA.

IGU (1976) World land use survey. Report of the Commission to the General Assembly of the International Geographic Union, *Geographica Helvetica*, 1: 1–28.

INSPIRE Environmental Thematic Coordination Group, (2002) Environmental thematic user needs – position paper, version 2, European Environmental Agency, Copenhagen.

Jansen, L.J.M., Carrai, G., Morandini, L., Cerutti, P.O. and Spisni, A. (2005) Analysis of the spatio-temporal and semantic aspects of land-cover/use change dynamics 1991–2001 in Albania at national and district levels, *Environmental Monitoring and Assessment*, 119: 107–136.

Jansen, L.J.M. and Di Gregorio, A. (1998) Land-use and land cover characterisation and classification: the need for baseline data sets to assess future planning, in *Proceedings of the IGBP-DIS and IGBP/IHDP-LUCC Data Gathering and Compilation Workshop*, 18–20 November 1998, Institute for Geography of Catalonia, Barcelona, Spain.

Jansen, L.J.M. and Di Gregorio, A. (2002) Parametric land cover and land-use classifications as tools for environmental change detection, *Agriculture, Ecosystems & Environment*, 91(1–3): 89-100.

Jansen, L.J.M. and Di Gregorio, A. (2003) Land-use data collection using the "Land Cover Classification System" (LCCS): results from a case study in Kenya, *Land Use Policy*, 20(2): 131–148.

Jansen, L.J.M. and Di Gregorio, A. (2004) Obtaining land-use information from a remotely sensed land cover map: results from a case study in Lebanon, *International Journal of Applied Earth Observation & Geoinformation*, 5: 141–157.

Lambin, E.F., Rounsevell, M.D.A. and Geist, H.J. (2000) Are agricultural land-use models able to predict changes in land-use intensity? *Agriculture, Ecosystems & Environment*, 82: 321–331.

Lambin, E.F., Geist, H.J. and Lepers, E. (2003) Dynamics of land-use and land-cover change in tropical regions, *Annual Reviews of Environmental Resources*, 28: 205–241.

Lombardo, S., Pecori, S. and Santucci, A. (2002) The competition for land between urban and rural systems investigated by G.I.S. and data-mining techniques, *Decision Making in Urban and Civil Engineering*, Vol. 1 and 3, London.

Meyer, W.B. and Turner, B.L. (eds) (1994) *Changes in Land Use and Land Cover: A Global Perspective*, Cambridge University Press, Cambridge, UK.

Rozelle, S. and Swinnen, J.F.M. (2005) Success and failure of reform: insights from the transition of agriculture, *Journal of Economic Literature* (in press).

Swinnen, J.F.M. (1999) The political economy of land reform choices in Central and Eastern Europe, *Economics of Transition*, 7(3): 637–664.

Swinnen, J.F.M. (2000) Political and economic aspects of land reform and privatization in Central and Eastern Europe, Paper prepared for the UDMS Conference on Land markets and Land Consolidation in Central Europe, UDMS, Delft.

UN (1989) *International Standard Classification of All Economic Activities (ISIC)*, Third revision, United Nations Statistics Division, Statistical Classifications Section, New York.

UN (1998) *Central Product Classification (CPC). Version 1.0*, United Nations Statistics Division, Statistical Classifications Section, New York.

UNEP/FAO (1994) Report of the UNEP/FAO expert meeting on harmonizing land cover and land use classifications, 23–25 November 1993, Geneva. *GEMS Report Series* No. 25.

van Dijk, T. (2003) Scenarios of Central European land fragmentation, *Land Use Policy*, 20(2): 149–158.

Veldkamp, A., Kok, K., De Koning, G.H.J., Verburg, P.H. and Bergsma, A.R. (2001) The need for multi-scale approaches in spatial specific land-use change modeling, *Environmental Modeling & Assessment*, 6: 111–121.

Verburg, P.H., De Groot, W.T. and Veldkamp, A. (2003) Methodology for multi-scale land-use change modelling - concepts and challenges, in Dolman, A.J., Verhagen, A., and Rovers, C.A. (eds) *Global Environmental Change and Land-use*, Kluwer Academic Publishers, Dordrecht. pp. 17–51.

Walker, B., Steffen, W., Canadell, J. and Ingram, J. (eds) (1997) *The Terrestrial Biosphere and Global Change: Implications for Natural and Managed Ecosystems*, Synthesis volume, IGBP Book Series 4, Cambridge University Press, Cambridge.

Walker, B. (1998) GCTE and LUCC – a natural and timely partnership, *LUCC Newsletter*, 3: 3–4.

Chapter 3

DRIVING FORCES OF LANDSCAPE CHANGE IN THE URBANIZING LIMMAT VALLEY, SWITZERLAND

A.M. Hersperger and M. Bürgi
Research Unit of Land Use Dynamics, Swiss Federal Research Institute (WSL), Birmensdorf, Switzerland

Abstract: This research aims to identify the driving forces that changed the Limmat Valley west of Zurich from a traditional agricultural valley in 1930 to a suburban region of the city of Zurich in 2000. The landscape changes are quantified based on the comparison of historical maps from 1930, 1956, 1976 and 2000. All individual changes, relating to new or disappeared landscape elements, were linked to a set of one to fourteen critical driving forces. Such an analytical and systematic study of driving forces, and specifically the clear link between changes of landscape elements and specific sets of driving forces and jurisdictional levels, is essential for the development of innovative land-use change models.

Key words: Greening; landscape ecology; land-cover change; peri-urban; rural-urban interface; urbanisation; agricultural intensification.

1.   INTRODUCTION

Over the past decades, urban sprawl and agricultural intensification have changed the traditional cultural landscape of the Swiss lowlands (Bundesamt für Raumentwicklung und Bundesamt für Umwelt Wald und Landschaft, 2001) and many other regions in Central Europe (Antrop, 2004) enormously. Many changes are associated with loss of biodiversity and quality of life and are therefore ecologically and socially undesirable. Since these developments occurred in Switzerland despite a well organized planning

E. Koomen et al. (eds.), Modelling Land-Use Change, 45–60.
© 2007 *Springer*.

apparatus in place, the question arises as to what extent the spatial development of the landscape was actually directed and controlled. In order to better understand the potential of planning and policy interventions for landscape development, we need to understand the underlying causes, i.e. driving forces, of landscape change.

Driving forces are the forces that cause landscape changes (Bürgi *et al.*, 2004), i.e. they influence the trajectories of landscape development. The study of driving forces of landscape change has a long tradition in geography and landscape research (Wood and Handley, 2001) and is gaining increasing attention in landscape-change research. Indeed, driving forces have been identified as one of the six core concepts for the modelling of land-use change (Verburg *et al.*, 2004a). The driving forces form a complex system of dependencies and interactions and affect a whole range of temporal and spatial levels. It is therefore difficult to analyse and represent them adequately. Debates on land-use change and their consequences are often confused and predictions misleading, because the relevant levels and important factors are often not specified or known at all (Baudry *et al.*, 1999).

We use an analytical framework in order to systematically approach the study of driving forces (Figure 3-1). Five major types of driving forces are identified: political, economic, cultural, technological, and natural driving forces, including driving forces that derive from the spatial configuration (this type is referred to as natural/spatial configuration) (Bürgi *et al.*, 2004). The economic driving forces include consumer demands, market structure and structural changes, as well as governmental subsidies and incentives. Political driving forces range from infrastructure policy to policies on nature protection and defense. Cultural driving forces include way of life, demography and the past development of society. Striking examples of how technology is shaping the landscape are the consequences of mechanization of agriculture. The term technology comprises man-made artifacts, knowledge of their production and use, but also social and organizational know-how and techniques (Grübler, 1994). For the natural/spatial configuration driving forces, we distinguish between site factors, such as spatial configuration (existing land uses and transportation networks), topography and soil conditions, and natural disturbances such as global change and mudslides. Calamities or extraordinarily grave events have also been proposed as driving forces (Antrop, 2005). We put natural disasters into the group of natural/spatial configuration driving forces and consider other disasters as rarely big enough to affect an entire landscape.

The five groups of driving forces hold distinctive positions within the framework. The cultural driving forces set the societal framework. The natural/spatial configuration driving forces set the physical background for the other driving forces. Individual actors of landscape change can rarely modify these two groups of driving forces. Political and economic driving forces are strongly interlinked since economic needs and pressures are reflected in political programs and economic instruments are used to implement political driving forces. Technological driving forces become relevant for the landscape in the context of political and economic driving forces.

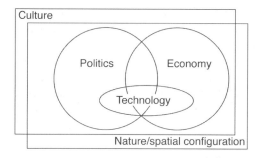

Figure 3-1. Conceptual framework for the five groups of driving forces.

Many studies simplify reality and only focus on a small number of driving forces. This clearly holds true for modelling approaches where statistical methods may limit the number of independent variables (e.g. Alig *et al.*, 2004). But also many descriptive approaches and case studies focus on a few driving forces. This approach can be appropriate if one is interested in the effect of, for example, a specific policy over time (e.g. Bürgi and Schuler, 2003). However, since we are interested in the causes of complex peri-urban landscape changes, we choose a comprehensive approach and include probable driving forces from the five types described above. In order to reduce complexity, we limit our study to the primary driving forces, i.e. driving forces acting directly upon landscape features. The main objectives of this research are:

1. to describe the landscape change from 1930-2000;
2. to identify the most important driving forces of landscape change; and
3. to assess to which type and jurisdictional levels the most important driving forces belong.

2.   STUDY AREA

The study area includes five municipalities of the Limmat Valley, namely Dietikon, Geroldswil, Oetwil an der Limmat, Spreitenbach, and Würenlos (Figure 3-2). The study area is part of the peri-urban area of the city of Zurich and is located approximately 13 kilometres from the centre of Zurich. The study area covers 31.6 km^2 with roughly 42,300 inhabitants (Census 2000). Today, the region is dominated by transportation infrastructure such as national and regional railway tracks, a national highway and a large regional freight train station (Koch *et al.*, 2003). In 1930, four of the five municipalities were small rural villages (Haller, 1957), with some industry in one municipality. The total population was 9,063. Since the 1930s, the study area experienced tremendous changes in land cover and land use. Only the forest cover stayed about the same, due to the strong legal protection of all forested land in Switzerland.

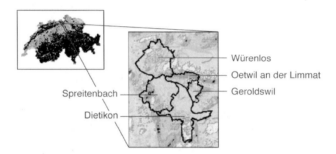

Figure 3-2. The study area of five municipalities located in the Swiss lowlands.

The study area, like the entire Swiss lowlands, has been subject to three main trends of landscape change (or lines of development): urbanisation, agricultural intensification and greening. Urbanisation is a complex process that transforms the rural landscape into an urban and industrial one. Characteristic is the increase in number of buildings and in transportation infrastructure. The trend of agricultural intensification finds its expression in a general decline of many elements of the traditional agricultural landscape such as hedgerows, orchards, solitary trees and stone walls. An opposite trend, greening, encompasses the reverse development in which new landscape elements appear in the agricultural landscape. The parallel existence of these two trends in the Swiss landscape is also reflected in official statistical reports (Bundesamt für Raumentwicklung und Bundesamt für Umwelt Wald und Landschaft, 2001).

3. METHOD

3.1 Data collection

To reduce time costs of data collection, we did not record the changes for the entire study area, but we selected 32 sample plots of 500 metres x 500 metres within the study area with a stratified random sampling scheme. As additional selection criteria, the plots had to be located completely within a single municipality and the proportion in forest had to be less than 50%.

We split the time from 1930 to 2000 into three periods of roughly 25 years and studied the changes for 1930-1956, 1957-1976 and 1977-2000. The limits of the periods were determined by the date of publication of the analysed maps. We studied all landscape elements of the Swiss topographical map on the scale of 1:25 000 (so-called 'Siegfriedkarte' for the years 1930/1931, National Maps 'LK 25' for the years 1956, 1976, and 2000). The analysis included point features such as buildings, ponds and solitary trees; line features such as highways, hedgerows and power lines; and area features such as vineyards and gravel pits.

Changes of the physical landscape elements in the sample plots were analysed based on a manual pair-wise comparison of the maps (comparing the maps issued in 1956 with the maps issued in 1930, 1976 with 1956, and 2000 with 1976). For each landscape element and each period, change was reported in one of the following categories: 'element new', 'element lost with built-up' and 'element lost without built-up'. Changes for line and area features were counted but not quantified in length and area.

Based on general information about the study area and the major landscape changes, a comprehensive list of the driving forces acting in peri-urban areas of the Swiss lowlands was compiled by the authors and verified and completed by experts with local expertise and experts in the relevant subjects (Hersperger and Bürgi, submitted). Expert opinions were used as a source of evidence according to Yin (2003). The driving forces were assigned to one of the following groups: political, economic, cultural, technological and natural/spatial configuration (Bürgi *et al.*, 2004). The driving forces were also assigned to a jurisdictional level, according to their origin, namely local, cantonal, national and international. Therefore, national transportation and infrastructure policy and cantonal transportation and infrastructure policy, for example, were treated as two distinct driving forces. Driving forces such as policies, taxes and subsidies, clearly originate from a jurisdiction because they are created within a legal framework of a jurisdiction. Other driving forces, such as ecological awareness and technological modernization, are not created within a legal framework of a jurisdiction. Nevertheless, they were assigned to a jurisdictional level since jurisdictional levels also represent a territory and its population. Seventy-three driving forces have been identified as relevant for landscape change in the Swiss urbanizing lowlands. Of the 73 driving forces, 52 were found to be relevant primary driving forces (acting directly upon the landscape feature) for the observed landscape changes in the Limmat valley. The remaining 21 driving forces act as secondary and tertiary driving forces. For example, international trade agreements affect national agricultural policy and the aesthetic quality of certain locations influences local land-use planning. Such secondary and tertiary driving forces have been excluded from the analysis.

3.2   Analysis

A two-step procedure was chosen for the analysis. First, with the help of experts (scientists and representatives of the local and cantonal administration), we assigned the relevant driving forces to each recorded landscape change. In order to accomplish that, we sorted all changes into 55 subsystems. The subsystems are characterized by a unique set of driving forces acting upon them. Examples of subsystems and driving forces, 'lost with built-up' (acted upon by 11 driving forces), 'new building before zoning was in effect' (acted upon by seven driving forces), and 'new connecter to foster mobility' (acted upon by eight driving forces). Since we were able to unambiguously assign all recorded landscape changes to a subsystem and each subsystem to one of the three major trends of change, each change recorded from the maps with its associated driving forces was directly assigned to one of these three major trends.

Second, the data was aggregated. The importance value of every driving force was determined for each of the three trends. The importance value of a driving force, IV(d), was calculated as:

$$\mathrm{IV}(d) = \sum_{l=1}^{L} \frac{1}{\sum_{d=1}^{D} I_{dl}} * I_{dl} \tag{1}$$

where D is the number of driving forces, L the number of landscape changes, and the impact $I_{dl} = 1$ if d affects landscape change l, and $I_{dl} = 0$ if d does not affect l. The importance value was adjusted for the number of years per period. In order to compare the results among periods, all importance values were transformed into a percentage of the importance value of the most significant driving force of a given period and trend. Driving forces with value <20% were considered marginal and therefore dropped from further analysis and discussion.

4. RESULTS

For the whole study period of 1930 to 2000, a total of 2,746 changes were recorded in the 32 sample plots (Figure 3-3). Of these changes, 2,087 can be attributed to urbanisation, 445 to agricultural intensification, and 214 to greening. In all three periods, urbanisation was the most important trend in absolute and relative terms, followed by agricultural intensification and greening. Urbanisation, namely the addition of new buildings and roads as well as the upgrading of existing roads, was comparatively slow in 1930-1956, strongly increased in 1957-1976, and slowed down since then. However, the total amount of change per year in the sampled 8 km^2 (32 plots of 0.25 km^2 each) was still considerable, i.e. reaching more than 30 changes per year in the last period. The trend of agricultural intensification was strongest in 1930-1956 and weakest in 1957-1976. However, the absolute number of changes contributing to this trend was about the same in all three periods. The trend of greening was steadily gaining importance and in the last period it reached the same values as agricultural intensification.

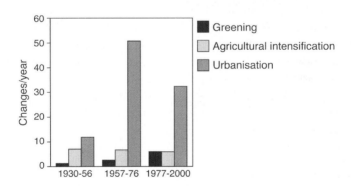

Figure 3-3. Trends of landscape change in 1930-1956, 1957-1976 and 1977-2000.

4.1 Urbanisation

In every period, up to 34 driving forces contributed to urbanisation. However, it is important to note that since 1957, a group of 11 driving forces clearly dominated urbanisation. These 11 most important driving forces contributed together 82% (1930-1956), 91% (1957-1976) and 88% (1977-2000) to the changes. The most important driving forces for urbanisation in all three periods were way of life, transportation and infrastructure policy, and the spatial configuration (Table 3-1). In the period 1977-2000, 12 driving forces were important (>20% of the most important driving force), in the period 1957-1976, 11 were important and in the period 1930-1956, 17 were identified. All 11 important driving forces during the period 1957-1976 were important in the other two periods, stressing their overall importance for urbanisation. Important driving forces originated from local to national levels. However, no international driving forces were among the most important ones. A broad range of political, economic, cultural and natural/spatial configuration driving forces were listed, but no technological driving force made the list.

Table 3-1. Most important driving forces for urbanisation

Driving force	Level	Type	1930-1956	1957-1976	1977-2000
Way of life	n	c	98	98	100
Transportation and infrastructure policy	c	p	100	100	98
Spatial configuration	l	n	96	98	97
Spatial configuration	c	n	96	98	97
Demography	c	c	94	97	92
Structural change in industry, commerce, trade and service	n	e	93	96	92
Taxes and subsidies	l	e	72	87	83
Property market	c	e	72	87	83
Taxes and subsidies	c	e	72	87	83
Local land use planning	l	p	33	85	82
Subsidies in agriculture and forestry	c	e	28	76	71
Demand for recreational, cultural and tourism facilities	l	c	50		25
Property market	l	e	44		
Regional competition	l	p	27		
Topography	c	n	24		
Topography	l	n	24		
Regional competition	c	p	24		

The importance of a driving force is indicated in percent of the most important driving forces
Level: n = national, c = cantonal, l = local. Type: p = political, e = economic, c = cultural, n = natural/spatial configuration

4.2   Agricultural intensification

In every period, 11 to 13 driving forces contributed to agricultural intensification. There was a more or less steady decline in importance from the most important to the least important driving force. However, the most important eight driving forces contributed 92% (1930-1956), 94% (1957-1976) and 93% (1977-2000) to agricultural intensification. In the period 1977-2000, nine driving forces were important (>20% of the most important driving force), in 1957-1976, ten and in 1930-1956, nine (Table 3-2). Seven driving forces were important in all three periods, a fact that stresses their overall importance for agricultural intensification. Clearly, the most important driving force for agricultural intensification was the technological modernization in agriculture. In 1930-1956, technological modernization was followed by the economic driving forces, i.e. prices/price relations and consumer demand. Since then, technological modernization in agriculture was followed by the political driving force of agricultural policy. Important driving forces originated from cantonal to international levels. However, no local driving forces were listed. A broad range of political, economic and technological driving forces were listed - whilst natural/spatial configuration driving forces did not make the list of driving forces responsible for agricultural intensification.

Table 3-2. Most important driving forces for agricultural intensification

Driving force	Level	Type	1930-1956	1957-1976	1977-2000
Technological modernization in agriculture	i	t	100	100	100
Agricultural policy	c	p	74	78	84
Agricultural policy	n	p	74	78	84
Structural change in agriculture and forestry	n	e		42	47
Prices/price relations	n	e	78	48	32
Consumer demand	n	e	78	48	32
Forestry policy	n	p	31	23	31
Prices on the world market	i	e	69	42	31
Forestry policy	c	p			16
Rural technologies	n	t	21		
Rural technologies	i	t	21		

The importance of a driving force is indicated in percent of the most important driving forces
Level: i = international, n = national, c = cantonal. Type: p = political, e = economic, t = technological

4.3 Greening

In every period, 11 to 18 driving forces contributed to greening. However, the most important seven driving forces contributed 94% (1930-1956), 89% (1957-1976) and 94% (1977-2000) to greening. From 1930-1956, the two most important driving forces contributed 69%, from 1957-1976, seven driving forces contributed 89%, and from 1977-2000, three driving forces contributed 78%. In 1977-2000, five driving forces were important (>20% of the most important driving force), in 1957-1976, seven and in 1930-1956, four (Table 3-3). Whereas in urbanisation and agricultural intensification, the vast majority of the essential driving forces remained constant, this did not hold true for greening. Technological modernization in agriculture and way of life were the only two driving forces important in all three periods. Between 1930-1956, only these two driving forces were relevant. These driving forces were still important in the following period but in 1976-2000 they were clearly topped by ecological awareness and agricultural policy. Important driving forces originated from cantonal to international levels. However, no local driving forces did make the list. A broad range of political, cultural, and technological driving forces were listed - economic and natural/spatial driving forces did not make the list.

Table 3-3. Most important driving forces for greening

Driving force	Level	Type	1930-1956	1957-1976	1977-2000
Ecological awareness	n	c		74	100
Agricultural policy	c	p		68	97
Agricultural policy	n	p		68	97
Technical modernization of society	i	t	100	100	24
Way of life	n	c	100	84	24
Transportation and infrastructure policy	n	p	23	87	
Transportation and infrastructure policy	c	p	23	71	

The importance of a driving force is indicated in percent of the most important driving forces
Level: i = international, n = national, c = cantonal. Type: p = political, c = cultural,
t = technological

5.   DISCUSSION

Urbanisation has been caused by local, cantonal and national driving forces, whereas agricultural intensification and greening have been caused by cantonal, national, and international driving forces. The strong impact of local driving forces on urbanisation is not surprising since the Swiss political system provides maximum autonomy to the municipalities in planning their built environment. The strong municipal federalism has been identified as one of the major causes of the sprawling growth patterns in Switzerland (Bundesamt für Raumentwicklung, 2005; Diener *et al.*, 2005). Indeed, the municipalities compete with each other in attracting businesses and good taxpayers with low taxes, high subsidies and attractive land-use planning (all identified as important driving forces for urbanisation). Local driving forces are shown to be unimportant for landscape changes associated with agricultural intensification and greening in the past 70 years. However, this might change in future. Demands for recreation, conservation, aesthetic experiences, locally produced fruits and vegetables *et cetera* already have increased. Therefore, local political and economic influence on the agricultural landscape might increase. Such a development would reflect the increasing multifunctionality of open landscapes such as the urbanizing Limmat Valley, where in addition to agricultural production a whole range of uses gains importance (van Huylenbroek, 2003).

5.1   Urbanisation

Our research shows that various driving forces have been important for urbanisation. Many of them, sometimes in some variations, have been found to be important in other studies. For example, Antrop (2005) and Verburg *et al.* (2004b) identified accessibility, and especially the road network, as a major driving force for the development of cities. In our study, transportation and infrastructure policy as well as the local and regional spatial configuration of the landscape (existing land uses and transportation networks) describe accessibility. Spatial configuration also refers to neighbourhood interactions. Neighbourhood interactions were found to contribute to explaining the location of further development (Verburg *et al.*, 2004b). These interactions are important because some existing land-use configurations in a neighbourhood attract further development, whereas others discourage development (Hersperger, submitted). Neighbourhood interactions therefore should be kept in mind for the analysis of land-use change and the planning of future developments. Way of life refers to changes in lifestyle which have occurred since 1930, and especially since 1950. These changes encompass, for example, the increasing demand of living space per person and the shrinking number of people per household. Specific, easily measurable socioeconomic factors included in this driving force have been identified in other studies (e.g. Nikodemus *et al.*, 2005; Alig *et al.*, 2004). The construction of new buildings and roads has many effects on the natural and the socioeconomic environment. Habitat loss and alteration, energy consumption, pollution, public health, social equity, aesthetics and sense of community are crucial issues (Gillham, 2002). In peri-urban landscapes, roads cause the removal and dissection of scarce natural habitat, the disruption of species movement and the reduction of biodiversity through traffic noise levels (Forman *et al.*, 2002, Forman and Hersperger, 1996).

5.2   Agricultural intensification

Since 1930, technological modernization in agriculture was the single most important driving force for agricultural intensification. The importance of this driving force has also been established, for example, by Schneeberger *et al.* (submitted). Agricultural policy on the cantonal and national level recently became more important while economic driving forces lost importance. Since the 1930s, Swiss agricultural policy has largely been driven by the goal to secure the capacity for Swiss food production for times of limited import and by the goal to improve the situation for farmers in climatically and topographically disadvantaged regions (Brugger, 1992). Since the Swiss Government slowly tries to open up the agricultural market to international access, the importance of the political driving forces for agricultural intensification might decline in future. Interestingly, we found that natural driving forces, i.e. topography and soil conditions, were unimportant for agricultural intensification in the Limmat Valley. Our result is consistent with the fact that agricultural intensification occurred in a consistent manner in all agriculturally used areas throughout the Swiss lowlands. This effect might be due to the strong influence of agricultural policy and a comparatively weak influence of economic factors on Swiss agriculture.

5.3   Greening

The crucial sets of driving forces for urbanisation and for agricultural intensification remain fairly constant. In greening, however, the crucial set of driving forces changes over time. The critical driving forces for 1930-1956 are different from those in 1977-2000. In the intermediate period from 1957-1976 we find the driving forces of the previous and the following period, i.e. we see a major shift in the relevant driving forces over the course of our study period. Before 1956, the changes in greening were primarily a result of a declining intensity of agriculture on comparatively marginal sites, especially at steep slopes. Since the 1980s, the changes in greening are primarily the result of conscious efforts to improve the ecological value of the landscape. They are implemented with various agri-environment schemes (Herzog, 2005). Between 1957 and 1976, we see an overlap of both these trends. This change in the set of important driving forces raises an important issue for modelling future landscape change. Often modelling is based on an analysis of driving forces of past land-use change. Consequently, forecasts will be correct only if one deals with a set of driving forces which remains stable over a long time period. This supports the request by Verburg *et al.* (2004a), to pay attention to temporal dynamics in a new generation of land-use models.

5.4   Methodological issues

The framework used in our study (Bürgi *et al.*, 2004) proved to be very useful to go beyond simply documenting the change in urbanisation, agricultural intensification and greening, and to systematically investigate the causes of recorded changes. Grouping the individual changes in subsystems characterized by a clearly defined set of driving forces, and relating these changes to three major trends of landscape change known to stand for core processes of changes in similar regions, enables us to interpret and to discuss individual changes in a more general context. In all studies of land-use change, it is pivotal to distinguish between correlation and causality (Bürgi and Russell, 2001). In our study, the causal relationships between driving forces and landscape change have been established as we determined the most important driving forces that act upon the subsystems (together with experts). This approach proved to be very suitable. So far, expert knowledge to quantify the relations between driving forces and land-use change has primarily been used in models that use cellular automata (Verburg *et al.*, 2004a). Our experience suggests that the use of expert knowledge might have a larger potential. We chose five municipalities of 31.6 km^2 in total to analyse the driving forces of landscape change. Such a small study area is rather unusual in land-use and land-cover change studies that generally focus on extensive regions (e.g. Schneider and Pontius, 2001), entire nations (e.g. Verburg *et al.*, 2004b), or even larger units (e.g. Klein Goldewijk and Ramankutty, 2004). However, local differences in drivers and processes strongly determine what has happened and what will happen (Leemans and Serneels, 2004). Since our study covered only a small area, our findings are rather explicit and therefore especially interesting for (local) politicians and planners. Moreover, we expect that our main conclusions hold true for other areas of the urbanizing Swiss lowlands with a similar economic, political, and cultural structure.

6.   CONCLUSION

The different trends of landscape change are characterized by quite distinct sets of driving forces. For example, the set of important driving forces for urbanisation and agricultural intensification are mutually exclusive. Greening, however, shares driving forces with the other two trends. Our findings support the statement of Verburg *et al.* (2004a) that models integrating the analysis of different trends of land-use change need a larger set of driving forces than for specific single-trend analyses. Our study also revealed that some trends of landscape change are caused by a specific and temporally constant set of driving forces (e.g. urbanisation), whereas others trends reveal a change in relevant driving forces over time (e.g. greening). We suggest that such changes in relevant driving forces over time indicate paradigmatic shifts. The existence of trend-specific paradigmatic shifts in landscape change in the course of only a few decades does stress the importance of thorough historical analysis of landscape changes, and has to be carefully considered in the design and validation of models of land-use change. We are convinced that the analytical and systematic approach to study driving forces, and specifically the clear link of changes of landscape elements to specific sets of driving forces, is essential for the development of innovative land-use change models.

ACKNOWLEDGEMENTS

This paper is a contribution to the WSL-Research focus 'Land Resources Management in Peri-Urban Environments' (www.wsl.ch/programme/peri-urban). Many thanks to Irmi Seidl, Nina Schneeberger and Werner Spillmann and various experts from the WSL-Research focus for their support.

REFERENCES

Alig, R.J., Kline, J.D. and Lichtenstein, M. (2004) Urbanisation on the US landscape: looking ahead in the 21st century, *Landscape and Urban Planning*, 69: 219–234.

Antrop, M. (2004) Landscape change and the urbanisation process in Europe, *Landscape and Urban Planning*, 67: 9–26.

Antrop, M. (2005) Why landscapes of the past are important for the future, *Landscape and Urban Planning*, 70: 21–34.

Baudry, J., Laurent, C., Thenail, C., Denis, D. and Burel, F. (1999) Driving factors of land-use diversity and landscape pattern at multiple scales - a case study in Normandy, France, in Krönert, R. (ed) *Land-use Changes and Their Environmental Impact in Rural Areas in Europe*, UNESCO, pp. 103–119.

Brugger, H. (1992) *Agrarpolitik des Bundes seit 1914*, Huber, Frauenfeld.

Bundesamt für Raumentwicklung (2005) *Raumentwicklungsbericht*, Bundesamt für Raumentwicklung (ARE), Switzerland.

Bundesamt für Raumentwicklung und Bundesamt für Umwelt Wald und Landschaft (2001) *Landschaft unter Druck: 2. Fortschreibung*, Bundesamt für Raumentwicklung und Bundesamt für Umwelt Wald und Landschaft, Switzerland.

Bürgi, M., Hersperger, A.M. and Schneeberger, N. (2004) Driving forces of landscape change - current and new directions, *Landscape Ecology*, 19: 857–868.

Bürgi, M. and Russel, E.W.B. (2001) Integrative methods to study landscape changes, *Land Use Policy*, 18: 9–16.

Bürgi, M. and Schuler, A. (2003) Driving forces of forest management - an analysis of regeneration practices in the forests of the Swiss Central Plateau during the 19th and 20th century, *Forest Ecology and Management*, 176(1–3): 173–183.

Diener, R., Herzog, J., Meili, M., de Meuron, P. and Schmid, C. (2005) *Die Schweiz: Ein städtebauliches Portrait*, Birkhäuser.

Forman, R.T.T. and Hersperger, A.M. (1996) Road ecology and road density in different landscapes, with international planning and mitigation solutions, in Evink, G. L., Garrett, P., Zeigler, D. and Berry, J. (eds) *Trends in Addressing Transportation Related Wildlife Mortality*, State of Florida, Department of Transportation, pp. 1–22.

Forman, R.T.T., Reineking, B. and Hersperger, A.M. (2002) Road traffic and nearby grassland bird patterns in a suburbanizing landscape, *Environmental Management*, 29: 782–800.

Gillham, O. (2002) *The Limitless City: A Primer on the Urban Sprawl Debate*, Island Press, Washington DC.

Grübler, A. (1994) Technology, in Meyer, W.B. and Turner II, B.L. (eds) *Changes in Land Use and Land Cover: A Global Perspective*, Cambridge University Press, pp. 287–328.

Haller. (1957) *Zur Geographie der Region zwischen Zürich und Baden*, Buchdruckerei J. Lerchmüller-Müri.

Hersperger, A.M. (2006) Spatial adjacencies and interactions: neighborhood mosaics for landscape ecological planning, *Landscape and Urban Planning*, 77(3): 227–239.

Hersperger, A.M. and Bürgi, M. (submitted) Going beyond landscape change description: a case study of analyzing driving forces of landscape change in Central Europe.

Herzog, F. (2005) Agri-environmental schemes as landscape experiments, *Agriculture, Ecosystems & Environment*, 108: 175–177.

Klein Goldewijk, K. and Ramankutty, N. (2004) Land cover change over the last three centuries due to human activities: the availability of new global data sets, *GeoJournal*, 61: 335–344.

Koch, M., Schröder, M., Schumacher, M. and Schubarth, C. (2003) Zürich/Limmattal, in Eisinger, A. and Schneider, M. (eds) *Stadtland Schweiz*, Birkhäuser, pp. 236–271.

Leemans, R. and Serneels, S. (2004) Understanding land-use change to reconstruct, describe or predict changes in land cover, *GeoJournal*, 61: 305–397.

Nikodemus, O., Bell, S., Ineta, G. and Ingus, L. (2005) The impact of economic, social and political factors on the landscape structure of the Vidzeme Uplands in Latvia, *Landscape and Urban Planning*, 70(1–2): 57–67.

Schneeberger, N., Bürgi, M., Hersperger, A.M. and Ewald, K.C. (submitted) Driving forces and rates of landscape change as a promising combination for landscape change research - An application on the northern fringe of the Swiss Alps, *Land use policy*.

Schneider, L. and Pontius Jr., R.G. (2001) Modeling land-use change in the Ipswitch watershed, Massachusetts, USA, *Agriculture, Ecosystems & Environment*, 85: 83–94.

Van Huylenbroek G. (ed) (2003) *Multifunctional Agriculture: A New Paradigm for European Agriculture and Rural Development*, Ashgate, Aldershot.

Verburg, P.H., Schot, P., Dijst, M. and Veldkamp, A. (2004a) Land use change modelling: current practice and reseach priorities, *GeoJournal*, 61: 309–324.

Verburg, P.H., Ritsema van Eck, J.R., de Nijs, T.C.M., Dijst, M.J. and Schot, P. (2004b) Determinants of land-use change patterns in the Netherlands, *Environment and Planning B*, 32: 125–150.

Wood, R. and Handley, J. (2001) Landscape dynamics and the management of change, *Landscape Research*, 26: 45–54.

Yin, R.K. (2003) *Case Study Research: Design and Methods*, Sage, Thousand Oaks, CA.

Chapter 4

LANDSCAPE CHANGES IN THE ISRAELI CARMEL AREA

An application of matrix land-use analysis

M. Sonis[1], M. Shoshany[2] and N. Goldshlager[3]

[1]*Department of Geography, Bar-Ilan University, Ramat-Gan, Israel;* [2]*Department of Transportation and Geoinformation Engineering, Technion, Israel Institute of Technology, Haifa, Israel;* [3]*Soil Erosion Station, Israeli Ministry of Agriculture, Emek-Hefer, Israel*

Abstract: The spatial redistribution of land uses can be measured by conventional remote sensing methods in the form of matrices of land-use redistributions within a given set of regions in a given time period. Two new methods of analysis of such land-use redistribution matrices are proposed: the 'superposition principle' and the 'minimum information' approach. For an empirical validation of these new methods, a settlement in the vicinity of the Haifa Carmel area is chosen. For different time intervals, the main trends in land-use redistribution are identified together with their minimum information artificial landscapes.

Key words: Matrix land-use analysis; superposition principle; minimum information approach; land-use change; Israeli Carmel area.

1.   INTRODUCTION

Over the last 50 years, Mediterranean landscapes have undergone major changes, mainly because of population growth and economic and social development. The existence of natural, agricultural and historical landscapes (Naveh and Kutiel, 1990; Perevolotsky *et al.*, 1992; Grossman *et al.*, 1993) is severely endangered due to these processes. Further complication results from a combination of factors not least associated with Israel being the focal point of intensive democratic and political change during the last 100 years, and its location in the transition zone between arid and Mediterranean climates.

E. Koomen et al. (eds.), Modelling Land-Use Change, 61–82.
© 2007 *Springer.*

The Carmel area (including Haifa's periphery) represents a system of varied Mediterranean landscapes, differentiated by soil conditions and vegetation, and by the anthropogenic activities that have taken place over the last 100 years. The main anthropogenic process influencing the rural system is accelerated urbanisation, in addition to agricultural regression due to national and global transformation. These processes have been in conflict with nature preservation efforts including legislation relating to reserves and parks. The current research objective is to provide a quantitative description of this conflict evolution from 1940 to 1990. It assesses rates of landscape transition in general and that of vegetation change in particular for a typical area in the Mount Carmel region.

The methodology employed here is a combination of multidate aerial photographic interpretation with analysis of temporal change using geographical information systems (GIS) and matrix land-use analysis as was developed by Sonis (1980) for migration matrix analysis and economic input-output analysis (Sonis *et al.*, 2000). Matrix land-use analysis utilizes the matrices of land-use redistribution within a given set of regions in a given time period. This type of analysis includes two new approaches to the analysis of land-use redistribution matrices: the 'superposition principle' and the 'minimum information' approach.

The first approach represents the geometric and analytical algorithm of decomposition of the land-use redistribution matrix into the convex combination of the land-use matrices which represent the main tendencies of land-use redistribution in a given set of regions in a given time period. Thus, each empirically given land-use redistribution can be presented as a superposition of the land-use redistributions connected with the optimal solutions of some extreme land-use redistribution corresponding to the parsimonious behaviour of land users in a given set of regions in a given time period.

The second approach represents the construction of an artificial land-use landscape corresponding to the minimum information land-use redistribution with fixed initial and final land-use distributions. The comparison of the empirical landscape with the artificial one represents the spatial specifics and preferences of an actual land-use redistribution connected with different parsimonious behaviour of the land users themselves.

For an empirical validation of these new methods we have selected a typical site in the Mount Carmel area: Zichron Ya'acov. This urban settlement within the urban-rural fringe of Haifa has experienced similar developments to other sub-areas in the Haifa periphery. For the selected site, the main tendencies of land-use redistribution for the 1944-1990 period are identified together with their minimum information artificial landscapes. Before we arrive at the description of these results we will first introduce the study area and research methodology. The chapter finishes with a discussion of the most important land-use transitions and the merits of the proposed methodology.

2.   STUDY AREA

The Carmel Mountain (240 km^2) is a triangle-shaped mountain. Its apex is Haifa and its base is along the Yoqne'am-Zichron Ya'acov road. The region consists of a mix of urban and agricultural areas (but more natural vegetation landscapes prevail), the latter comprising partly of nature reserves, national parks and forest plantation areas, mainly owned by the Keren Kayemeth LeIsrael (KKL).

This research is concerned mainly with the landscape dynamics occurring in composite areas representing the major conflicting trends of landscape evolution. The historical dimension is essential for understanding these dynamics in general and vegetation recovery and disturbance in particular. It is important to note several major events: the declaration of the area as Forest Reserve by the British administration Forest Act of 1926; the war of independence (1948) which marked a major decrease in the human disturbance to the natural vegetation due to the abandonment of most of the Palestinian villages; and the legislation on nature reserves and national parks in 1966 by an Act of the Israeli Knesset (Parliament). Another important event was the Six Day War (1967), representing a turning point in the processes of land-use change and in the economic and social structure of Israel.

We restrict our consideration to the analysis of the land-use transitions in Zichron Ya'acov: a Jewish settlement experiencing an advanced process of urbanisation. It was established in 1882 at the beginning of the First Aliya (wave of Jewish immigration) and, until the 1970s, the villagers were mostly engaged in the grape-wine growing. In 1990, it had a population of 6,220, but agriculture is now no longer the major economic activity. Most of the inhabitants commute to Haifa, where they are employed in various jobs, but some are able to benefit from the emerging local tourist industry. The Zichron Ya'acov site includes all principal types of anthropogenic activities in the region of the Carmel Zone. This enables us to understand the complexity of the landscape processes throughout the Carmel Zone.

3.   METHODOLOGY

The research methods employed in this study for the construction of an empirical database are based upon the concepts and methodology of GIS, combined with research methods of historical and settlement geography and ecology. This study uses quantitative and periodic analyses for gathering the data. In order to employ these means, two preliminary stages were carried out: identifying and mapping the landscape units using air-photos (similar to the work of Gavish and Sonis, 1979) and establishing a geographic data base. The research methodology consists of the following main steps that are described in more detail in the subsequent sections:

- data collection and processing;
- construction of sum products matrices (SPM) and artificial land-use transition landscapes; and
- decomposition and assemblage analyses.

3.1   Data collection and processing

Air photograph availability was one of the principal criteria for choosing the research sites, delineating their extents and determining the dates included. The only air photographs available from the 1940s were those taken by the British Government in 1944-1945. This survey provides the first full set of air photographs of Israel. The second set of air photographs from 1956 represents the stage of stabilization of settlement activity after the major waves of immigration which entered Israel following the establishment of the State. The air photographs from 1970 were chosen because of their proximity to the period just following the Six Day War when the economy of Israel was reshaped and restructured, and to the time when steps were taken to preserve the Carmel by declaring it a national park and a nature reserve. The last set of air photographs was taken in 1990-1992. Since the various air photographs were taken along different flight paths, the study area size and its location in relation to the settlements were determined according to the overlap between the flight strips. An area of some 8 km^2 was found to represent the average site size, although the site outline is considerably amorphic; it basically formed a circle with a diameter of three kilometres. In these areas, most of the possible types of landscape units were present. The scale of the selected air photographs ranged from 1:10,000 to 1:20,000. The landscape units were defined according to their interpretability from a mirror stereoscope. Table 4-1 presents an overview of the landscape categories that were defined.

Table 4-1. Landscape units description

Unit	Name	Additional remarks
1	Dense vegetation coverage	Includes natural pine forest, oakwood and shrubs
2	Medium vegetation coverage	Includes open oak and pine forests, scattered shrubs
3	Light vegetation coverage	Includes bare areas, grassy meadows and isolated Trees and shrubs
4	Orchards, olive groves	
5	Cultivated fields	
6	Recently abandoned fields	Identification was based upon presence of abandoned irrigation canals and terraces and the existence of randomly scattered sparse bush vegetation
7	Old abandoned fields	Representing the characteristics of previous category, but with heavier vegetation cover coexisting with remnants of historical agricultural systems
8	Built-up areas	High and medium building densities, including commercial shopping centres - functional zones in rural settlements and surrounding urban areas
9	Sparsely distributed buildings	Also with single houses on outskirts of urban areas

The boundaries of the landscape units have been drawn onto a transparency. They were scanned, encoded and georeferenced to form a layer representing the landscape at a single point in time. Combining the layers of the different dates for each site formed a multi-temporal database that provides a clear description of the basic land-use changes over time. The main aggregated average tendencies in land use (Table 4-2) are that natural vegetation (I) is covering about 60% of Zichron Ya'acov area; agricultural uses (II) gradually decrease from nearly 40% to 20%; and built-up area (III) strongly increases from 4% to nearly 30%.

Table 4-2. Temporal change in aggregated land-use shares for Zichron Ya´acov

	1944	1956	1970	1990
I Vegetation (1, 2, 3)	58.6	59.5	63.5	53.7
II Agricultural uses (4, 5, 6, 7)	37.5	29.0	21.6	16.9
III Built-up areas (8, 9)	3.9	11.5	14.9	29.4

3.2   Matrix land-use analysis basics

The statistical data for matrix land-use analysis is presented in the form of matrices of land-use transitions of the following form:

$$M = \left[p_{ij} \right], \; i, j = 1, 2, ..., K \tag{1}$$

Here the relative land-use transition rates p_{ij} possess the following properties of the probabilistic vectors:

$$0 \le p_{ij} \le 1; \quad \sum_{i,j=1}^{K} p_{ij} = 1 \tag{2}$$

where p_{ij} is the relative frequency in percentage of the area changes from landscape category i to category j and K is the total number of landscape units in a given time period.

The row sums $S_{i\bullet}$ give the initial distribution of land-use ratios (*ID*) at the beginning of the time period. The column sums $S_{\bullet j}$ give the final distribution of land-use ratios (*FD*) in the end of the time period:

$$\begin{aligned} ID : S_{i\bullet} &= \sum_{j=1}^{K} p_{ij}, \quad i = 1,2,...,K \\ FD : S_{\bullet j} &= \sum_{i=1}^{K} p_{ij}, \quad j = 1,2,...,K \end{aligned} \tag{3}$$

It is obvious that the vectors *ID* and *FD* of initial and final distributions are probabilistic vectors.

3.2.1   Sum products matrix (SPM) and artificial land-use transition landscape

The land-use transition matrices M and their row and column sums $S_{i\bullet}, S_{\bullet j}$ (Eq. 3) can be used for the calculation of sum products matrices (SPM):

$$S = \left(S_{\bullet i} S_{j \bullet} \right) = \begin{pmatrix} S_{1\bullet} \\ S_{2\bullet} \\ \vdots \\ S_{n\bullet} \end{pmatrix} \left(S_{\bullet 1}, S_{\bullet 2}, \cdots S_{\bullet n} \right) \quad \left(= \left[s_{ij} \right] \right) \tag{4}$$

It is important to emphasise that the column and row multipliers of the SPM are the same as those of the land-use transaction matrix M. The SPM provides a visual representation (artificial landscape) of the structure of land use, giving a basis for the comparison of structures of different land-use transitions in the same area over time. The definition (Eq. 4) defines a specific cross structure of the SPM which will be presented below. First of all, the largest component of the SPM is the product of the largest column and row sums:

$$\max_{ij} s_{ij} = (\max_i S_{i\bullet})(\max_j S_{\bullet j}) = S_{i_I\bullet} S_{\bullet j_I} \tag{5}$$

Moreover, all rows of the SPM are proportional to the row of column sums and the i_I th row, corresponding to the largest row sum $S_{i_I\bullet}$, is the 'biggest' row with the maximal components in each column. Analogously, all columns of the SPM are proportional to the column of the row sums and the j_I th column, corresponding to the largest column sum $S_{\bullet j_I}$, is the 'largest' column with maximal components in each row. These proportionality properties imply that the largest components of the SPM are included in the cross (i_I, j_I) generated by the i_I th row and j_I th column in such a way that for each column (row) of the SPM the largest element lies in the i_I th row (j_I th column). The largest component of the SPM is located in the centre of this cross. Furthermore, if the cross (i_I, j_I) is excluded from the SPM, then the next cross (i_{II}, j_{II}) will include the largest remaining elements; the same property holds for the succeeding crosses (i_{III}, j_{III}) , (i_{IV}, j_{IV}) ,...., (i_N, j_N).

This cross-structure of the SPM is essential for the visualization of the land-use transition structure with the help of artificial structural landscapes. Essentially, SPM is presented as a three-dimensional picture of land-use transitions; by corresponding manipulation of the row and column ordering, it is possible to directly compare the land-use transition structure of several areas in different time periods. For the construction of these landscapes, one can reorganize the location of rows and columns of the SPM in such a way that the descending sequence of the centres of crosses appears on the main diagonal. This rearrangement also reveals the descending rank-size hierarchies of row and column sums. Moreover, we can consider the rank-size sequences of the components of the SPM and replace the entries with their ranks. On the basis of the rearranged SPM, the three-dimensional diagram of descending economic landscape can be drawn, where the two-dimensional plane represents the hierarchy of column and row sums, and the third dimension - the height of the bars - represents the volume of the products of the column and the row sums. It is important to stress that the construction of artificial landscapes for different regions, or for the same region at different time periods, creates the possibility for the creation of taxonomy of the land-use transitions on the basis of visual representation of the similarities and differences in the structure of transitions.

3.2.2   Maximum entropy property of SPM

Consider all land-use transition matrices $N = \left[r_{ij} \right]$ with the property that the row and column sums are equal to the row and column sums of the concrete land-use transition matrix $M = \left[p_{ij} \right]$:

$$\sum_j r_{ij} = \sum_j p_{ij} = S_{i\bullet}, \; \sum_i r_{ij} = \sum_i p_{ij} = S_{\bullet j} \qquad (6)$$

We can attribute to each positive matrix N the Shannon entropy (here we apply the usual assumption $0 \log 0 = 0$):

$$EntN = -\sum_{i,j} r_{ij} \log r_{ij} \qquad (7)$$

The sum product matrix S has a maximum entropy property (Sonis, 1968; Sonis *et al.*, 2000):

$$EntN = -\sum_{i,j} r_{ij} \log r_{ij} \; \leq -\sum_{i,j} S_{i\bullet} S_{\bullet j} \log S_{i\bullet} S_{\bullet j} = EntS = E_{max} \qquad (8)$$

The proof that this statement is possible, is obtained by direct calculation from the well-known Shannon information inequality (Shannon and Weaver, 1964, p. 51). The SPM matrix S may be considered to present the most homogeneous distribution of the components of the column and row sums of the land-use shares M. Thus, while the SPM does not take into account the specifics of the land-use transformations, it does provide the aggregate representation of land-use equalization tendencies in the spatial interactions between land uses. To emphasise this, let us note that if the land-use transitions matrix M has equal column and row sums, then the artificial land-use landscape will be a flat, horizontal plane.

3.3   Decomposition and assemblage of land-use change

The land-use transformations in the given area during some time interval can be considered from the viewpoints of 'decomposition' and 'assemblage'. Decomposition means the division of a sub-area under some definite land use at the beginning of the time interval into the set of sub-areas under different land uses at the end of the time interval; assemblage means the bringing together all sub-areas under the same land use at the end of time interval into the unified area under one type of land use. This section introduces the topic of spatial representation of decomposition and assemblage in real land-use transformations. We restrict ourselves to a detailed discussion of the analysis of decomposition, since the scheme of assemblage analysis can be considered analogously. The application of those methods is presented in Sections 5 and 6.

3.3.1   Convex polyhedron of admissible land-use transformations

Let us consider the land-use transformations on K different types of land use in a given geographical area in a given time interval. These transformations can be statistically described by the transformation matrix. An initial land-use distribution for decomposition analysis is:

$$S_{i\bullet} = \sum_{j=1}^{K} p_{ij}, \quad i = 1,2,...,K \tag{9}$$

These data allow for the incorporation of the actual state of the land-use system, M, into the polyhedron of admissible states. For the decomposition analysis, the convex polyhedron of admissible states includes the transition matrices $X = \left[x_{ij} \right]$, satisfying the following system of linear constraints:

$$\begin{cases} x_{ij} \geq 0, \quad i,j = 1,2,...,n \\ \displaystyle\sum_{i,j=1}^{K} x_{ij} = 1 \\ \displaystyle\sum_{j=1}^{K} x_{ij} = S_{i\bullet}, \quad i = 1,2,...,K \end{cases} \tag{10}$$

3.3.2   Normalized unit cube of admissible land-use transitions

The description of the polyhedron of admissible land-use transitions (Eq. 10) can be simplified by compressing them into a many-dimensional unit cube of stochastic matrices $R = \left[r_{ij} \right] = \left[x_{ij} / S_{i\bullet} \right]$ for decomposition analysis:

$$r_{ij} \geq 0, \quad i,j = 1,2,...,n$$
$$\sum_{i=1}^{n} r_{ij} = 1, \quad j = 1,2,...,n \tag{11}$$

The correspondence between the matrices X of admissible transition matrices from the polyhedron (Eq. 10) and the stochastic matrices R from the normalized polyhedron (Eq. 11) is one to one; transfer from matrix R to X is easily done by multiplication of rows of the matrix R on the sums $S_{i\bullet}$. The unit cube (Eq. 11) of the stochastic matrices is generated by the vertices V which are 0-1 stochastic matrices with only one non-zero component 1 in each row. The matrix $M = \left[p_{ij} \right]$ of actual land-use transitions is converted into the stochastic matrix $R_0 = \left[p_{ij} / S_{i\bullet} \right]$ within the unit cube (Eq. 11) of all stochastic matrices; thus, the procedure of the decomposition can be applied for analysis of a normalized transition matrix. Moreover, because of the 0-1 structure of the vertices V, each vertex-matrix V presents the extreme tendency of transfer of land uses only to the one type of land use. Thus the vertices of the normalized unit cube are defined by the rule: 'everything or nothing' – each row of the vertex-matrix V includes only one non-zero coordinate.

3.3.3   Superposition principle and definition of main tendencies

The superposition approach decomposes the actual land-use transition matrix R_0 into the weighted sum of matrices V_k representing the action of the extreme transition tendencies:

$$R_0 = p_1 V_1 + p_2 V_2 + ... + p_m V_m \tag{12}$$

where: $1 \geq p_s \geq 0$ and $p_1 + p_2 + ... + p_m = 1$

The complete expressions of these extreme tendencies define the set of vertex-matrices V_s. Each extreme transition matrix V_s enters the actual transition matrix R_0 with the weight $p_s \leq 1$, and the sum of weights is equal to 1. The procedure of the decomposition analysis consists of the successive extraction from an actual transition matrix of the shares corresponding to the constructed set of extreme tendencies. At the beginning, we construct an extreme vertex-matrix V_1, which is the complete expression of the main extreme tendency of land-use transition tendency, and determine its share (weight) in the actual transition matrix, and simultaneously determine the residual of the actual transition after the extraction of the action of the main extreme tendency V_1. In this residual R_1, we choose the next extreme tendency V_2, and so forth. The most significant fact is that the set of residuals R_s corresponds to the meaningful set of the 'bottlenecks', corresponding to those parts of the actual transition process where the action of environmental factors compels the actual transition to diverge from the extreme transition. These transition bottlenecks determine the weights of the extreme transitions V_s in the actual transition matrix R_0.

4. INITIAL MATRIX ANALYSIS RESULTS

In this chapter, we are using the matrices of landscape transition rates in the area of Zichron Ya'acov, in 1944-1956, 1956-1970, and 1970-1990 for a total of nine landscape units. To provide an initial overview of the results of this analysis, the row and column sums (reflecting the initial and final distributions) are presented in Table 4-3 for the three studied periods. To better represent the temporal changes in land-use shares, we will use the ranking of the shares according to their size.

Table 4-3. Rank-size sequences of initial and final land-use distributions

1944-1956				1956-1970				1970-1990			
Initial		Final		Initial		Final		Initial		Final	
3	29.3	2	28.7	2	25.7	2	23.2	2	25.3	8	26.0
5	25.2	3	23.7	3	24.7	3	22.6	3	22.4	1	22.5
2	19.8	5	15.5	5	16.8	1	16.0	1	17.4	2	19.6
1	9.5	4	10.2	4	10.9	4	11.5	4	12.0	3	11.6
4	8.9	1	8.5	1	7.6	8	10.5	8	10.1	5	10.4
8	3.5	9	7.0	9	6.6	5	10.1	5	9.6	4	6.5
7	2.6	8	4.8	8	4.7	9	6.1	9	3.2	9	3.4
6	0.8	6	1.6	6	3.0	6	0.0	6	0.0	6	0.0
9	0.4	7	0.0	7	0.0	7	0.0	7	0.0	7	0.0

The figures in this table relate to landscape unit and share (%) respectively; the dominant landscape unit in 1944 (3) thus covered 29.3% of the total study area

It is possible to see that land-use shares at the end of a time period do not coincide quantitatively with the shares in the beginning of the next time period. This can be explained by the fact that the air photographs in each period are not identical in scale and angle of view and, therefore, the practical measurement the land-use shares gives the measurement deviations which in our case do not exceed 3% in any of the time periods. Nevertheless, the rank-size sequences coincide qualitatively, i.e. the ranking of land uses is the same at the end of each time period and at the beginning of the next one. This fact supports significantly the robustness of our method of measurement of land uses. The dynamics of redistribution of land-use shares reveals the following tendencies of change:

- light vegetation coverage (3) loses its magnitude (from 29% to 12%), descending from first place in the ranking in 1944 to fourth place in 1990;
- medium vegetation coverage (2) occupies in 1944 and in 1990 about the same size (about 20%) and takes first place in the ranking from 1956 until 1970 (covering about 25%);
- heavy and dense vegetation (1) climbs up from fourth place in 1944 and fifth place in 1956 (about 9%) to the second place in the ranking in 1990 (23%);

- orchards and olive plantations (4) stay in all time periods on fourth-fifth place (magnitude declines from about 10% to 7%);
- cultivated fields (5) descend from second place in 1944 (about 25%) to fifth place in 1990 (about 10%);
- abandoned fields (6,7), which occupy about 3% in 1944, disappear in 1970;
- dense built-up areas (8) are growing strongly from six-seventh place in 1944-56 (about 5%) to first place in the ranking in 1990 (26%); and
- sparse urban built-up areas (9), which reached six place in 1956 (7%), are last in the ranking in 1990 with 3%.

4.1   Sum Products Matrix

In a next step, the Sum Products Matrix (SPM) was calculated to visualise the land-use transitions structure with the help of artificial structural landscapes. Table 4-4 presents the rank-size hierarchies of row and column sums of the land-use transitions matrix for Zichron Ya'acov area and the corresponding maximum entropy SPM. This matrix is calculated with the help of Eq. (4) on the basis of the land-use transition matrix. The corresponding cross-structure is presented graphically in Figure 4-1, where the order of the rows provides the hierarchy of rows in the initial distribution of land uses while the order for the columns provides a similar structure for the column sums in the final distribution of land uses. The land uses are arranged in such a way that the northwest quadrant provides the highest elevation and the artificial land-use transitions landscape slopes towards the east and south. At the top of the hierarchy (8.4 %) is the transition of land use 3 (light vegetation coverage) to 2 (medium vegetation coverage).

Table 4-4. Rank-size hierarchies of row and column sums and corresponding maximum entropy SPM, for Zichron Ya'acov area, 1944-1956

	7	6	8	9	1	4	5	3	2	ID
3	0.00	0.47	1.41	2.05	2.49	2.99	4.54	6.94	8.41	29.3
5	0.00	0.40	1.21	1.76	2.14	2.57	3.91	5.97	7.23	25.2
2	0.00	0.32	0.95	1.39	1.68	2.02	3.07	4.69	5.68	19.8
1	0.00	0.15	0.46	0.67	0.81	0.97	1.47	2.25	2.73	9.5
4	0.00	0.14	0.43	0.62	0.76	0.91	1.38	2.11	2.55	8.9
8	0.00	0.06	0.17	0.25	0.30	0.36	0.54	0.83	1.00	3.5
7	0.00	0.04	0.12	0.18	0.22	0.27	0.4	0.62	0.75	2.6
6	0.00	0.01	0.04	0.06	0.07	0.08	0.12	0.19	0.23	0.8
9	0.00	0.01	0.02	0.03	0.03	0.04	0.06	0.09	0.11	0.4
FD	0.0	1.6	4.8	7.0	8.5	10.2	15.5	23.7	28.7	

The figures in this table relate to landscape unit and share (%) respectively; the table for example shows that the transition of unit 3 to unit 2 covers 8.41% of the total study area

The preferences of land uses existing in the actual land-use matrix can be revaluated by calculating the difference matrix $M - S$. This matrix, for 1944-1956, reveals that preferable land uses in the Zichon Ya'acov area were: 1, 2, 3 (vegetation coverage), 4 (orchard and olives plantation) and 5 (cultivated fields). In 1956-1970, the land-use preferences were 1, 2, 3 (vegetation coverage), 5 (cultivated fields) and new land use 9 (sparse built-up area). In 1970-1990, the land-use preferences were 1, 2, 3 (vegetation coverage), 5 (cultivated fields) and land use 9 (sparse built-up area) was exchanged with land use 8 (dense built-up area). It is important to note that in 1956-1970 and 1970-1990, the artificial land-use landscapes became more flat.

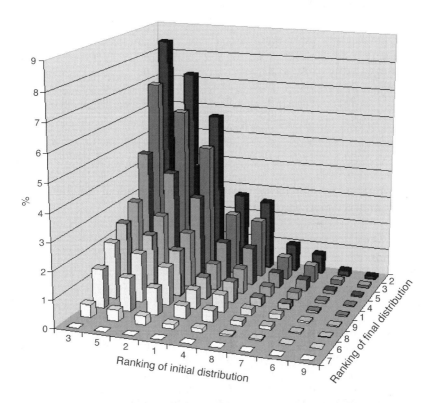

Figure 4-1. Artificial land-use transition landscape showing the share (%) of the total area claimed by each transition and the rank-size hierarchies of the initial and final distribution of land-use types for Zichron Ya'acov, 1944-1956.

5.   DECOMPOSITION ANALYSIS

The decomposition and assemblage analysis will be illustrated in this and the following section starting from the 1944-1956 land-use transitions in the Zichron Ya´acov area (Table 4-5). The starting-point for both approaches is matrix M with the corresponding initial and final distribution (see below). The initial distribution (*ID* or row sums $S_{i\bullet}$) is used in the decomposition analysis, while the assemblage analysis starts from the final distribution (*FD* or column sums $S_{\bullet j}$).

Table 4-5. Matrix M for the Zichron Ya´acov area in 1944-1956, also indicating Initial and Final Distribution (*ID* and *FD*)

	1	2	3	4	5	6	7	8	9	ID
1	4.3	3.5	0.8	0	0	0	0	0	0.9	9.5
2	2.1	15.4	0	0	0	0	0	0	2.3	19.8
3	1.8	6.6	14.2	1.3	2.6	0	0	1.3	1.5	29.3
4	0	1.1	0.6	5.6	1.1	0	0	0	0.5	8.9
5	0.3	2.1	5.8	2.8	11.8	1.6	0	0	0.8	25.2
6	0	0	0.8	0	0	0	0	0	0	0.8
7	0	0	1.5	0.5	0	0	0	0	0.6	2.6
8	0	0	0	0	0	0	0	3.5	0	3.5
9	0	0	0	0	0	0	0	0	0.4	0.4
FD	8.5	28.7	23.7	10.2	15.5	1.6	0	4.8	7	

The normalized stochastic matrix R_0 corresponding to the land-use transition matrix M has the form:

Table 4-6. Normalized stochastic matrix R_0 , with related main tendency V_1

	1	2	3	4	5	6	7	8	9	V_1
1	0.45	0.37	0.08	0	0	0	0	0	0.09	1
2	0.11	0.78	0	0	0	0	0	0	0.12	2
3	0.06	0.23	0.48	0.04	0.09	0	0	0.04	0.05	3
4	0	0.12	0.07	0.63	0.12	0	0	0	0.06	4
5	0.01	0.08	0.23	0.11	0.47	0.06	0	0	0.03	5
6	0	0	1	0	0	0	0	0	0	3
7	0	0	0.58	0.19	0	0	0	0	0.23	3
8	0	0	0	0	0	0	0	1	0	8
9	0	0	0	0	0	0	0	0	1	9

This matrix is calculated by dividing the rows of the matrix M on the row sums $S_{i\bullet}$. The shaded coefficients of this stochastic matrix represent the maximal elements of each row. Based on this classification we can construct a 0-1 stochastic matrix, V_1 , by replacing each shaded coefficient with a value of 1. The other coefficients then receive a value of 0. This 0-1 matrix can be compressed to reflect main tendency of land-use change, by noting for each row (initial land use) the relating most common final land use indicated with a 1. This main tendency (V_1) is also indicated in Table 4-6 and shows a strong preference for preservation for most land-use types. Vegetation coverage (1, 2, 3 agricultural uses (4, 5) and built-up areas (8, 9) tend to preserve. The abandoned fields (6, 7) are transferred to sparse vegetation coverage (3). The weight of this tendency is 0.45. This value presents the minimum from the set of all shaded coefficients in matrix R_0 and also defined the first bottleneck problem: the interdiction to the preservation of sparse vegetation coverage (3).

The following decomposition holds:

$$R_0 = 0.45V_1 + 0.55R_1 \tag{13}$$

where R_1 is the first remainder. Remember that V_1 has the form of a 0-1 matrix. Equation (13) implies that:

$$R_1 = 1.8182R_0 - 0.8182V_1 \tag{14}$$

This remainder is presented in the table below together with the compressed form of the second extreme tendency.

Table 4-7. The first remainder matrix R_1 with related main tendency V_2

	1	2	3	4	5	6	7	8	9	V_2
1	0	0.67	0.15	0	0	0	0	0	0.17	2
2	0.19	0.6	0	0	0	0	0	0	0.21	2
3	0.11	0.41	0.06	0.08	0.16	0	0	0.08	0.09	2
4	0	0.22	0.12	0.33	0.22	0	0	0	0.1	4
5	0.02	0.15	0.42	0.2	0.03	0.12	0	0	0.06	3
6	0	0	1	0	0	0	0	0	0	3
7	0	0	0.23	0.35	0	0	0	0	0.42	9
8	0	0	0	0	0	0	0	1	0	8
9	0	0	0	0	0	0	0	0	1	9

The extraction of the next extreme tendency V_2 (lowest maximum value for any of the rows; also presented in Table 4-7) from the first remainder, R_1, gives:

$$R_1 = 0.33V_2 + 0.67R_2 \tag{15}$$

Here, the second bottleneck problem prevents the preservation of orchard and olives plantations (4). Equation (15) implies:

$$R_2 = 1.4925R_1 - 0.4925V_2 \tag{16}$$

The substitution of Eq. 15 into Eq. 13 gives the following decomposition:

$$R_0 = 0.45V_1 + 0.18V_2 + 0.37R_2 \tag{17}$$

The aggregated weight of two extreme tendencies equals 0.63 and the second remainder R_2 calculated with the help of Eq. 17 is given in Table 4-8.

Table 4-8. The second remainder matrix R_2

R_2	1	2	3	4	5	6	7	8	9
1	0	0.51	0.24	0	0	0	0	0	0.26
2	0.29	0.4	0	0	0	0	0	0	0.32
3	0.17	0.12	0.09	0.12	0.24	0	0	0.12	0.14
4	0	0.34	0.18	0	0.33	0	0	0	0.15
5	0.03	0.23	0.13	0.3	0.05	0.17	0	0	0.09
6	0	0	1	0	0	0	0	0	0
7	0	0	0.34	0.52	0	0	0	0	0.13
8	0	0	0	0	0	0	0	1	0
9	0	0	0	0	0	0	0	0	1

The sequential analysis of remainders can be continued further by including additional extreme tendencies with preset average weight. In our analysis, we choose the value of the preset average weight to be about 0.80. For 1956-1970, the normalized stochastic matrix of the land-use shares, R_0, has the following decomposition into four extreme tendencies:

$$R_0 = 0.39V_1 + 0.25V_2 + 0.10V_3 + 0.07V_4 + 0.19R_4 \tag{18}$$

Equation (18) has an aggregated weight 0.81 and its extreme tendencies have the following compressed form:

Table 4-9. Compressed form of the extreme tendencies for 1956-1970 and 1970-1990

	1956-1970 period				1970-1990 period			
	V_1	V_2	V_3	V_4	V_1	V_2	V_3	V_4
1	1	2	1	1	1	1	2	8
2	2	1	2	1	1	2	8	1
3	3	2	3	3	3	2	8	2
4	4	5	3	4	4	5	8	2
5	5	4	3	3	5	5	9	4
6	3	3	4	3	n.a	n.a	n.a	n.a
7	n.a	n.a	n.a	n.a	n.a	n.a	n.a	n.a
8	8	8	8	8	8	8	8	8
9	9	9	9	9	9	8	8	1

We can see that in 1956-1970 the old abandoned fields (7) disappeared. For 1970-1990, the normalized stochastic matrix of the land-use shares, R_0, has the following decomposition into four extreme tendencies:

$$R_0 = 0.32V_1 + 0.21V_2 + 0.13V_3 + 0.08V_4 + 0.26R_5 \qquad (19)$$

Equation (19) has an aggregated weight of 0.74 and delivers the extreme tendencies as presented in Table 4-9. In 1970-90, the recent and old abandoned fields (6 and 7) disappeared. The extreme tendency V_3 represents the strong urbanisation process in this decade.

6.   ASSEMBLAGE ANALYSIS

Analogous to the analysis of decomposition of land-use coverage, we will consider first the main tendencies of assemblage of land-use transitions in the 1944-1956 period, as presented in Table 4-5. We start from the final distribution (*FD*) of shares of different land uses at the end of 1956. The normalized Markovian matrix Q_0 corresponding to the land-use transition matrix *M* is presented in Table 4-10. Please note that in a Markovian matrix, the column sums of coefficients always equal to 1.

Table 4-10. Markovian matrix Q_0 , corresponding to land-use transition matrix *M* for 1944-1956

Q_0	1	2	3	4	5	6	7	8	9
1	0.51	0.12	0.03	0	0	0	0	0	0.13
2	0.25	0.54	0	0	0	0	0	0	0.33
3	0.21	0.23	0.60	0.13	0.17	0	0	0.27	0.21
4	0	0.04	0.03	0.55	0.07	0	0	0	0.07
5	0.03	0.07	0.24	0.28	0.76	1	0	0	0.11
6	0	0	0.03	0	0	0	0	0	0
7	0	0	0.07	0.04	0	0	0	0	0.09
8	0	0	0	0	0	0	0	0.73	0
9	0	0	0	0	0	0	0	0	0.06

Analogous to Eq. 19, the following superposition of four extreme tendencies of the assemblage process with the aggregated weight 0.77 can be derived (Eq.20). The compressed form of the sequential extreme tendencies is presented in Table 4-11.

$$Q_0 = 0.32W_1 + 0.21W_2 + 0.13W_3 + 0.11W_4 + 0.23Q_5 \qquad (20)$$

Table 4-11. The four extreme tendencies of the assemblage process for 1944-1956

	1	2	3	4	5	6	7	8	9
W_1	1	2	3	4	5	5	n.a	8	2
W_2	2	2	3	5	5	5	n.a	8	3
W_3	3	3	5	4	5	5	n.a	3	1
W_4	1	1	5	3	3	5	n.a	8	5

The first extreme tendency W_1 presents the fact that almost all (1-8) land-use types are the biggest suppliers of coverage to themselves; only the sparse urban areas coverage (9) is supported by transition from medium vegetation coverage area (2) and this transition contains the bottleneck. The weight of this first tendency is 0.32. It is interesting to note that on the level of aggregated land uses of the types I, II, III, other tendencies are similar with bottlenecks interdicting the transition to the same sparse built-up area (9). Analogous assemblages can be constructed for 1956-70 and 1970-1990. These present the continuation of self-support of land-use coverage together with the enlargement of the built-up areas, which contains all bottleneck problems interdicting the assemblage of urban land uses (8, 9).

7.   DISCUSSION

In this chapter, we considered the dynamics of land-use transition in the Zichron Ya´acov site of the Israeli Carmel area. The two most important trends in the 1944-1990 period were: a loss in agricultural area and transitions within areas of more natural vegetation. We will first discuss these trends and then finish with some remarks relating to the applied methodology.

7.1   Loss in agricultural area

Our analysis of the Zichron Ya´acov site shows a sharp decline (from 38% to 17%) in the total agriculture area for the period. Historical records of the British Government land-use survey (Village statistics, 1945) reveal that agricultural land uses covered almost the total area of the villages in the Mount Carmel region in the 1940s. Built-up areas, on the other hand, have expanded and became a dominant land use in most of the sites. It is important to note that the mountain agriculture was almost completely abandoned or decreased radically while cultivated fields and orchards were less affected in the valleys. In broad terms, this land transformation can be divided into two phases: in the first phase, the rate of decreasing agriculture is higher than that of increasing built-up area, while an inverse relationship exists in the second phase. The replacement of agricultural land by built-up areas is well known around the world in general and in Mediterranean countries in particular (see, for example, Frandez Ales *et al.*, 1992 and Barbero *et al.*, 1990). However, a distinction must be made between endogenic processes of built-up area growth on agricultural lands taking place in core rural areas in general and in the Carmel in particular, and exogenic processes where agricultural land is lost due to the expansion of urban areas into the rural zone, as is the case in the periphery of the Tel Aviv Metropolitan Zone and Haifa (Gavish and Sonis, 1979; Amiran, 1996).

7.2   Vegetation transitions

The general picture of changes in these areas is the increase in vegetation density as a result of the forced reduction of grazing pressures and woodcutting mainly because of the enforcement of the legislation on natural landscape preservation (National Parks and Nature Reserves Law of 1963 and the Law of the Protection of Vegetation of 1950). The general average of sparsely vegetated areas decreased from 32% in 1944 to 10% in 1990, while areas with a high vegetative cover increased from 14% to 44% in the same period of time. By tracing the trends of vegetation change in the different study locations for the three categories of coverage we have charted four different types of vegetation processes:

Natural recovery: this process is represented by two parameters: the expansion of areas of heavy vegetation coverage and the transformation of areas of moderate vegetation cover into thick cover. One should note that these results correlate with findings of Kutiel (1994) and Broide *et al.* (1996) who examined the process of vegetation renewal after fires.

Recovery linked to afforestation: vegetation recovery rates were enhanced due to afforestation activities mainly by the KKL. These activities were widespread throughout the Carmel region and mainly during the 1960s and the beginning of the 1970s.

Disturbance due to grazing and woodcutting: settlements having a significant proportion of agricultural land use are showing a delay in the vegetation development rate in open areas.

Disturbance due to the expansion of built-up areas: vegetation recovery since the 1940s is most prominent in the open areas of the Mount Carmel region. The main threat to this positive phenomenon is from the expansion of built-up areas. Controlled grazing in these areas may help in preserving the ecological values of this landscape and will also allow the continuation of some characteristics of the local population's traditional culture. Another

advantage concerned improving the forest structure and, by that, a reduction
of fire threats (Perevolotsky *et al.*, 1992).

7.3   Methodological considerations

This study presents the theoretical principles, the methodology, and the
use of matrix analysis of landscape transitions by means of a GIS, and
employs, for the first time in Israel, a quantitative method for measuring
changes over a long period of time. The relationships between man and the
environment have been presented in the past as being reciprocal ones, and as
two separate systems - nature and man. The latter is an 'external' force,
creating disturbances and stopping them (Naveh and Lieberman, 1984).
Only for the past few years have we begun to see studies pertaining to
landscape transitions in the world as slow quantitative changes. This subject
is still at its very beginning. In none of these studies was there a combination
of a number of landscape methods over a relatively long period of time for a
good number of locations. Studies which pertained to an analysis of the
landscape of Israel were concerned with descriptions of the landscape and
settlement processes in a qualitative manner (Ben Artzi, 1986; 1996;
Grossman *et al.*, 1993).

Because of the rise of awareness of the topic of open areas, studies have
begun to appear dealing with an empirical analysis of the landscape
(Feitleson, 1995). Our study presents an integral landscape system in which
man lives and works and influences nature 'from the inside'. Even though
the general trends of landscape development on the Carmel were well-
known to Israeli scientists, the dynamics and rate of the processes were not
studied quantitatively using methods of remote sensing. The scientific
importance of this type of work results from the opportunity to use the same
methods in different places and to compare the results despite differences in
structure, characteristics and function.

The use of landscape transformation matrices is a basic instrument for
this purpose. The matrices were constructed by building layers of data in a
regional GIS and by overlaying each pair of successive layers. The rates of
landscape transition were calculated by these matrices. They pointed to rapid
processes of change on the Carmel - urban encroachment at the expense of
agricultural areas and natural vegetation. The continuation of these trends
will undoubtedly bring about sharp conflicts between the needs for land for
agriculture and building and the desire to preserve natural landscapes, the
Carmel National Park, and the valley bottom landscapes which have been
renewed on the periphery of Haifa.

The new methodology of matrix land-use analysis presents an analytical
computerized approach to the study of land use. It is hoped that it will

broaden and deepen the study of landscape systems in Israel and enable its comparison with other Mediterranean landscapes.

ACKNOWLEDGEMENTS

Thanks to the editors and Tintu Thomas for help in reducing the size of the original manuscript.

REFERENCES

Amiran, D.H.K. (1996) Preservation and future land use in agriculture, the mosaic of Israeli geography, The Negev Press, Ben Gurion, pp. 107–112.

Barbero, G., Bonin, G., Loiesel, R. and Quezel, P. (1990) Changes and distribution of western part of ecosystem caused by human activities in the Mediterranean basin, *Vegetation,* 87: 151–173.

Ben Artzi, J. (1986) *The Settlement of the Carmel and its Coastal Plain in the Modern Period, in The Carmel - Man and Settlements*, Society for the Preservation of Nature, Jerusalem (in Hebrew).

Ben Artzi, J. (1996) *From Germany to Eretz Israel - The Settlement of the Templars in Eretz Israel*, Yad Ben Zvi, Jerusalem (in Hebrew).

Broide, H., Kaplan, M. and Perevolotsky, A. (1996) The development of woody vegetation in the Ramat Hanadiv Park and the impact of fire: a historical survey based on aerial photos, *Ecology and Environment*, 2(1–2): 127–131 (in Hebrew).

Feitleson, E. (1995) Open space protection in Israel at turning point, *Horizons in Geography*, 42–43: 7–27 (in Hebrew).

Frandez Ales, R., Martin, A., Ortega, F. and Ales, E.E. (1992) Recent change in land use structure and function in a Mediterranean region of SW Spain (1959–1984), *Landscape Ecology*, 7(1): 3–18.

Gavish, D. and Sonis M. (1979) *Urban Expansion of Tel-Aviv into the Rural Area, Karka (land)*, Jerusalem, pp. 15–26 (in Hebrew).

Grossman, D., Shoshany, M. and Kutiel, P. (1993) Landscape Changes in the Carmel Ridge in the Modern Era – Process and Methodology, A Geographic Perspective on the Social and Economic Restructuring of Rural Areas, International Geographical Congress, Kansas State University.

Kutiel, P. (1994) Fire and ecosystem heterogeneity: a Mediterranean case study, *Earth Surface Processes and Landforms*, 129: 187–194.

Naveh, Z. and Kutiel, P. (1990) Changes in the Mediterranean vegetation in response to human habitation and land use, in Woodwell, G.M. (ed) *The Earth in Transition Patterns of Biotic Impoverishment*, Cambridge, pp. 259–299.

Naveh, Z. and Lieberman, A.S. (1984) *Landscape Ecology Theory and Application*, Springer Verlag, New York.

Perevolotsky, A., Lachman, E. and Follak, G. (1992) The human impact on Mediterranean vegetation, in *The Mediterranean Scrubland, A Literature Review*, Ramat Hanadiv Project, Research Reports Series, no 8, pp. 89–130.

Shannon, C.E. and Weaver, W. (1964). *The Mathematical Theory of Communications*, University of Illinois Press. Urbana, IL.

Sonis, M. (1968) Significance of entropy measures of homogeneity for the analysis of population redistributions, *Geographical Problems. Mathematics in Human Geography*, 77: 44–63, (in Russian).

Sonis, M. (1980) Locational push-pull analysis of migration streams, *Geographical Analysis*, 12(1): 80–97.

Sonis, M., Hewings, G.J.D. and Guo, J.M. (2000) A new image of classical key sector analysis: minimum information decomposition of the Leontief inverse, *Economic Systems Analysis*, 12(3): 401–423.

Chapter 5

NEW LAND-USE DEVELOPMENT PROCESSES ASSOCIATED WITH THE ACCELERATION OF URBANISATION IN CHINA
Case study of the Pearl River Delta metropolis

Z.-G. Wu[1], S.-H. Zhou[2] and C.-C. Feng[1]
[1]Center of Real Estate Research and Appraisal, Peking University, Beijing, China; [2]School of Geography and Planning, Zhongshan University, Guangzhou, China

Abstract: In recent years, new processes for land-use development have appeared that are not entirely in accord with state law in China, especially in areas of rapid urbanisation. This situation results from conflicts between land use and the rapid development of society and economy in these areas. Although urbanisation contradicts the national law of land use in various ways, it reveals the irrationality of the current land institutions, which have to be reformed in order to achieve urban sustainability. After summarizing and analysing some typical processes of land-use development for the rapidly urbanizing areas in the Pearl River Delta metropolis, this chapter presents some conclusions and suggestions for land institution reform in the future.

Key words: Urbanisation; processes of land-use development; Pearl River Delta of China.

1.   INTRODUCTION

The revision of the Chinese Land Administration Law, given the background of the new market economic system, is now a hot topic for debate by scholars and government officials. This theme is of particular interest for the analysis and simulation of land-use change in China since it relates to various processes, such as land-use management (Liu and Zhang, 2001; Wang, 1994), cooperation between different actors in land-use development (Tang, 1995; Zhang, 1996) and organizational and financial modelling of land-use dynamics (Li and Chen, 2001; Li and Liu, 2004).

E. Koomen et al. (eds.), Modelling Land-Use Change, 83–94.
© 2007 *Springer.*

This chapter focuses on recent land-use changes in a rapidly urbanising part of China and describes five new land-use development processes based on an extensive inventory of development projects in the past 10 years. The newly distinguished development processes are then compared with the legal approach. This comparison reveals some irrationality in the current law and some possible reforms in the land-conversion process are therefore suggested.

2.   STUDY AREA

The Pearl River Delta (PRD) comprises the low-lying areas alongside the Pearl River estuary near the South China Sea. It is one of the most important economic regions and manufacturing centres in mainland China. The metropolitan PRD region in the Guangdong Province consists of eight municipalities, namely Guangzhou, Shenzhen, Zhuhai, Dongguan, Zhongshan, Foshan, Huizhou and Jiangmen (Figure 5-1). The eastern part of the PRD, near Hong Kong, is the most economically developed. New transport links between Hong Kong, Macau and Zhuhai, such as the proposed 29 kilometres long Pearl River Bridge, are expected to open up new areas for development within the region.

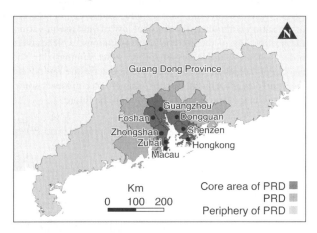

Figure 5-1. Pearl River Delta study area.

The transformation from a predominantly rural and agricultural region to one of the leading economic centres followed the economic reform program that started in 1979. The region's impressive growth was fuelled by foreign investment, initially coming largely from Hong Kong manufacturers that moved their operations into the PRD. During the 1980-2000 period, the average real rate of GDP growth in the PRD Economic Zone exceeded 16%, well above the impressive national average of close to 10%. According to the 2000 Census, the Economic Zone has a population of 40.8 million people, equivalent to 3.2% of the total population of mainland China, on only 0.4% of the land area. Yet it accounted for 8.7% of the GDP, 32.8% of the total trade, and 29.2% of utilised foreign capital in mainland China in 2001. The Greater Pearl River Delta is an internationally oriented manufacturing centre of global importance: nearly five percent of the world's goods were produced here in 2001. There has been a rapid development of privately-owned enterprises in the PRD Economic Zone. These have played an increasingly important role in the regional economy, particularly after 2000 when the environment for private enterprise was greatly relaxed.

The abundance of job opportunities created a reserve of middle-income, professional consumers with an annual per capita income putting them among China's wealthiest groups. The PRD has also been a pioneer in reform and openness. Two of the original four Special Economic Zones (Shenzhen and Zhuhai) are located in the PRD, and have been instrumental in the transition from a planned to a market economy.

With the growth of the economy and acceleration of urbanisation, the model of land development has been a serious problem for the sustainability and coordination for the development of the society, economy and environment in the PRD. The latter is notoriously polluted, with sewage and industrial waste facilities unable to keep pace with population and industrial growth. Frequent smog and pollution threaten people and ecosystems alike. The key to avoid the future risks is to ensure intensive and sustainable land development. First of all, active models of land development have to be made clear for the readjustment and assembly of land.

3.   CURRENT LAND ADMINISTRATION LAW

3.1   Ownership of land

According to the law, land in the inner urban areas of cities is owned by the State; land in rural and suburban areas is owned by peasant collectives, except for those portions of land which belong to the State as indicated by law; housing sites and private plots of cropland and hilly land are also owned by peasant collectives. Thus, there are two types of ownership of land in China: land owned by the State and land owned by peasant collectives.

Development or construction of non-agricultural projects, including real estate and industrial buildings on land owned by peasant collectives, is forbidden without legal permission. Only State-owned land can be used for real estate development or any other profit-making projects. There is a legal procedure through which the land owned by a collective can be requisitioned by the State. During the conversion, the unit or individual has to pay compensation to the organisation associated with the peasant collective (Figure 5-2).

Figure 5-2. Legal procedure for land ownership conversion.

3.2   Right to use of land

The right to use land owned by the State can be granted to the State, a collective or a private unit, in which case the user has to pay land revenues to the Government. The access time is 70, 50 or 40 years from the time that the current land function of residence, industry or commerce respectively is put in place, according to the current land administration law of China, enacted in 1998. Rights to the use of land owned by the State can involve conveyancing, transactions and real estate development.

All units and individuals that need land for construction purposes shall, in accordance with law, apply to use State-owned land. The exceptions are twofold. Firstly, there are those that have lawfully obtained approval for using the land owned by peasant collectives to build townships or town enterprises or to build houses for villagers. Secondly, there are those who have obtained approval for using the land owned by peasant collectives to build public utilities or public welfare undertakings in a township (town) or village. The State-owned land mentioned in the preceding paragraph includes land owned by the State and land originally owned by peasant collectives but requisitioned by the State.

3.3   Legal process for land-use development

If there is a construction project scheduled to run, the right to use of the land owned by the State should be available to the developer first of all. The Government monopolizes the first-hand bargaining of land-use rights. During this bargaining, the land is transferred from natural land to 'prepared' land that can be auctioned in public. Those who want to develop a real estate project have to buy the prepared land-use rights in a public auction.

The land resource agency authorized by the Government is responsible for the first-hand bargaining for land-use rights. This agency usually requisitions land from the peasant collective land, idle land and replacement land, taking these as the land resources for the public land-rights auction. It has been authorized for land requisitioning but not for real estate development. Real estate developers obtain the land-use rights in a public land auction and can then have the land developed to build houses or other buildings for sale or rent. This process is known as land-use development. The legal process for land-use development in accordance with the law is described in Table 5-1.

According to Article 63 of the Land Administration Law of China, the rights to the use of land owned by peasant collectives may not be assigned, transferred or leased for non-agricultural construction. The exceptions are enterprises that have lawfully obtained land for construction in conformity with the overall plan for land utilization but have to transfer, according to law, their land-use rights because of bankruptcy, merger or for other reasons.

Table 5-1. The legal process for land-use development in accordance with law

Types of land	Main participating body	Lawful process for conversion	Financial flow	Market
Natural land	Owned by State or peasant collective economic organization	Land ownership turned to State through requisition	Compensate the original land owner for the preparation cost	First hand land market
Prepared land	Conserved by Government or authorized land resource agency	Lease and land auction	Infrastructure investment	Second hand land market
Land for development	Accessed from public land auction by any developer (e.g. company, unit or individual)	Lease of the right to use the land accessed in public land auction	Pay land revenue to Government for land-use right	Second hand land market

4. NEW LAND DEVELOPMENT PROCESSES

We investigated more than 20 cases including projects such as real estate development, industrial park and downtown regeneration in areas of the rapidly developing PRD counties of Nanhai, Panyu and Shunde. All the selected projects were carried out between 1995 and 2004. Cadastral data are taken as the basic information for the investigation. In total, 900 questionnaires were distributed, of which 855 were returned and 831 proved valid. In addition, 25 interviews were undertaken with local government officers and stakeholders. This extensive investigation gave us a clear insight in the current land-use change processes in the research area. The applied methodology is summarised in Figure 5-3.

Based on our investigation of land-use change processes and several external consultation and discussion sessions, we distinguish a number of new models for land development. These are characterised by different structures, processes for the conversion of land and financial flows which conflict with current land legislation and may produce a loss of land revenue. All of these processes play an active role in economic growth generation but also bring the potential risk of socioeconomic development because of an extensive, irrational and unsustainable land use. The characteristics of the new land development processes are described below. A discussion of the forces that drive these land development changes and of possible improvements in this process follows in Section 3.

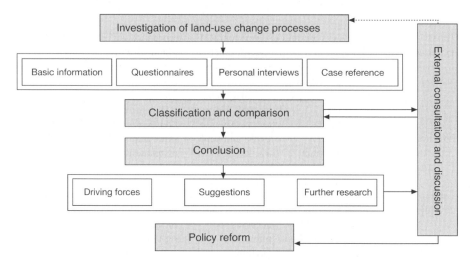

Figure 5-3. Research framework.

4.1   Characteristics of new land development processes

Besides the traditional legal process, there are five typical types of land development process appearing in urban China with the acceleration of urbanisation.

- Process 1: where land is developed by various bodies which may be the land resource agency, the peasant collective economic organization or the enterprise, and in which part of the peasant collective's land has been converted into State-owned land;

- Process 2: where the peasant collective economic organization is the only main body to develop the land;
- Process 3: where the land has been developed after being requisitioned in a special economic development zone;
- Process 4: where land has been assembled in the special economic zone and leased by the land resource agency; and
- Process 5: where the lots of State-owned land used by different parties have been assembled to be developed as a whole.

4.1.1 Process 1: Several main bodies collaborate in development of land with dual ownership

In this instance, the main bodies of land development are respectively the land resource agencies authorized by the Government, the peasant collective economic organizations and some other enterprises managing the development. The land resource agency acts as a coordinator amongst the local peasant collective economic organizations and enterprises. This agency assembles the plots of land from the group of villagers and negotiates with the Government. During the development process, part of the land is requisitioned as State-owned land for use in the land market and real estate market directly, while another part of the land is kept in collective ownership. In this process, the land ownership is of two types: part of the land will be State-owned, and the other part will be collective-owned.

4.1.2 Process 2: Peasant collective economic organization as the main body for land development

According to the current Land Administration Law of China, the land owned by peasant collectives without lawful procedure for requisitioning is forbidden to be used for the construction of non-agricultural projects including real estate and industrial buildings. But, in this instance, the peasant collective economic organization has the land developed directly for the use as industrial buildings or real estate development. During this process, the ownership of the developed land remains with the collective.

4.1.3   Process 3: Requisition of collective-owned land for development in special economic development zones

In this process, land is identified as being in a special economic development zone, requisitioned and developed by the land resource agency authorized by the local government. After the construction of infrastructure, land is auctioned on the open market, and the land-use rights will be transferred to the enterprises and developers involved. The land ownership is changed from being owned by the collective to that of being owned by the State according to a legal process of requisition.

4.1.4   Process 4: Assembly of land for development or lease in special economic development zones

In this process, land is assigned by the peasant collective organization to be within an economic development zone and leased by the land resource agency. After the necessary construction of infrastructure, some of the land will be subleased directly, and the remainder will be used to construct some buildings, such as standard workshops, which will be leased thereafter. Part of the benefits are taken by the land resource agency in management fees whilst the remainder are returned to the peasant collective organization and subsequently returned to the villagers. The land ownership does not change; it remains with the collective throughout the development process.

4.1.5   Process 5: Several bodies join State-owned land development

Some plots of State-owned land are assembled by the local government. After negotiation over the development and distribution of the benefits, the users will sign a contract, and the Government will entrust an enterprise to manage the development. The benefit will be distributed according to the contract. Most steps in this process correspond to the national land policy.

4.2   Comparison of the processes

Table 5-2 summarises the most important characteristics of the newly distinguished land development processes. The characteristics relate to aspects such as: the main participating body, land ownership, the process of land conversion and related financial flows. For the sake of comparison the characteristics of the legal process have been added to the table, indicating that the processes 3 and 5 come closest to the envisaged legal process. The following section discusses the forces that drive the changes in the land development process and proposes a number of possible improvements in this process.

Table 5-2. Comparison of the five new processes for land development with the legal process

Aspect	Characteristic	1	2	3	4	5	Legal
Land source	State-owned land			●		●	●
	Peasant collective-owned land	●	●	●	●		●
Main body of land use development	Government			●		●	
	Peasant collective organization	●	●				
	Enterprise	●				●	●
	Land resource agency	●		●	●		
	Individuals						●
Approach to obtain prepared land	Lease	●	●		●		
	Requisition			●			●
	Assembly of State-owned land			●		●	
	Land returned from requisition to collective organization	●					
Infrastructure investment by	Government	●	●	●	●	●	●
	Collective organization		●				
	Land resource agency	●		●	●		
	Individuals	●				●	
Types of land development	Land hiring			●		●	
	Industrial house hiring			●		●	
	Right transfer	●					
	Right lease			●		●	●
	Contract			●	●		
Land ownership	State			●		●	●
	Peasant collective		●		●		
	Both State and peasant collective	●					
Compensation	Original landowners			●	●	●	●
	Land resource agency	●		●	●		
Land revenue	Government	●		●		●	●
	Collective organization	●	●	●	●		

5. DISCUSSION

The conflict between the agricultural demand for land and that of urbanisation and industrialisation is becoming more and more significant. Excessive cultivation of grasslands, over-grazing, utilization of land by village and township enterprises, soil pollution and erosion and rapid urbanisation have caused serious damage to, and deterioration of, land resources.

The per capita usable land resources of China are very low and will continue to decrease both in quantity and quality. Due to increasing population, industrialisation and urbanisation, the demand for land resources has increased. A shortage of land resources has become a major limitation to the sustainable economic and social development of China. The Chinese Government has realized that an efficient and powerful management and legal system for land resources is vital for not only the present but also the future of China.

5.1   Forces driving the new forms of land development

With the acceleration of urbanisation and growth of industrialisation in the PRD of China, more and more land resources have been transferred into non-agriculture use for more profit via legal or illegal processes. The new processes for land development mentioned above have their origin in a combination of economic and social driving forces, as is discussed below.

5.1.1   Economic driving forces

The direct value of cultivated land is much lower than the value of the land for real estate development or industrial use. In the market economy system, all resources including land would seek a higher profit and this is one of the main reasons why more and more land resources have been converted into construction use, extensively and irrationally.

5.1.2   Rapid urbanisation and industrialisation

With the rapid urbanisation and industrialisation, growth of the urban population and increasing investment in industries, demand for land has shot up quickly. Given the strict land administration laws, the development of land becomes increasingly difficult and more and more time is required to obtain and develop the land through legal processes. Some approaches have emerged for experimental land development in those areas of rapid urbanisation and industrialisation in China.

5.1.3   Policy on the conversion of rights to the use of land owned by a peasant collective

The sluggish policy associated with the conversion of rights to the use of land owned by peasant collectives has made this a serious problem. Sometimes there are models adapted to economic demand but these tend to conflict with State land law. This inconsistency between policy and market practice may cause unsustainable development in the future.

5.2   Measures and suggestions

The processes for land development mentioned above may cause the loss of land revenue and potentially the unsustainable development of the national economy. It is necessary to take these problems seriously now. Some measure should be taken to solve these problems as follows:

- revise some articles of the Land Administration Law of China regarding the conversion of the land owned by the peasant collective;
- strengthen the land resource control and the implementation of the Overall Plan for Land; and
- encourage experiments using models of sustainable land development

5.3   Further research

All the research above is preliminary. Firstly, we have learnt about the problems which have emerged in many places especially in the rapid urbanisation and industrialisation areas of China, such as the PRD. We have recognized that the conflict between the lack of land resources and the demands for economic growth have been increasingly significant. How to obtain intensive, rational, sustainable land development is the key for producing a sustainable society, economy and environment. Further research might usefully focus on the following:

- the adjustment and assembly of land resources;
- the possibility of using methods of the marketing the conversion of lands owned by the peasant collective economic organizations; and
- a process for sustainable land development in areas of rapid urbanisation.

ACKNOWLEDGEMENTS

This chapter is supported by the National Natural Science Foundation of China grants 40401019 and 40471051, and Natural Science Foundation of Guangdong Province grant 04300547.

REFERENCES

Li, D.P. and Chen, D.G. (2001) Model for management of land development and assembly, *China Land Science*, 1:43–48 (in Chinese).
Li, Y.F. and Liu, Q.Q. (2004) New model for land assembly finance, *Forum on Economy*, 12: 92–94 (in Chinese).

Liu, W.X. and Zhang, H. (2001) Urban management, land storage and land finance, *China Real-Estate Financing*, 10: 6–10 (in Chinese).

Tang, W.M. (1995) Consummating the urban plan to guide the land remise, *City Planning Review*, 6: 51–53 (in Chinese).

Wang, L.P. (1994) Reviewing on the urban planning from the view of urban land institution, *Urban Planning Review*, 8: 8–10 (in Chinese).

Zhang, X.K. (1996) Discussion on the Relationship between Land Development and Management, *Urban Planning Review*, 2: 39–42 (in Chinese).

PART II: EXPLANATORY MODELS OF LAND-USE CHANGE

Chapter 6

DRIVING FORCES OF LAND-USE CHANGE IN A CULTURAL LANDSCAPE OF SPAIN
A preliminary assessment of the human-mediated influences

J. Peña[1], A. Bonet[1], J. Bellot[1], J.R. Sánchez[1], D. Eisenhuth[1], S. Hallett[2] and A. Aledo[3]

[1]*Departamento de Ecología, Universidad de Alicante, Spain;* [2]*National Soil Resources Institute, Cranfield University at Silsoe, UK;* [3]*Departamento de Sociología y Teoría de la Educación, Universidad de Alicante, Spain*

Abstract: The aim of this chapter is to examine the processes of change in land cover and land use over the last 44 years, at regional scale, in a traditional, rural south-eastern Spanish catchment. Land use has changed dramatically over recent decades throughout the Mediterranean. Much of this change has been driven by shifts in agricultural and socioeconomic policy. Analysis of aerial photography for the Marina Baixa catchment has revealed a significant decline in traditional agriculture and conversion to forestry or intensive croplands. The consequences of economic globalisation are reflected here in a shift from traditional to intensive agriculture and in human migration from rural to urban areas, as well as in the development of tourism. Land-use changes are correlated with socioeconomic structural forces in order to demonstrate how these changes affect the basic resources of the area and to provide a clearer understanding of possible future trends.

Key words: Landscape change; land-use and land-cover change; driving forces; agricultural abandonment; agricultural intensification; urbanisation.

1.   INTRODUCTION

Land-cover (the biophysical attributes of the earth's surface) and land-use change (human purpose or intent as applied to these attributes) play an important role in current global change phenomena (Turner *et al.*, 1990; Vitousek, 1992). Changes in landscape structure represent some of the most

E. Koomen et al. (eds.), Modelling Land-Use Change, 97–115.
© 2007 *Springer.*

prevalent and important impacts of land-use change (Forman and Godron, 1986). For instance, habitat fragmentation and environmental stress have profound consequences on ecosystem processes and are found to affect species diversity (Chapin *et al.*, 1997).

During the last few decades, many parts of the planet have experienced an impressive human-mediated switch (Wilson and King, 1995). Not all areas, however, are undergoing similar types of land-use changes. In developed countries, agriculture is now concentrated in the most productive lands. More marginal areas have been abandoned or subjected to less intensive land uses and to afforestation. Conversely, in less developed countries, forestland is being cleared and converted for agricultural use (Lambin *et al.*, 2001).

Comprehension of landscape change requires a sound understanding of the underlying processes. The driving forces are those underlying elements that trigger landscape changes. Consequently, these elements represent influential processes in the evolutionary trajectory of the landscape (Bürgi *et al.*, 2004). The main driving forces for these changes include economic, social and territorial planning, and these are therefore the key elements for decision makers. Recently, land managers have begun to realise that ecosystems and landscapes are dynamic, and that disturbance and succession processes operate over many scales, maintaining ecosystems and landscapes in a constant state of flux (Bonet *et al.*, 2004). This dynamism is essential in preserving biodiversity.

Cultural landscapes reflect the long-term interactions between people and their natural environment (Farina, 1998), indicating that landscapes have been shaped over time in an interactive process that links human needs with natural resources and features in a specific topographical and spatial setting (Turner, 1990). As a dynamic system, the present landscape is the result of past processes and provides the basis for the formation of future landscapes (Peña *et al.*, 2005c). Traditional land-use activities are at least partly responsible for maintaining the high levels of ecological quality found in Mediterranean landscapes (Blondel and Aronson, 1999). Accordingly, landscapes change in a somewhat chaotic way, while human intervention assists to control this evolution regularly with planned actions which are seldom realised as they are intended (Antrop, 1998).

It is important to emphasize that the study area presented is located in a semi-arid Mediterranean climate. The sensitivity of the semi-arid zones to climatic fluctuations has been dramatically illustrated with periods of severe drought in 1981-1984 (Quereda-Sala *et al.*, 2000) and in 1993-1994 (Llamas, 2000) and also in more recent times. These circumstances must be reflected in land-planning strategies aimed at deciding which land practices should be applied during especially dry periods, based upon the understanding that the environment is in a state of fragile equilibrium (Bellot *et al.*, 1999) in which both socio-cultural activities and biological diversity and ecosystem functions trigger the susceptibility of this land to desertification (Palutikof *et al.*, 1996).

Historical analysis is the basis of landscape evaluation (Marcucci, 2000). It is not possible to assess the present conditions of a landscape mosaic without at least knowing its recent history. It is only by considering the evolution of a landscape that it is possible to understand the level of reaction to different types of perturbation (Moreira *et al.*, 2001).

The analysis of variation in land cover and land use over time, as sources of information and geographical diagnosis at a regional scale, is central to improving knowledge of land-cover and land-use change in Mediterranean habitats (Bonet *et al.*, 2001; Fernández-Alés *et al.*, 1992; Pan *et al.*, 1999). In fact, changes in land use and the way in which such changes occur can be detected primarily by using land-cover maps handled by geographical information systems (GIS). In this chapter, we emphasize the importance of GIS representation and photographic interpretation (Dunn *et al.*, 1991), because this provides the principal methods with which to obtain evidence of land-use change. A high-precision land-use and land-cover integrated mapping system has been established in order to improve the knowledge relating to change at the regional level over the last 44 years in Marina Baixa (Peña *et al.*, 2005a).

This study establishes a diachronic cultural landscape information system at a regional scale that can serve the following goals:

- analysis of the evolution of land cover and land use over time;
- examination of the spatial transitions between different land-use categories;
- construction of a stochastic Markov model, with the assistance of the associated attribute database; and
- support for strategic decision-making in landscape planning and nature conservation management in Spain.

2.   STUDY AREA

The Marina Baixa (MB) catchment is located in SE Spain (Figure 6-1). Its boundaries roughly correspond to the county of the same name in the Alicante province of the autonomous region of Valencia. The catchment has a surface of 641 km^2, with a complex topography, ranging from sea level to 1,558 metres above sea level. It has a semi-arid to sub-wet Mediterranean climate gradient with hot summers and mild winters, an annual average temperature from 9 to 18°C and a mean annual precipitation from 300 to 800 mm

The parent geological material is mainly calcareous in composition with marls, and the soils are classified following the soil taxonomy system as entisols or inceptisols, being shallow and poorly developed (Quereda-Sala, 1978).

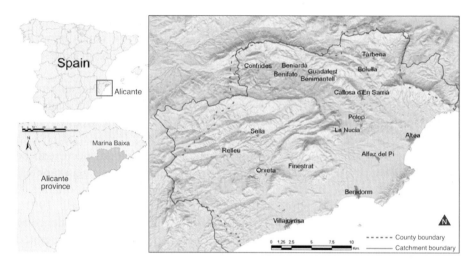

Figure 6-1. The study area.

This Mediterranean region is located in a rural-coastal gradient with a traditionally high diversity of agriculture, resulting in a complex cultural landscape (Solanas and Crespo, 2001). The landscape is characterised by concentrated land-use patterns of irrigated crops, dry crops, urbanisation, and sequential abandonment during the last century, as well as Mediterranean shrublands and woodlands (garrigue). Low and tall shrublands, mainly formed by *Quercus coccifera*, *Pistacia lentiscus*, *Ulex parviflorus*, *Stipa tenacissima*, *Arbutus unedo*, are the dominant vegetation evolution after land abandonment. Some minor parts are also covered by oak trees (*Quercus ilex* sbsp. *rotundifolia*) and principally by non-cropped pine trees (*Pinus halepensis*, *Pinus pinea* and *Pinus nigra*).

MB is a landscape mosaic comprised of 18 municipalities. Before the 1950s, the MB catchment was considered almost as an island, being isolated by abrupt topography. In those times, due to the topography, the most common means of transport and communication with other regions was by sea. Since the 1950s, the socio-economic structure of MB has undergone radical change due to the fact that today over 60% of the Valencian tourist activity is concentrated here (mostly located in Benidorm). Prior to the tourism boom in the 1960s, agriculture and fishing provided the main sources of employment (Quereda-Sala, 1978).

3.   METHODOLOGY

3.1   Data acquisition

Land use in the MB basin was mapped manually using aerial photographs and image processing techniques (Peña *et al.*, 2005a). Stereo-pairs of photographs for 1956 were provided by the Universidad de Alicante at a scale of 1:33,000 and for 1978 at a scale of 1:18,000. Stereo-pairs of photographs of 1998, 1999 and 2000 were provided by the Diputación de Alicante at a scale of 1:25,000. Through the use of advanced photo-scanning technologies, the aerial images were captured and processed together at the improved scale of 1:10,000. These data have been incorporated into ArcGIS 9 and used to obtain subsequent land-use maps in polygonal vector format. The resulting maps were later converted to a raster format to facilitate the cell-based transition matrix analysis of land-use change.

The selected thematic mapping units capture elements of both land use and land cover. For convenience, these will be referred to as land-use types. The selected approach can be typified as being physiognomic since it is based on the features of the landscape that are observable from the aerial photographs. The chosen land-use nomenclature is based on the CORINE and LUCC projects whose classification is hierarchical (Peña *et al.*, 2005a); differentiating 31 categories in the second level of detail. A legend of seven general classes was adopted at the first level, as it was found far more appropriate for analysis of the main changes at the regional level.

3.2   Analysis of land-use changes

Land-use dynamics were analysed during the 1956-2000 period by means of cross-tabulation tools handled by ArcGIS 9 and spreadsheet software. A matrix of land-use changes was created in order to determine substitution patterns during the time period studied. This used a Markov chain procedure, a stochastic model that analyses a pair of land-use images and outputs a transition probability matrix (Peña *et al.*, 2005c). The resultant matrix records the probability of any land-use category changing to any other category. This analysis establishes the quantity of anticipated land-use changes from each existing category to that of each of the other categories over the time period. Consequently, it is possible to gain an insight into how the study area might change in overall terms, but not in specific terms as to where these changes could occur.

Social and environmental forces driving land-use changes were identified subsequently by an exhaustive study of the leading cadastral statistics of the region. This was supported by a series of detailed interviews with farmers, managers of local irrigation systems, managers of local councils and other key actors and stakeholders.

4.   RESULTS

4.1   Land-use dynamics

The study of the land-use evolution in the MB catchment over time (1956-2000) reveals an increase in the presence of all artificial surfaces, mostly near the coastline (Figure 6-2). However, there has also been a significant growth in natural areas due to the abandonment of dry crops due to low productivity. Irrigated croplands were seen to increase in the catchment area in 1978, but then to decrease again near the coastline in 2000.

The MB region has undergone enormous socioeconomic change over recent decades, which can be attributed to both the development of tourism, and the intensification of agricultural activity (Peña *et al.*, 2005b). The driving forces have transformed the hydrological resources and the coastline. The change attractors can be described as coastal proximity (tourism) and water availability (irrigated crops).

Throughout the time period considered the MB landscape was dominated by shrub land/herbaceous cover (39-38% of the total area), followed by non-irrigated arable land (36-14%), woodland (17-30%), irrigated arable land (7-10%) and urban areas (0.3-6.5%). The area of non-irrigated crops registered a 61% decline over 44 years, being most pronounced during the 1956-1978 period. The shrub land/herbaceous coverage remains stationary in time, with woodland cover increasing to 30%. Urban areas experienced the most significant growth, with the current coverage area being some 21 times bigger than in 1956.

The present situation can be seen as representing an increasing polarisation between the more inland, rural and mountainous areas on the one hand, and the coastal region on the other. The different land uses are clearly in competition with one another, resulting in the generation of a series of conflicts (Meyer and Turner, 1994). In fact, this evolution may be understood from the wider perspective of growth concentrating along much of the Mediterranean coastal region. This has produced saturation problems due to the intense competition for space (land occupation) and pressure on the resources of the area (and of these especially water).

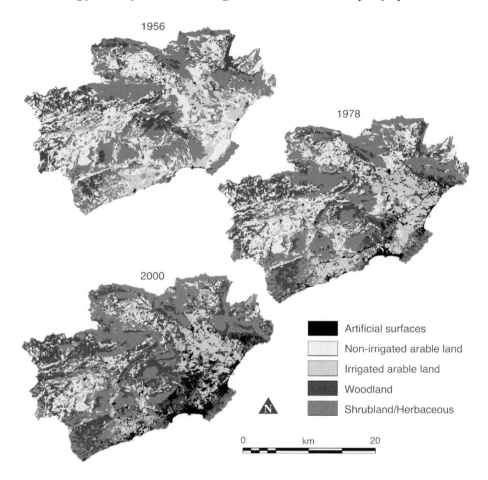

Figure 6-2. Land-use chronosequence in the Marina Baixa catchment, 1956-2000.

4.2   Land-use transitions

A simple transition-matrix model was developed to explore changes in land use over the time period 1956-2000, in order to predict future land-use composition (Baker, 1989). Transition matrices have been used quite commonly to model and explore landscape dynamics and change (Shugart, 1998).

Such models have at their heart a transition matrix (A) that describes the probability of a cell changing from state i to state j (for all classes) in some discrete time step, and a vector (x_t) containing the abundance of each class (absolute or proportional) at time t (Pastor and Johnston, 1992). Multiplying A by x_t gives a new vector (x_{t+1}) that describes landscape composition one time step into the future. If this iterative process is repeated sufficiently, then the stable stage distribution for the landscape will be reached. Thus:

$$x_{t+n} = x_t \cdot A_n \qquad\qquad\qquad (1)$$

A key assumption of transition models is stationarity (i.e. that transition rates do not change over time). We have overlaid 1956 and 2000 land-use maps in seven categories in order to calculate landscape changes. The outcome indicates that 49.5% of the total area has changed and the remainder has not changed for 44 years.

The first step was to calculate the land-use transition using cross-tables. To use Markov chain analysis, it is necessary to convert data from spatial units to probabilities held in a transition matrix. Table 6-1 identifies the combination of land-use maps from 1956 to 2000 and the correspondence to each possible combination of land uses that change from one use to other.

Table 6-1. Transition matrix of land-use change, 1956-2000

		Land-use map 2000						
		Water	Bare s.	Urban	Non-irrig.	Irrigated	Woodland	Shrubland
Land-use map 1956	Water	45.90	0.73	1.99	1.89	31.89	6.40	11.20
	Bare soil	0.13	73.83	21.43	0	0.13	0.27	4.20
	Urban	0.06	0.53	99.30	0	0.02	0.01	0.08
	Non-irrig.	0.38	1.27	6.84	31.50	10.06	25.25	24.70
	Irrigated	1.32	1.45	19.49	1.88	59.07	3.97	12.82
	Woodland	0.20	0.52	0.95	6.50	1.51	62.49	27.83
	Shrubland	0.21	1.27	5.28	3.70	3.57	25.43	60.54

Transition matrices illustrate the origin of land uses in MB for the study period. Each row indicates the proportion (%) of the original land use that changed into other land uses by the end of the period. Diagonal elements are the retention frequencies. Codes for land-use categories are: continental water bodies (water), bare soil (bare s.), artificial sealed surfaces (urban), non-irrigated arable land (non-irrig.), irrigated arable land (irrigated), woodland (woodland) and shrub land/herbaceous (shrubland).

In order to create a future model, it is fundamental to address first the requirements necessary for creating a suitable transition matrix (Lipschutz, 1968). For instance, the summation of the probabilities of the rows must all equal unity, or 100%, for each land-use category of the study area. Therefore, this matrix could be suitable for further projections, taking into account the initial state and change probabilities in that state.

Markov analysis offers a powerful modelling technique with strong applications in time-based reliability and availability analysis. The reliability behaviour of the MB landscape system was represented using a transition matrix consisting of seven discrete states that the system could be in, with speeds defined for the rates at which transitions or flows between those states take place.

Markov models provide great flexibility in modelling the timing of events (Dale *et al.*, 2002). The Markov model is analyzed in order to determine comprehensive representations of future trends at a given point in time. In Figure 6-3, we can observe real data (from 1956 to 2000), as well as estimated data (from 2000 to 2044) using Markov analysis using the transition matrix of 1956-2000 (Peña *et al.*, 2005c).

From the empirical results noted, we can propose a future continuation of the trends observed within the period: a decline in non-irrigated and irrigated croplands due to land abandonment and an impressive growth in urban development, which in 2022 is expected to be the third major type of land use in MB. However, we cannot anticipate the specific socio-economic changes responsible for the magnitude of land-use changes.

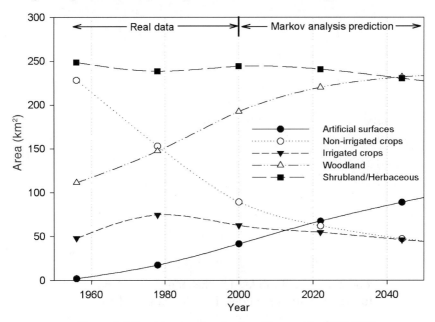

Figure 6-3. Land-use trends using transition matrix, 1956-2000.

The model presented here is unable to predict long-term trends accurately as it is based upon empirical initial input conditions. The small differences in the initial set of key parameters could lead to bigger differences in the development of trends through time, due to the complexity of the feedback and accumulative processes that characterise both the socio-economic and natural components of the system.

4.3   Land-use change processes

It is often difficult to elucidate the different types of processes that contribute to the landscape change progression. Thus, it is necessary to simplify reality and to focus only on a small number of processes. This approach can prove appropriate in order to summarise the complexity of the system. The analysis of the different processes was made by means of a spatial query built by cross-tabulating features from 1956 and 2000 land-use maps, in a manner similar to the construction of the transition matrix. The results can be divided in two different groups (Figure 6-4).

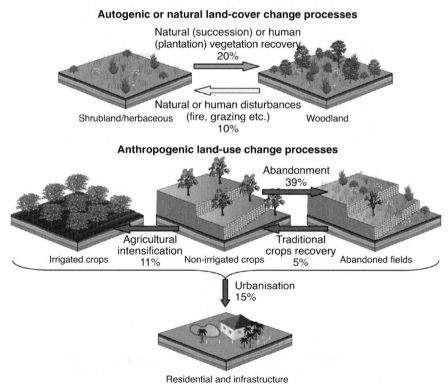

Figure 6-4. Land-cover/use change processes in the Marina Baixa catchment, 1956-2000. (See also Plate 2 in the Colour Plate Section)

Autogenic or natural land-cover change processes explain changes in the natural or human indirect environment, indicating that these processes are caused by natural forces. They represent 30% of the changes in MB and are composed of two opposite processes. Natural (succession) or human (plantation) vegetation recovery (20%) is the transition from shrubland or herbaceous to woodland (pine forest). On the other hand, natural or human disturbances (fire, grazing, etc.) convert woodland to shrubland or herbaceous cover (10%).

Anthropogenic or human-mediated land-use change processes include human land-use shifts. These are determined by socio-economic and political driving forces and are enabled by technological changes. Abandonment ('greening') contributes to increases in natural areas by the set aside of irrigated and non-irrigated crops (39%). Agricultural intensification consists of the modernisation of agriculture, which is typically the modification of non-irrigated to irrigated crops with a correspondingly higher demand for water (11%). Traditional crop recovery represents the inverse flow to that of abandonment, that is, to reversion of abandoned fields to dry crops. This phenomenon, known as 'fallowing', was quite common in the past (5%). Finally, urbanisation presents an irreversible change that converts irrigated, non-irrigated and abandoned fields to residential and infrastructural areas with sealed soil surfaces (15%). Urban expansion always takes place in fertile agricultural areas (mostly abandoned).

It is important to highlight how predominantly three opposite processes have shaped the MB landscape: *land abandonment, agricultural intensification* and *urbanisation* (Peña *et al.*, 2005b). These processes have profound consequences on the structure and functioning of landscapes and related ecosystem services. The intensification of agriculture and urban settlements has severe consequences on the water cycle, nutrients and pollutant contamination. On the other hand, the expansion of natural land cover (361.6 km^2 to 433.6 km^2) from a state of land abandonment indicates a decrease of human pressure in the more rural inland areas. It should be noted that this situation also implies an increase in wildfire risk.

Land abandonment: The abandonment of terraced dry crops can be widely observed within the study area, but it is concentrated in areas that have steep slopes or are far from settlements and roads. Some small areas of dry crops (typically almond, olive and carob) were still under management in rural areas, but almost all the cropland near the coast was abandoned between 1956 and 1978. The socioeconomic history of MB has generated a cultural landscape characterised by small properties, interspersed with shrubland and forest fragments, which facilitates the availability of seed and vegetal establishment. These locations do not suffer the effects of intensive desertification as the terraces there, even in an abandoned state, offer good growing conditions (good soil content, flat slopes), enabling spontaneous afforestation after abandonment (Bonet, 2004). Land abandonment (greening) was always the most important trend in absolute and relative terms, and is also in magnitude the event that masks other changes, especially so since 1978.

Agricultural intensification: Agricultural intensification is defined as a transition to systems having higher levels of inputs (water, fertilizer, pesticide) and increased output (in quantity or value) of cultivated or reared products per unit area and time. Monoculture (mostly medlar and citrus) is more productive economically than traditional dry crops (olive, almond or carob). Agricultural transformation from dry to irrigated crops is only possible in suitable sites in order to maximise outputs and is preferred in locations with low altitude, good soil conditions and proximity to settlements and roads. However, agricultural intensification requires major water consumption and soil requirements, and is viewed sceptically by observers contemplating the future of a stressed system in which the water resources are inadequate to maintain the actual pace of change. Most of the new irrigated crops belong to large companies and co-operatives related to agribusiness or commercial activities, housing or speculative operations. This type of activity is totally different to the traditional dry crop agricultural patterns. The trend of agricultural intensification was strongest in the period 1956-1978 and weakest between 1978 and 2000. However, the trend is important as it has remained relatively constant.

Urbanisation: Changes in the area of urban land *per se*, appear to be central to land-cover change in tourist areas. Urbanisation is a complex process that transforms the rural or natural landscape into urban, industrial and infrastructure areas. Urban area as land cover, in the form of built-up, sealed or paved-over areas, occupied some 6.5% in MB in 2000, although this had grown from 0.3% in 1956. Large-scale urban agglomerations and extended peri-urban settlements fragment the landscapes of such large areas, threatening various existing ecosystems. Ecosystem fragmentation, however, in peri-urban areas may be compensated by urban-led demands for conservation and recreational land uses. Coastal cities in MB attract a significant proportion of the local and national rural population by way of permanent and circulatory migration, which is just the opposite trend to that occurring in the inland municipalities which have suffered an ongoing decline in population since 1960. The trend of urbanisation is steadily becoming more important (as measured by the changes per year contributing to that trend) and has in the period 1978-2000 actually become more significant than agricultural intensification.

4.4   Landscape change driving forces

It is well known at the regional landscape level that environmental variables influence both human and natural patterns of landscape change. However, the major contributions are inevitably human-mediated and are promoted by several underlying processes or driving forces, influential in determining the evolutionary trajectory of the landscape change. The driving forces form a complex system of dependencies, interactions, and feedback loops, affecting several temporal and spatial levels. It is therefore difficult to analyse and represent them adequately (Blaikie, 1985).

An attempt in this direction, however, is made in Figure 6-5. This representation is a chronosequence of the major types of driving forces, identified as being: *political, technological, natural, cultural* and *socioeconomic* (Brandt *et al.*, 1999). It is based on an exhaustive study of the leading cadastral statistics of the region and a series of detailed interviews with farmers, managers of local irrigation systems, managers of local councils and other key actors and stakeholders. A discussion of the main driving forces is provided below.

Political driving forces are strongly interlinked with economic driving forces, due to economic needs and pressures. These are expressed and reflected in political programmes, laws and policy. The role of governance at national, regional and local scale is always to support increasing economic development based in territorial planning. Most of the laws and policies are directly connected to assessments of urban planning and water management. The principal policy of land-use change at municipal level is the drawing up of a 'PGOU' (a General Urban Plan), which serves as a tool to administer the land. It defines, for example, the areas disposed to change from cropland to urban. This kind of land reclassification is in accordance with political ordinances, but this also sometimes represents a quick method of enrichment and speculation as the price of land can and does change dramatically from before to after a revision of the plan. Benidorm, the largest city of MB County, approved its own General Urban Plan in 1956. It was the second municipality in Spain to do this (preceded by Barcelona). The Plan consisted of drawing up a rectilinear design of streets whose main axes would be two great avenues, that would cross to define the extension of this part of the city, resulting in the so-called 'matchbox strategy'. This planning model established the present layout of Benidorm, although at that time it was impossible to predict the numbers of tourists who now visit the city. This city is a model of tourist development and continues to grow in order to maintain its position as one of the premier tourist resorts in the world.

Technological driving forces maintain advances in the development of civilisation and these driving forces have contributed to shape the landscape enormously. Principally, technology has been used to improve resource management, land productivity and the quality of life. For instance, water is the most restrictive resource in MB; therefore technical advancements are evidenced in the building of dams, recycling by means of sewage plants or by desalinization plant construction to obtain fresh water. Technological modernisation, such as industrialised agriculture, has influenced some social changes. Farming trends have conditioned the shift of many fields from traditional dry crops to drip-irrigated systems, with the associated intensified use of fertilizers, pesticides, and the introduction of foreign crop varieties in orchards.

Natural driving forces include site factors (spatial configuration, topography, and soil conditions) as well as natural disturbances such as drought, wildfires and floods which induce long term global change (Antrop, 2005). The major natural attractiveness of the MB catchment for tourism is the weather, which provides more than 300 days of sun, as well as the sea, with numerous services and infrastructure for leisure and relaxation activities (e.g. open beaches, water sports). The Mediterranean semi-arid conditions in MB are characterized by drought periods of several years and torrential rainfalls in autumn. The environment has adapted to these circumstances accordingly. However, human pressure over the natural environment has caused an amplification of disturbances, i.e. during drought periods, it is more vulnerable to wildfires. Also, torrential rainfalls, known locally as *'gota fría'*, provoke major damage to infrastructure due to the intensity of the rains over short periods of time. To place this in context, in some cases more than half the annual rainfall may occur in one day. These events may trigger substantial flash floods in susceptible urban areas, particularly those that are built near or over drainage channels.

Cultural driving forces shape landscapes and these are then interpreted by individual landowners. In turn, people shape landscapes according to their beliefs, in order to achieve a good quality of life. The evolution of cultural perspectives through which the territory is managed is strongly interlinked with associated *socioeconomic* drivers. Traditional agriculture as the principal source of income was challenged during the 1960s as the industrial model emerged. This in turn never came to fruition due to strong competition from agricultural intensification and from tourism. Today, this spatial conflict among different wider land uses has all but ended, and practically all endeavours are now concentrated on urban or touristic development. Urban cadastral prices increase strongly due to sustained high demand. Equally, irrigated and non-irrigated land values have declined during the same period. The estimated cadastral value for a hectare of terraced dry cropland is 2,721 €/ha; for irrigated crops it is 5,723 €/ha and for urban it is 716,041 €/ha (Alcázar *et al.*, 1998). These extreme inequalities between land-use prices are designated by the General Urban Plan and determine land-use change patterns.

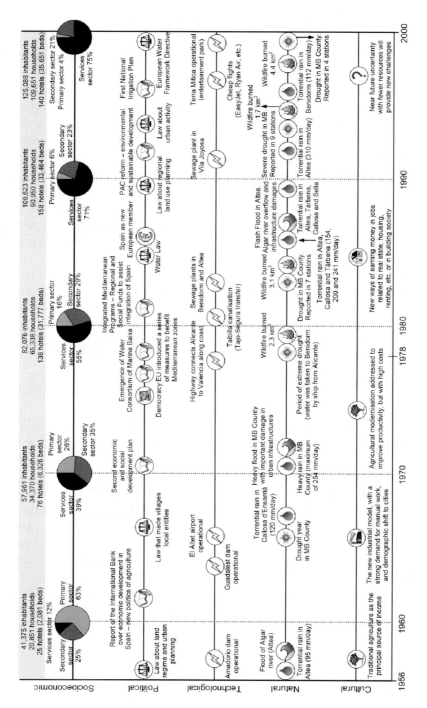

Figure 6-5. Timeline for the five major types of driving forces in Marina Baixa, 1956-2000.

5.   DISCUSSION

Two opposite processes can be observed in the MB landscape: firstly, intensification of agriculture and urbanisation, and secondly, land abandonment. Both processes have exerted profound consequences on the structure and functioning of landscapes. The intensification of agriculture and urban settlements has had severe impacts upon the hydrological cycle, nutrient loadings and pollutant contamination. Conversely, the expansion in natural land cover from abandoned land clearly indicates a decrease in human pressure in inland mountainous and rural areas. This implies an increasing and unrestrained wildfire risk but also represents a modification of the natural resources (e.g. increasing biodiversity, decreasing water availability and aquifer recharge). In addition, two further phenomena are observed, namely, habitat fragmentation in urban areas (in the coastal zone) and aggregation of new forestry cover (in the rural zone).

The pattern of expansion of urban and irrigated arable lands has led to a number of associated difficulties, an example of which has been the availability of secure water resources. However, the Government aims to counter its water problems by increasing availability through policy and seeks to solve supply deficits through the use of water transfers. This situation has failed to restrain non-sustainable growth in tourism, as well as in agriculture, and tends to promote continued growth and increasing water demands. Plans for future water transfers can only serve to stimulate the future non-sustainable development in MB.

There is a lack of balanced economic growth, as agricultural sector earnings are unstable and there is increasing pressure for more land to be made available to support the tourism sector. Equally, natural resources are often used unwisely, and conservation of the environment is over-looked. For example, aquifers are depleted, desertification problems abound and protected natural spaces are threatened by the expansion of agriculture and tourism. The irony is that the tourists are attracted by many factors, importantly including the quality of the environment at their destination.

Without sustainable development being adopted as a governing priority for policymakers, and with the continuance of current trends, future scenarios can be developed for MB in which the landscape will likely become a degraded landscape because of resource exhaustion, tourists will leave for other Mediterranean destinations and agriculture will be developed in the North of Africa or other areas of Spain and Eastern Europe. Predicted climate changes are likely to exacerbate these circumstances.

One of the main causes of these problems is the lack of planning and the lack of control of tourism and agriculture. Administrations are focused upon solving water demand problems and often seek to satisfy these increases in demand by means of technological advances (e.g. water transfers from the Tajo-Segura, desalinisation plants and sewage wastewater treatment plants). However, this type of solution fails to control the non-sustainable increase of tourism and agricultural activity and therefore the increase in demand for water is unlikely to cease in the foreseeable future. There is no easy panacea evident to address these issues, but certainly one important means to address this situation is the implementation of an integrated process of sustainable development and modelling approaches such as those evidenced in this research have an important role to play in helping the policymakers explore a range of alternative futures resulting from their land-use policies.

REFERENCES

Alcázar, M., Gilabert, M. and López, M. (1998) El Catastro en España, *Universidad Politécnica de Valencia*, Valencia.

Antrop, M. (1998) Landscape change: plan or chaos? *Landscape and Urban Planning*, 41: 155–161.

Antrop, M. (2005) Why landscapes of the past are important for the future, *Landscape and Urban Planning*, 70: 21–34.

Baker, W.L. (1989) A review of models of landscape change, *Landscape Ecology*, 2: 111–135.

Bellot, J., Sanchez, J.R., Bonet, A., Chirino, E., Martinez, J.M., Hernandez, N. and Abdelli, F. (1999) Effect of different vegetation type cover effects on the soil water balance in a semi-arid areas of south eastern Spain, *Physics and Chemistry of the Earth (B)*, 24: 353–357.

Blaikie, P. (1985) *The Political Economy of Soil Erosion*, Longman, London.

Blondel, J. and Aronson, J. (1999) Biology and Wildlife in the Mediterranean Region, Oxford University Press, Oxford.

Bonet, A., Bellot, J. and Peña, J. (2004) Landscape dynamics in a semiarid Mediterranean catchment (SE Spain), in Mazzoleni, S., Di Pasquale, G., de Martino, P. and Rego, F. (eds) *Recent Dynamics of Mediterranean Vegetation Landscape*, Wiley, London, p. 256.

Bonet, A., Peña J., Bellot, J., Cremades, M. and Sánchez, J.R. (2001) Changing vegetation structure and landscape patterns in semi-arid Spain, in Villacampa, Y., Brebbia, C.A. and Usó, J.L. (eds) *Ecosystems and Sustainable Development III*, WIT Press, Southampton, pp. 377–386.

Bonet, A. (2004) Secondary succession on semi-arid Mediterranean old-fields in South-eastern Spain: Insights for conservation and restoration of degraded lands, *Journal of Arid Environments*, 56: 213–233.

Brandt, J., Primdahl, J. and Reenberg, A. (1999) Rural land-use and dynamic forces - analysis of "driving forces" in space and time, in Krönert, R., Baudry, J., Bowler, I.R. and Reenberg, A. (eds) *Land-use Changes and their Environmental Impact in Rural Areas in Europe*, UNESCO, pp. 81–102.

Bürgi, M., Hersperger A.M. and Schneeberger, N. (2004) Driving forces of landscape change - current and new directions, *Landscape Ecology*, 19: 857–868.

Chapin, F.S., Walker, B.H., Hobbs, R.J. et al. (1997) Biotic control over the functioning of ecosystems, *Science*, 277: 500–504.

Dale, M, Dale, P. and Edgoose, T. (2002) Using Markov models to incorporate serial dependence in studies of vegetation change, *Acta Oecologica*, 23: 261–269.

Dunn, C.P., Sharpe, D.M., Guntenspergen, G.R., Stearns, F. and Yang, Z. (1991) Methods for analyzing temporal changes in landscape pattern, in Turner, M.G. and Gardner, R.H. (eds) *Quantitative Methods in Landscape Ecology: The Analysis and Interpretation of Landscape Heterogeneity*, Springer-Verlag, New York.

Farina, A. (1998) *Principles and Methods in Landscape Ecology*, Chapman and Hall, London.

Fernández-Alés, R., Martín, A., Ortega, F. and Alés, E.E. (1992) Recent changes in landscape structure and function in a mediterranean region of SW Spain (1950–1984), *Landscape Ecology*, 7(1): 3–19.

Forman, R.T.T. and Godron M. (1986) *Landscape Ecology*, Wiley, New York.

Lambin, E.F., Turner, B.L., Geist, H.J., Agbola, S.B., Angelsen, A., Bruce, J.W., Coomes, O.T., Dirzo, R., Fischer, G., Folke, C., George, P.S., Homewood, K., Imbernon, J., Leemans, R., Li, X., Moran, E.F., Mortimore, M., Ramakrishnan, P.S., Richards, J.F., Skanes, H., Steffen, W., Stone, G.D., Svedin, U., Veldkamp, T.A., Vogel, C. and Xu, J. (2001) The causes of land-use and land-cover change: moving beyond the myths, *Global Environmental Change*, 11: 261–269.

Lipschutz, S. (1968) *Theory and Problems of Probability*, McGraw-Hill Book Company, New York.

Llamas, M.R. (2000) Some lessons learnt during the drought of 1991–1995 in Spain, in Vogt, J.V. and Somma, F. (eds) *Drought and Drought Mitigation in Europe*, Kluwer Academic Publishers, Dordrecht, pp. 253–264.

Marcucci, D. (2000) Landscape history as a planning tool, *Landscape and Urban Planning*, 49: 67–81.

Meyer, W.B. and Turner, B.L. (1994) *Changes in Land Use and Land Cover: A Global Perspective*, Cambridge University Press, Cambridge.

Moreira, F., Rego, F. and Ferreira, P. (2001) Temporal (1958–1995) pattern of change in a cultural landscape of northwestern Portugal: implications for fire occurrence, *Landscape Ecology*, 16: 557–567.

Palutikof, J.P., Conte, M., Casimiro, Mendes, J., Goodess, C.M. and Espirito Santo, F. (1996) Climate and climatic change, in Brandt, C.J. and Thornes, J.B. (eds) *Mediterranean Desertification and Land Use*, Wiley, Chichester, pp. 43–86.

Pan, D., Domon, G., De Blois, S. and Bouchard, A. (1999) Temporal (1958–1993) and spatial patterns of land use changes in Haut-Saint-Laurent (Quebec, Canada) and their relation to landscape physical attributes, *Landscape Ecology*, 14: 35–52.

Pastor, J. and Johnston, C.A. (1992) Using simulation models and geographic information systems to integrate ecosystem and landscape ecology, in Naiman R.J. (ed) *Integrated Watershed Management*, Springer-Verlag, New York, pp. 324–346.

Peña, J., Martínez, R. M., Bonet, A., Bellot, J. and Escarré, A. (2005a) Cartografía de las coberturas y usos del suelo de la Marina Baixa (Alicante) para 1956, 1978 y 2000, *Investigaciones Geográficas 37*, Universidad de Alicante.

Peña, J., Bonet, A. and Bellot, J. (2005b) Historical land cover and land use changes in Marina Baixa from 1956 to 2000, *European IALE congress "Landscape Ecology in the Mediterranean: Inside and Outside Approaches'*, Faro (Portugal), 29 March-2 April.

Peña, J., Bonet, A., Bellot, J. and Sánchez, J.R. (2005c) Trends and driving factors in land use changes (1956–2000) in Marina Baixa, SE Spain, Paper presented at the 45th Congress of the European Regional Science Association on Land Use and Water Management in a Sustainable Network Society. Amsterdam, 23–27 August.

Quereda-Sala, J.J. (1978) Comarca de la Marina. Estudio de Geografía Regional. *Excma.* Diputación Provincial de Alicante, Alicante.

Quereda-Sala, J.J., Gil-Olcina, A., Pérez-Cueva, A., Olcina-Cantos, J., Rico-Amorós, A. and Montón-Chiva, E. (2000) Climatic warming in the Spanish Mediterranean: natural trend or urban effect, *Climatic Change*, 46(4): 473–483.

Solanas, J.L. and Crespo, M.B. (2001) Medi físic i flora de la Marina Baixa, Universitat d'Alacant.

Shugart, H.H. (1998) *Terrestrial Ecosystems in Changing Environments*, Cambridge University Press, Cambridge.

Turner II, B.L., Clark, W.C. Kates, R.W., Richards, J.F., Mathews, J.T. and Meyer, W.B. (eds) (1990) *The Earth as Transformed by Human Action*, Cambridge University Press, Cambridge.

Turner, M.G. (1990) Spatial and temporal analysis of landscape patterns, *Landscape Ecology*, 4: 21–30.

Vitousek, P.M. (1992) Global environmental change: an introduction, in Fautin, D.G., Futuyma, D.J. and James, F.C. (eds) *Annual Review of Ecology and Systematics*, 23: 1–14.

Wilson, J.B. and King, W.M. (1995) Human-mediated vegetation switches as processes in landscape ecology, *Landscape Ecology*, 10: 191–196.

Chapter 7

EMPIRICALLY DERIVED PROBABILITY MAPS TO DOWNSCALE AGGREGATED LAND-USE DATA

N. Dendoncker[1], P. Bogaert[2] and M. Rounsevell[1]

[1]*Département de Géographie, Université Catholique de Louvain, Belgium;* [2]*Département d'Agronomie, Université Catholique de Louvain, Belgium*

Abstract: Land-use simulation results are often provided at spatial resolutions that are too coarse to establish links with local or regional studies that, for example, deal with the physical or ecological impacts of land-use change. This chapter aims to use novel spatial statistical techniques to derive representations of land-use patterns at a resolution of 250 metres based on aggregate land-use change simulations. The proposed statistical downscaling method combines multinomial autologistic regression and an iterative procedure using Bayes' theorem. Based on these methods, a set of probability maps of land-use presence is developed at two time steps. The method's low data requirements (only land-use datasets are used) make it easily replicable, allowing application over a wide geographic area. The potential of the method to downscale land-use change scenarios is shown for a small area in Belgium using the CORINE land-cover dataset.

Keywords: Downscaling; multinomial logistic regression; Bayes' theorem; suitability maps.

1. INTRODUCTION

The ATEAM (Advanced Terrestrial Ecosystem Analysis and Modelling) project developed land-use scenario maps of Europe for the years 2020, 2050 and 2080 (Rounsevell *et al.*, 2006; Schröter *et al.*, 2005). The ATEAM land-use maps give land-use shares (in percentages) for each cell over a 10 minute longitude/latitude grid. This resolution does not allow the

E. Koomen et al. (eds.), Modelling Land-Use Change, 117–131.
© 2007 *Springer.*

identification of land-use change effects at the landscape level and is insufficient to establish a link with local case studies (Verburg *et al.*, 2006). The creation of pan-European, fine-resolution land-use change scenarios would be useful for various purposes. For example, land-use patterns have been shown to affect numerous ecological processes (Parker and Meretsky, 2004). Since land use is a decisive factor in modelling community and species distribution, it is nowadays also being integrated into ecological modelling approaches (Peppler-lisbach, 2003). Animal ecologists need to know the precise location of land-use change. This is especially important for migratory bird species whose population dynamics can be strongly influenced by land-use change over wide areas (Gauthier *et al.*, 2005). Downscaling is also essential to better assess the land-use change impacts on the biodiversity of natural areas, which depend not only on the quantity but also on the spatial configuration of natural areas, determining the relative connectivity or isolation of species and habitat (Wimberly and Ohmann, 2004). Soil scientists study changes in soil organic carbon stocks which directly depend on land use (Lettens *et al.*, 2004).

A simple land-use allocation procedure disaggregating coarse-resolution land-use data and the ATEAM data in particular, would be useful in such studies. Therefore, the main objective of this chapter is to set the basis for a statistically consistent downscaling procedure based on: (1) a multinomial autologistic regression model to generate maps of the probability of land-use presence based on the European-wide CORINE land-cover raster dataset (European Commission, 1993) and (2) a methodology to update these probability maps using Bayes' theorem and the ATEAM land-use change scenarios. Whilst the work presented here is based on an application that uses land-use change scenarios, the methodology is appropriate for the downscaling of other types of aggregated land-use data, e.g. at the level of administrative units.

2.   CONTEXT OF THE STUDY

2.1   Land-use drivers, spatial autocorrelation and neighbourhoods

Broadly speaking, two main types of variables are used within empirical land-use change models: non-neighbourhood based variables and neigh-bourhood-based variables. While the former category is included in virtually all land-use studies, only a limited number of studies explicitly deal with neighbourhood effects and spatial autocorrelation. Neighbourhood effects, reflecting centripetal forces, are known to have a role in the spatial structuring of land use and the landscape (Verburg *et al.*, 2004; White and Engelen, 1993). When regression is performed on spatial data, it is likely that spatial autocorrelation will remain in the residuals (Anselin, 2002). This will occur unless the regression is performed on a non-autocorrelated data sample, as discussed by Serneels and Lambin (2001) or Peppler-lisbach (2003), but this results in a loss of information. Alternatively, in spatial models, a part of the variance can be explained by neighbouring values (Overmars *et al.*, 2003).

2.2   Logistic regression in land-use modelling

Binomial logistic regression (BLR) has been widely used in the field of land-use change modelling to quantify the relationships between driving factors and land-use (change) patterns and to derive land-use probability maps (Veldkamp and Fresco, 1996; Mertens and Lambin, 1997; Hilferink and Rietveld, 1998; Serneels and Lambin, 2001; Peppler-lisbach, 2003; Verburg *et al.*, 2002). However, it is a time consuming process that requires the acquisition and treatment of a large number of datasets of the independent variables.

Conversely, multinomial logistic regression (MLR) models have not often been used in land-use modelling. However, MLR models are probably better than BLR if the aim is to obtain the best fit probability maps of land-use presence. This is because they consider each land use as a possible alternative amongst the whole set of land-use categories. Indeed, MLR models generate consistent probability maps of land-use presence, the sum of probabilities for a given pixel being equal to 1, which is not the case in BLR for which probabilities are computed for each land use independently of other land uses. In summary, MLR allows discrimination between land uses on a non-biased basis.

This study uses an autologistic MLR based on neighbourhood variables, which is described in the following section. Using only neighbourhood-based variables to build probability maps allows other variables or drivers to be ignored. This makes the statistical basis of the land-use allocation procedure simple and easily replicable. Some recent work has incorporated spatial dependencies into qualitative dependent variables and discrete choice models (e.g. Augustin *et al.*, 2001; Bhat and Guo, 2003; Ben-Akiva and Lerman, 1985; Mohammadian and Kanaroglou, 2003). However, examples of autologistic MLR in land-use studies are rare. An exception is offered by

McMillen (2001), who applied this framework to model land use in the urban fringe area of Chicago considering three land-use classes.

3.   DATA AND METHODS

The analysis presented here consists of two main parts: (1) deriving a statistically consistent autologistic MLR model to estimate baseline probability maps of land-use presence; and (2) updating these probability maps with an iterative procedure based on the Bayes' theorem and on land-use change scenario data. The CORINE (Coordinated Information on the European Environment) land-cover map serves as the baseline dataset to derive the probability maps of land-use presence whilst the ATEAM scenario data (Rounsevell *et al.*, 2006; Schröter *et al.*, 2005) provides the quantitative constraints to update the probability maps.

3.1   Data

3.1.1   Dependent variable: land-use categories

There are very few European wide land-use datasets. The two main datasets are the PELCOM land-cover dataset (Mücher *et al.*, 2000) and the CORINE land-cover dataset (European Commission, 1993). The CORINE dataset was used here because it has a finer spatial resolution (i.e. 250 metres instead of 1100 metres) and because it is generally assumed to be of better quality than PELCOM (Schmit *et al.*, 2006). The official classification accuracy of CORINE is about 87%.

3.1.2   Independent variables

To keep the autologistic MLR model as simple as possible, a neighbourhood variable was built by taking the percentage of the same land use within the eight cells immediately surrounding the cell (Moore neighbourhood) with that land use. No others variables were included.

3.1.3   Land-use change scenarios

The land-use classes given by the ATEAM scenarios of land-use change are urban, cropland, grassland, forest, biofuels (liquid, non-woody and woody) and surplus (i.e. abandoned land). The latter two are absent from the CORINE land-cover map. To match the baseline land-cover categories, liquid biofuels (e.g. oilseed rape) were grouped with cropland while non-woody and woody biofuels (e.g. willow plantations) were grouped with forest. These land-use change scenarios provide the basis to update the probabilities of land-use presence given by the logistic regressions.

To demonstrate the method, a subset of CORINE was arbitrarily chosen corresponding to a small area within southern Belgium, covering nine ATEAM cells or, approximately, 3000 km^2. The method is, however, applicable to larger geographic areas. The various land-cover classes present in the study area were reclassified into 'urban', 'cropland', 'grassland', and 'forests'. All minor land cover classes (<1 % of the total area) that did not match any of these categories (e.g. water courses and bare rock) were reclassified as 'others'.

3.2 Multinomial autologistic regression

The general form of MLR model can be written as:

$$p(c_i|N_\alpha) = \frac{\exp(\mu_i N_\alpha)}{\sum_j \exp(\mu_j N_\alpha)} \quad i = 1\ldots k \tag{1}$$

where $p(c_i|N_\alpha)$ is the probability that a CORINE cell α would take land-use class i from the set of k possible land-use categories conditional on the knowledge of the vector of characteristics N_α specific to the CORINE cell α. μ_i represents non-negative scale parameters. In the case of the autologistic MLR used in this study, only the eight immediate neighbours of each CORINE cell were taken into account to build each N_α. The CORINE cells contained in the nine eastern ATEAM cells were used as a training set to parameterise the MLR model while the CORINE cells contained in the nine western ATEAM cells were used as a validation set.

3.3 Updating the probabilities of land-use presence

Without loss of generality, let us consider that for any given CORINE cell, the eight immediate neighbours have been taken into account when estimating the probability of observing each possible category for the central cell at an initial time step t. If we denote N_α as a specific frequency of land use i in the neighbourhood (among the set of nine possible frequencies for land use i), the multinomial logit procedure provides us with the conditional probabilities $p(c_i|N_\alpha)$ (where $i = 1, \ldots, k$), i.e. the probabilities of each possible category of the central cell given the neighbourhood category frequencies N_α. Moreover, from the CORINE database, we are also able to obtain good estimates of the global theoretical frequencies (marginal probabilities) for each land use, that will be denoted as $p(ci)$. From Bayes' theorem, we can also write that:

$$p(c_i|N_\alpha) = \frac{p(N_\alpha|c_i)p(c_i)}{\sum\limits_{j=1}^{k} p(N_\alpha|c_i)p(c_i)} = \frac{1}{A}p(N_\alpha|c_i)p(c_i) \qquad (2)$$

where the denominator (A) plays the role of a normalization constant, so that we always have $\Sigma_i\, p(c_i|N_\alpha) = 1$. Based on this result we can state that, up to a multiplicative constant, the conditional probabilities can be computed as the product of marginal probabilities $p(c_i)$ – that may change according to possible global modifications of the land-use frequencies over time – and conditional probabilities $p(N_\alpha|c_i)$ that characterize the local spatial organization of the various land uses, assumed to be largely invariant over time. As we have explicitly specified, the $p(c_i|N_\alpha)$ values from the multinomial logit model as well as the $p(c_i)$ values from the whole CORINE database, these various $p(N_\alpha|c_i)$ values can be computed from the MLR model, with:

$$p(N_\alpha|c_i) = A\frac{p(c_i|N_\alpha)}{p(c_i)} \qquad (3)$$

This way of presenting the problem is particularly useful as it allows us to take into account the effect of the various ATEAM scenarios. Indeed, let us consider that at time t' after t the theoretical frequencies of land use become equal to the new values $p'(c_1), \ldots, p'(c_k)$. From Eq. (2), it is clear that a change of the marginal probability from $p(c_i)$ to $p'(c_i)$ will cause a change in the conditional probability from $p(c_i|N_\alpha)$ to a new value $p'(c_i|N_\alpha)$. However, conditional to the observed category c_i, one can assume that the probabilities values $p(N_\alpha|c_i)$ remain unchanged when moving from time t to t'. Stated in other words, conditionally to the fact that a given land-use category c_i is observed for the central cell in the Moore neighbourhood, the probability of observing a given frequency of identical pixels in this neighbourhood will remain unchanged over time, despite the fact that the probability that the central cell belongs to this category c_i may change over time. Using Eq. (3), we can thus write that subject to the constraint $\Sigma_i\, p'(c_i|N_\alpha) = 1$:

$$A \frac{p(c_i|N_\alpha)}{p(c_i)} = p(N_\alpha|c_i) = A' \frac{p'(c_i|N_\alpha)}{p'(c_i)} \Leftrightarrow p'(c_i|N_\alpha) \propto \frac{p'(c_i)}{p(c_i)} p(c_i|N_\alpha) \quad (4)$$

The last relationship in Eq. (4) is thus an updating rule for the computation of new conditional probabilities $p'(c_i|N_\alpha)$ when the marginal probabilities are changed from the set of values $\{p(c_1), \ldots, p(c_k)\}$ that apply at time t to the new set of values $\{p'(c_1), \ldots, p'(c_k)\}$ that apply later at time t'. In practice, these new marginal probabilities can be estimated from the land-use frequencies (i.e. land-use proportions f'_i) of the ATEAM cell within which the CORINE cell is included. For each CORINE cell, the outputs of Eq. (4) are normalized so that $\Sigma_i\ p'(c_i|N_\alpha) = 1$, as required for a valid probability distribution.

Although the above procedure provides us with a simple method for computing new conditional probabilities for any CORINE cell included within a given ATEAM cell, there is still a consistency issue. The marginal probabilities as directly specified by the corresponding ATEAM frequencies $\{f'_1, \ldots, f'_k\}$ may differ somewhat from the marginal probabilities $\{p'(c_1), \ldots, p'(c_k)\}$ that can also be estimated later from the new set of conditional probabilities using the formula:

$$\hat{p}'(c_i) = \frac{1}{n} \sum_{j=1}^{n} p'(c_i|N_{[j]}) \quad (5)$$

where $p'(c_i|N_{[j]})$ refers to the estimated conditional probability of category c_i for the jth CORINE cell. This consistency problem is an indirect consequence of the normalization step that involves a different normalization constant for each CORINE cell in general. Indeed, using Eq. (5) with the (non-normalized) ouputs of Eq. (4) will lead to estimated marginal probability values that are identical to the specified f_i ATEAM frequencies, but these outputs do not provide valid distributions (as in general these probabilities will not sum up to one for each CORINE cell). Conversely, the normalized outputs will respect this validity condition, but using Eq. (5) will no longer guarantee that the estimated marginal probabilities match the ATEAM frequencies in general. In other words, the above procedure does not allow the user to derive conditional probability values that respect at the same time the validity condition $\Sigma_i\ p'(c_i|N_\alpha) = 1$, and the specified frequencies f_i for the corresponding ATEAM cell. A way of avoiding this consistency problem is to use an iterative procedure that aims to modify the set of conditional probabilities in such a way that both conditions can be fulfilled at the same time.

If the procedure described above is applied when deriving the new set of marginal probabilities at time step t' using the ATEAM cells as defined in the initial data set, there will be a clear border effect. The same marginal probabilities will apply for all CORINE cells belonging to the same ATEAM cell, whatever their location within the cell. As a consequence, there will be a sudden shift of these marginal probabilities when one crosses the border between two adjacent ATEAM cells, thus creating maps with clear lines that are obvious artefacts. The validity of using the same set of marginal probabilities for all CORINE cells inside the same ATEAM cell is debatable, especially for cells located close to the borders; ATEAM land-use frequencies are artificially integrated values over an area of about 15x15 kilometres, whereas the land use itself is a continuous variable across space. To avoid this artefact, locally smoothed estimates of the marginal probabilities were obtained using an inverse distance weighting (IDW) procedure. More detail about this IDW procedure and the iterative procedure to modify the set of conditional probabilities are given in Dendoncker *et al.* (2005).

4.   RESULTS

4.1   Multinomial autologistic regression

The general fit of the model is very good (Mc Fadden's likelihood ratio index = 0.63) and the neighbourhood variable is highly significant (p value < 0.0001). This confirms the contention that using neighbourhood variables only to derive probability maps of land-use presence at the CORINE resolution was appropriate. The associated output probability table is the starting point for the rest of this study, as the main goal of this model is to derive baseline probability maps. This set of probabilities will be the reference database for the updating methodology described in Section 3.3.

The confusion matrix (or contingency table) comparing the original CORINE data with the predicted land use (i.e. the land-use category with the highest probability in the validation set) is given as Table 7-1. The overall accuracy (0.77) and the Kappa coefficient (0.72) (Richards and Jia, 1999; Pontius, 2000) are both good. To consider the similarity of location and the similarity of quantity independently, the Kappa statistic was partitioned into 'K-location' (as defined by Pontius, 2000) and 'K-quantity' (as defined by Hagen, 2002). The former equals 0.65 while the latter equals 0.98. This suggests that the model predicts quantity with more accuracy than location. Not surprisingly, the model does not predict the 'others' category well as only 6 of the 199 'others' pixels are predicted correctly. This can be explained by the under-representation of this artificially constructed category and the fact that it mainly consists of linear elements (rivers), that are not clustered. However, this category is not subject to change and will be masked when the allocation procedure is applied to real data. It was only included here for completeness, to have a consistent set of alternatives (ensuring that predicted probabilities add up to one) and to avoid holes in the gridded dataset. All other land-use categories are fairly well predicted.

Table 7-1. Confusion matrix of CORINE (reclassified) compared to the land use predicted by the MLR model at 250m resolution for the selected region

	PREDICTED LAND USE					
	Urban	Cropland	Grassland	Forest	Others	Total
Urban	2,979	961	303	373	116	4,732
Cropland	1,356	14,387	747	729	10	17,229
Grassland	261	537	1,496	220	5	2,519
Forest	323	1,035	263	5,415	62	7,098
Others	1	0	0	0	6	7
Total	4,920	16,920	2,809	6,737	199	31,585

Values represent number of pixels.

A spatial comparison of the observed CORINE data with the predicted land use for the western, validation part of the study area in southern Belgium is given in Figure 7-1. This shows that the model tends to further compact existing land-use clusters while isolated cells are lost. This leads to a more aggregated and less fragmented landscape pattern. These results are confirmed by some simple landscape metrics computed with FRAGSTAT (McGarigal *et al.*, 2002). The total number of patches decreases from 1,280 in the original CORINE land-use map to 881 in the predicted land-use map. Conversely, the mean patch size increases from 154 ha in the original CORINE map to 224 ha in the predicted map. Moreover, cells located at the fringes of land-use clusters are usually poorly predicted. This is the consequence of using a purely neighbourhood-based regression model.

It is worth noting that the result of the procedure is a map showing the most probable (in a statistical sense) land-use pattern, but not necessarily the most realistic one. Indeed, a maximum probability map does not include any measure of the uncertainty attached to it (e.g. for such a map, cells mapped as grassland may have conditional probabilities that are quite different from one place to another, even if grassland is the maximum probability land use for those cells). The map reflects only a small part of the full information that is made available from the corresponding conditional probability distribution for all possible land uses, as computed using the autologistic MLR model. It is, however, a convenient choice for visualization purposes.

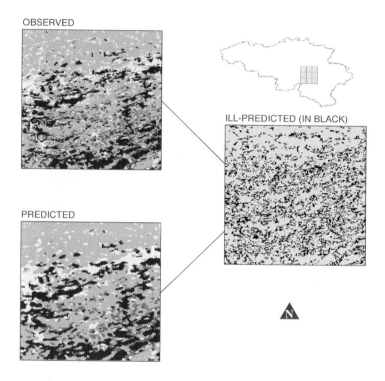

Figure 7-1. Observed versus predicted land use, with incorrectly predicted pixels shown in black for the study area in southern Belgium measuring approximately 50 km across.

4.2   Updated probabilities of land-use presence

Figure 7-2 shows the land-use maps resulting from the downscaling of the four ATEAM scenarios using the updated conditional probabilities for the year 2020 (time t'). The land use with the highest predicted conditional probability is represented. As expected from the baseline probability map resulting from the MLR procedure (Figure 7-1), the landscape pattern is more aggregated and less fragmented. However, no border effects can be seen.

All river cells from the baseline CORINE dataset were artificially replaced by different land-use categories. This is not considered to be a problem as the 'others' (i.e. river and bare rock) category will be restricted from change in a practical application of the method. Similarly, no indication is given as to which pixels will be abandoned from commercial agricultural management (i.e. the 'surplus' category from the ATEAM scenarios). However, other studies suggest that for agriculture these are likely to take place in less favoured areas (Verburg *et al.*, 2006).

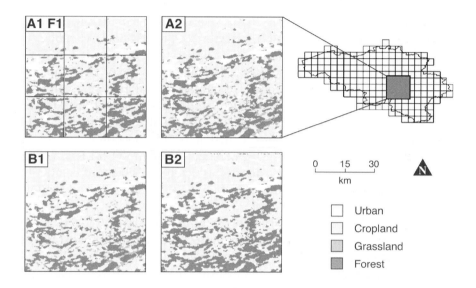

Figure 7-2. Updating the conditional probabilities: four ATEAM scenarios in 2020 (A1 FI refers to a fossil fuel intensive version of A1). (See also Plate 3 in the Colour Plate Section)

The main differences between the four scenarios of land-use change that can be observed in Figure 7-2 result from the differences in the land-use frequencies given by the ATEAM scenarios rather than from the probability updating methodology itself. For example, the A2 and B1 scenarios display very similar land-use patterns and quantities, both preserving most of their grassland cells although, on the whole, European grassland diminishes much more in A2 (Rounsevell *et al.*, 2006). In the A1 scenario, grassland diminishes mainly in the Least Favoured Areas (LFAs) represented by the southern cells in this example. In general, grassland is replaced by 'liquid' biofuels classified as cropland here. Finally, B2 shows the almost complete disappearance of grassland and its replacement by 'solid' biofuels here classified as forest.

5.   DISCUSSION

The presented methodology to downscale aggregate land-use data proved to be able to visualize detailed land-use patterns from the coarse ATEAM scenarios of land-use change. The maps, resulting from a combination of multinomial autologistic regression and a subsequent iterative procedure based on Bayes' theorem, give an appropriate representation of the land-use pattern despite a tendency to increase clustering and *a fortiori* lose part of the initial fragmentation. This is a consequence of using a purely autoregressive model. However, using neighbourhood variables is considered to be a real advantage as no ancillary data are needed to derive the statistically based probability maps. These datasets are often incomplete, are not always available at adequate resolution or spatial extent, can be of debatable quality or simply do not exist. Thus, finding appropriate datasets to derive optimal variables that represent land-use drivers is always a time and effort consuming process. The methodology developed in this chapter is simpler and quicker to implement. A final advantage of the MLR procedure is that it discriminates between land uses as complete vectors of conditional probabilities are derived (i.e. the probabilities of presence for all land-use classes within each cell sum to one).

The downscaled land-use scenarios presented in Figure 7-2 are 'mean' maps representing the most probable patterns (in a statistical sense) but not necessarily the most realistic. In order to derive such maps, it is necessary to adapt the calculated probability with decision rules that reflect the assumed changes in location preferences (Verburg *et al.*, 2006). In addition to the above comments, it should be noted that downscaling is merely the last step in the multi-scale land-use modelling process. The main differences between land-use models (especially in land-use quantities) will always arise from the interpretation of storylines and their translation into quantitative scenarios. Nevertheless, it is important to use optimal probability maps to underpin a rule-based procedure. Rules need to include elements such as policy constraints (e.g. designation of protected areas, precise delineation of LFAs). Some conversions should be made impossible (e.g. existing urban and river cells will always remain as such), others should only be possible after certain time lags. A simple conversion matrix as proposed by Verburg *et al.* (2006) would account for this. Finally, some rules could be specific to certain scenarios and some *a posteriori* decisions need to be made regarding the location of biofuels and abandoned land. When this rule-based approach is achieved, the resulting land-use scenario maps will be useful for a wide range of scientific disciplines. Possible users for the detailed land-use maps include soil scientists, who are concerned with the evolution of soil carbon related to land-use change, or for ecologists who are concerned with the evolution of the spatial patterns of land-use and their relationships with biodiversity. The statistical analysis presented here offers a first step in this direction.

ACKNOWLEDGEMENTS

This work was funded by the Framework 5 Programme (RTD priority 2.2.1: Ecosystem Vulnerability) of the European Commission via the project FRAGILE (Fragility of Arctic Goose habitat: Impacts of Land use, conservation and Elevated temperature). The authors appreciate the availability of the data of the ATEAM (Advanced Terrestrial Ecosystem Analysis and Modelling, No. EVK2-2000-00075) project.

REFERENCES

Anselin, L. (2002) Under the hood. Issues in the specification of spatial regression models, *Agricultural Economics*, 27: 247–267.

Augustin, N.H., Cummins, R.P. and French, D.D. (2001) Exploring spatial vegetation dynamics using logistic regression and a multinomial logit model, *Journal of Applied Ecology*, 38: 991–1006.

Ben-Akiva, M. and Lerman, S.R. (1985) *Discrete Choice Analysis: Theory and Applications to Travel Demand*, Cambridge, MA.

Bhat, C.R. and Guo, J. (2003) A mixed spatially correlated logit model: formulation and application to residential choice modelling, In IATBR conference, Lucerne, Switzerland.

Dendoncker, N., Bogaert, P. and Rounsevell, M. (2005) A statistical methodology to downscale aggregated land use data and scenarios, *Journal of Land Use Science* (Submitted).

European Commission (1993) Corine land cover map and technical guide, Technical report, European Union Directorate General Environment (Nuclear Safety and Civil Protection), Luxembourg.

Gauthier, G., Giroux, J.F., Reed, A., Brechet, A. and Belanger, L. (2005) Interactions between land use, habitat use, and population increase in greater snow geese: what are the consequences for natural wetlands? *Global Change Biology*, 11(6): 856–868.

Hagen, A. (2002) Map comparison – methods, Technical report, Riks, Maastricht.

Hilferink, M. and Rietveld, P. (1998) Land use scanner: an integrated GIS-based model for long term projections of land use in urban and rural areas, Technical report, Tinbergen Institute, Amsterdam.

Lettens, S., Van Orshoven, J., van Wesemael, B., Perrin, D. and Roelandt, C. (2004) The inventory-based approach for prediction of SOC change following land use change, *Biotechnology, Agronomy, Society and Environment*, 8(2): 141–146.

McGarigal, K., Cushman, S.A., Neel, M.C. and Ene, E. (2002) FRAGSTATS: Spatial pattern analysis program for categorical maps. Computer software program produced by the authors at the University of Massachussetts, Amherst. Available at the following website: www.umass.edu/landeco/research/fragstats/fragstats.html

McMillen, D. (2001) An empirical model of urban fringe land use, *Land Economics*, 65(2): 138–145.

Mertens, B. and Lambin, E.F. (1997) Spatial modelling of deforestation in Southern Cameroon. spatial disaggregation of diverse deforestation processes, *Applied Geography*, 17: 143–162.

Mohammadian, A. and Kanaroglou, P.S. (2003) Applications of spatial multinomial logit model to transportation planning. In: moving through nets: the physical and social dimensions of travel, Lucerne.

Mücher, C.A., Steinnocher, K.T., Kressler, F.P. and Heunks, C. (2000) Land cover characterization and change detection for environmental monitoring of pan-Europe, *International Journal of Remote Sensing*, 21(6): 1159–1181.

Overmars, K.P., de Koning, G.H.J. and Veldkamp, A. (2003) Spatial autocorrelation in multi-scale land use models, *Ecological Modelling*, 164: 257–270.

Parker, D.C. and Meretsky, V. (2004) Measuring pattern outcomes in an agent-based model of edge-effect externalities using spatial metrics, *Agriculture, Ecosystems and Environment*, 101(2–3): 233–250.

Peppler-lisbach, C. (2003) Predictive modelling of historical and recent land use patterns, *Phytocoenologia*, 33(4): 565–590.

Pontius, R.G. (2000) Quantification error versus location error in comparison of categorical maps, *Photogrammetric Engineering and Remote Sensing*, 66(8): 1011–1016.

Richards, J.A. and Jia, X. (1999) *Remote Sensing Digital Image Analysis*, An Introduction. Springer, Berlin.

Rounsevell, M.D.A., Reginster, I., Araujo, M.B., Carter, T.R., Dendoncker, N., Ewert, F., House, J.I., Kankaanpää, S., Leemans, R., Metzger, M., Schmit, C., Smith, P. and Tuck, G. (2006) A coherent set of future land use change scenarios for Europe, *Agriculture, Ecosystems and Environment*, 114: 57–68.

Schmit, C., Rounsevell, M.D.A. and La Jeunesse, I. (2006) The limitations of spatial land use data in environmental analysis and policy, *Environmental Science and Policy*, 9(2): 174–188.

Schröter, D., Cramer, W., Leemans, R., Prentice, C., Arnell, A.W., Araujo, M.B., Bondeau, A., Bugmann, H., Carter, T.R., de la Vega-Leinert, A.C., Erhard, M., Ewert, F., Fritsch, P., Friedlingstein, P., Glendining, M., Gracia, C.A., Hickler, T., House, J.I., Hulme, M., Kankaanpää, S., Klein, R.J.T., Lavorel, S., Lindner, M., Liski, J., Metzger, M., Meyer, J., Mitchell, T., Morales, P., Reidsma, P., Pla, E., Pluimers, J., Pussinen, A., Reginster, I., Rounsevell, M., Sanchez, A., Sabaté, S., Sitch, S., Smith, B., Smith, J., Smith, P., Stykes, M.T., Thonicke, K., Thuiller, W., Tuck, G., van der Werf, G., Vayreda, J., Wattenbach, M., Wilson, D.W., Woodward, F.I., Zaehle, S., Zierl, B., Zudin, S., Acosta-Michlik, L., Moreno, R., Espinera, G.Z., Mohren, F., Bakker, M., and Badeck, F. (2005) Ecosystem service supply and vulnerability to global change in Europe, *Science*, 310(5752): 1333–1337.

Serneels, S. and Lambin, E.F. (2001) Proximate causes of land-use change in Narok district, Kenya: a spatial statistical model, *Agriculture, Ecosystems and Environment*, 85: 65–81.

Veldkamp, A. and Fresco, L.O. (1996) Clue-CR: an integrated multiscale model to simulate land use change scenarios in Costa Rica, *Ecological Modelling*, 91: 231–248.

Verburg, P.H., Veldkamp, A. and Espaldon, R.L.V. (2002) Modeling the spatial dynamics of regional land use: the clue-s model, *Environmental Management*, 3: 391–405.

Verburg, P.H, de Nijs, T., van Eck, J.R., Visser, H. and de Jong, K. (2004) A method to analyse neighbourhood characteristics of land use patterns, *Computers, Environment and Urban Systems*, 28(6): 667–690.

Verburg, P.H, Schulp, C.J.E., Witte, N. and Veldkamp, A. (2006) Downscaling of land use change scenarios to assess the dynamics of European landscapes, *Agriculture, Ecosystems and Environment*, 114: 39–56.

White, R. and Engelen, G. (1993) Cellular automata and fractal urban form: a cellular modelling approach to the evolution of urban land use patterns, *Environment and Planning A*, 25: 1175–1199.

Wimberly, M.C. and Ohmann, J.L. (2004) A multi-scale assessment of human and environmental constraints on forest land cover change on the Oregon (USA) coast range, *Landscape Ecology*, 19: 631–646.

Chapter 8

A SPATIAL INTERACTION MODEL FOR AGRICULTURAL USES

An application to understand the historic land-use evolution of a small island

J. Gonçalves and T. Dentinho
Gabinete de Gestão e Conservação da Natureza, Departamento de Ciências, Agrárias, Universidade dos Açores, Angra do Heroísmo, Portugal

Abstract: A spatial interaction model is constructed to simulate the historic agricultural land-use evolution of Corvo Island, Azores. Basic ingredients for the model are the local attractiveness for the different agricultural land-use types and historic population counts from 1590 to 2001. The spatial interaction model is then used to distribute employment, residents and respective surface areas, taking into account local agricultural attractiveness and distances between the different zones. These are in turn used to generate past land-use patterns.

Key words: Planning; land use; GIS; spatial interaction model.

1. INTRODUCTION

The aim of this chapter is to explain the historical evolution of land use in small islands through the simulation of an economic and environmental spatial interaction model, calibrated to a small, insular community based on historical data about the island's population and export sector.

We start by introducing Corvo Island, the study area in the Portuguese Azores archipelago, in Section 2. Then we design a spatial interaction model with five different sectors in which employment can be closely related to surface area (Section 3). Section 4 describes an assessment of the attractiveness of the island territory for different uses based on agronomic analysis. Section 5 then presents the model simulations based on historical data on population and main export crops for the years 1590, 1820 and 2001.

E. Koomen et al. (eds.), Modelling Land-Use Change, 133–144.
© 2007 *Springer.*

From these simulations, it is possible to estimate historical land uses and identify the causes of past environmental disruptions connected with the lack of food, firewood or water, as will be discussed in the concluding section.

2.   STUDY AREA

Corvo is the most north-western island of the Azores group, lying half way between the Iberian Peninsula and Newfoundland. It is the smallest of the nine inhabited Azorean islands, with an area of 17 km² and a current population of around 440. The land is constituted by a single volcanic cone with a large crater. Although the cone was once symmetrical, erosion of its western and northern faces, caused by the prevailing winds and waves, has led to the formation of high cliffs (more than 600 meters high), giving the island a pear-like shape (Figure 8-1).

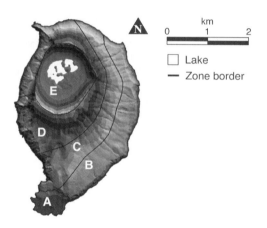

Figure 8-1. Shaded relief map of Corvo Island, including zonal borders. (See also Plate 4 in the Colour Plate Section)

Situated far away from any large land masses, Corvo is subject to the full force of North Atlantic gales, having a climate characterized by plentiful rain, frequent fogs, almost permanent winds and high air humidity. The sea around the island is very deep (more than 3,000 metres at 5 kilometres from the shore) and, due to a branch of the Gulf Stream, relatively warm (17°C in winter and 21°C in summer). The island is covered with grass, with occasional peat bogs and wind-stunted shrubs, giving it a rather high latitude appearance.

Today Corvo is one of the 19 Azorean municipalities, and constitutes also an electoral assembly or 'circle' for the Azorean Parliament. Since the Portuguese Constitution requires each circle to elect a minimum of two representatives, Corvo, despite its small population, elects two members of the Regional Parliament. Due to its status as a municipality and irrespective of its small size and number of inhabitants, Corvo has all the services one would expect in an urban area: a mayor; water and sanitation; two banking offices; an airport with an airline office and its part-time fire department; a health centre with attending doctor and nurse; a border and customs police office; postal service office; telecommunications exchange; a harbour with a harbour master; a fiscal delegation, and representative officers of several Azorean government agencies. With all these, Corvo is indeed one of the smallest humanized ecosystems. Today, the economy of Corvo, as the rest of the Azores, is dominated by cattle production for beef and external public support from the Portuguese Government and the European Union (Dentinho and Meneses, 1996).

The population of Corvo Island (Figure 8-2) evolved from 1590 to 1820 generally with a positive growth rate (Madeira, 1999). From then until today the growth rate became negative generally with a few oscillations (SREA, 2004). The population in 2001 (425) represents half of the population in 1820 (910), when it achieved its maximum.

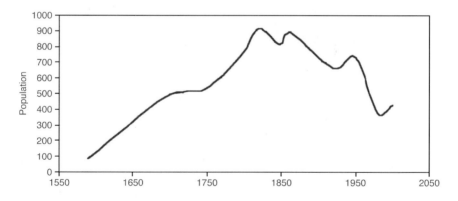

Figure 8-2. Population of Corvo from 1590 to 2001.
Source: Madeira, (1999); SREA, (2004)

During these centuries, land use also varied, influenced by the main exports and by the demands of the local population. Population growth in the beginning of the 18th century for example caused the clearing of all forest on the island.

3.   MODEL DEFINITION

A spatial interaction model distributes employment and residents by different zones of the region, taking into account the distances between those zones and their attractiveness (Dentinho and Meneses, 1996). In the application described in this chapter, it is assumed that residents and each type of employment generate land-use patterns based on coefficients of land use for each activity. The relationships between and within the external and the internal economic systems can be explained using the structure of a basic model (Hoyt, 1939; North, 1955; Tiebout, 1956) according to which exports, or basic activities, are the propulsive factors of the economy, demarcating not only its dimension but also the pattern of local production.

3.1   Mathematical formulation

The model is composed of the equations (1)-(4). The population that lives in each zone is dependent on the employment, basic and non-basic, which is established in all the other zones within a commuting range:

$$T_{ij} = E_{ki} \{r.W_j \, exp(-\alpha d_{ij})/\Sigma_j \, [r.W_j \, exp(-\alpha d_{ij})]\} \tag{1}$$

and

$$P_j = \Sigma_i \, T_{ij} \tag{2}$$

where:

T_{ij} is the population that lives in j and depends on the activity k in zone i;

E_{ki} is the employment in sector k in zone i;

r is the inverse of the activity rate, thus the ratio of population over employment;

W_j is the residential attractiveness of j;

α is the parameter that defines the friction produced by distance for the commuters;

d_{ij} is the distance between i and j; and

P_j are all the residents in j.

On the other hand, the activities generated for each zone serve the population that lives in all the other zones within a service range:

$$S_{ikj} = P_i \{sk.V_{kj} \exp(-\beta d_{ij})/\Sigma_j [sk.V_{kj} \exp(-\beta d_{ij})]\} \tag{3}$$

and

$$E_{kj} = \Sigma_i S_{ikj} \tag{4}$$

where:

S_{ikj} is the activity generated in sector k in zone j that serves the population in i;

V_{kj} is the activity attractiveness of sector k in zone j;

sk is the ratio of employment of non-basic activity k over population;

β is the parameter that defines the friction produced by distance for the people that look for activity services; and

d_{ij} is the distance between i and j.

Figure 8-3 explains the functioning of the spatial interaction model. The term basic employment is related to the employment oriented at, or supported by, external markets and/or instituitions. The non-basic employment is related to the local population. In the first instance, it is possible to estimate the population of the different zones dependent on the basic activity (exports and external supports) of various zones by multiplying its amount by the proportion of those dependent on the activity of zone *i* living in zone *j* following Eq. (1). Secondly, the existing population for each zone i induces the creation of non-basic activity (services to the population) in different zones following Eq. (3). Thirdly, the non-basic activity in the various zones generates more dependent population across the region, again taking into account Eq. (1). The second and third stages are repeated iteratively until the total employment and total population derived from the model both converge to actual levels. The endogenous variables (P_i, E_{kj}) can be obtained from the exogenous variable basic employment (Eb_{ik}) through the use of matrices *[A]* and *[B]*.

$$[E_{ik}] = \{I- [A] [B]\} [Eb_{ik}]; [P_i] = \{I-[A] [B]\} [Eb_{ik}] [A] \tag{5}$$

where:

$$[A] = [\{r.W_j \exp(-\alpha d_{ij})/\Sigma_j [r.W_j \exp(-\alpha d_{ij})]\}] \tag{6}$$

and

$$[B] = [\{sk.V_{kj}\ exp\ (-\beta d_{ij})\ /\ \Sigma_j\ [sk.V_{kj}\ exp\ (-\beta d_{ij})]\}]\tag{7}$$

However, there are also area constraints that must be fulfilled. The area occupied by the different activities (basic, non-basic, residential) in each zone should not exceed the total area of that zone as indicated in Eq. (8). Furthermore, the area occupied by the different activities for each suitability class and zone should not exceed the total area of that suitability class and zone as shown in Eq. (9).

$$\Sigma_k[\sigma_{ik}S_{ik}] + \rho_i P_i + \Sigma_{ik}[\sigma_{ik}Eb_{ik}] \le A_i \text{ (for all zones } i)\tag{8}$$

$$\Sigma_k[\sigma_{ik}S_{ik}] + \Sigma_{ik}[\sigma_{ik}Eb_{ik}] \le A_{ik} \text{ (for all zones } i \text{ and activities } k)\tag{9}$$

where:
σ_{ik} is the area occupied by one employee of sector k in zone i;
ρ_i is the area occupied by one resident in zone i; and
A_{ik} is the area available for sector k in zone i.

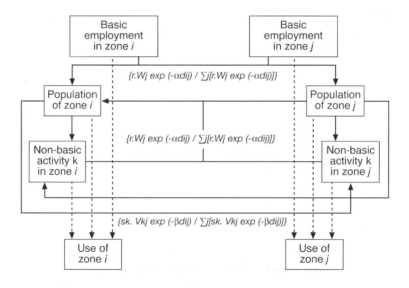

Figure 8-3. Spatial interaction model with land-use patterns.

The general structure of the system can be represented, in a simple way, by a model with three interrelated blocks (as shown in Figure 8-3). Firstly, there are the exports and external flows of money that are considered as the basic employment or the engine of small economies (Dommen and Hein, 1985). These are assumed to set the initial dimension of the five sectors considered. Secondly, there is the internal economic system that describes the relations between the local population and the various activities (k) that fulfil the demand of that population. This is the non-basic employment and includes also five sectors: urban, horticulture, agriculture, pasture and forest production. The third block focuses the use of natural resources, here represented by the use of surface area, which is crucial to analyze the sustainability of the whole system. Actually, when some demand cannot be fulfilled by the local supply, it must be fulfilled by external production which reveals the dependence of local population on external fundamental supplies from horticulture and agriculture, cattle production and forest services.

3.2   Model calibration

Population is the only quantitative data used to calibrate the model. Beyond that, only qualitative information on the type of exports can tell us that the main export was always cattle and some rent was paid with wheat. Using the population data for different years of the island's history, the model simulated the basic employment – for 'pasture', 'agriculture' and 'urban' sectors – and for the non-basic activity for each of the five sectors in each zone. The second step was to distribute spatially the activities within each zone; this was done using the attractiveness coefficients for each activity and using a hierarchy relating to the five sectors. This hierarchy was given assuming the following order of land-use allocation: first urban use, second horticulture, third agriculture, fourth pasture and finally forest are located closest to the residences.

The estimates for the distance inertia parameters, α and β, were obtained from available data on work-residence movements on Flores island, that has a similar structure in terms of employment. The coefficients (sk) and the inverse of the activity rate (r) were derived from the official statistics of Corvo relating to the year 1991. We assumed that all these parameters were stable throughout the centuries because no previous information was available.

4.   ASSESSMENT OF ATTRACTIVENESS

The attractiveness was determined through the assessment of the environmental conditions of each zone for the five sectors: urban, horticulture, arable farming, pasture and forest. The environmental features considered were: temperature, precipitation, elevation, slope, soil quality and exposure. The spatial climate data were obtained by the CIELO model (Azevedo, 1996). The general formula used is expressed as follows:

$$V_{kj} = \Sigma_m \left(\Pi_{qm} \, C_{jkqm}^{\;(1/Q)} \right) / M_j \tag{10}$$

where:

V_{kj} is the attractiveness of sector k for zone j;

Π_{qm} denotes that for each place (m) and each factor (q) the product of the following expression will be taken;

C_{jkqm} is the normalized value of factor (q) for sector (k) in place (m) of zone (j);

Q is the number of factors (q) considered; and

M_j is the number of spots (m) of 0.0625 hectare in zone j.

Due to lack of data, it was not possible to assess the relative importance of each factor and therefore all the factors took the same weight ($1/Q$). A discussion of individual attractiveness maps is provided below.

The attractiveness for *urban use* (Figure 8-4A) was achieved through the analysis of three parameters: slope, exposure and elevation (Gonçalves and Dentinho, 2005). Slope was divided into three classes: 0-7, 7-25 and >25%. Slopes greater than 25% are not suitable for construction, and the interval between 0 and 7 was preferable. The north and west exposures were excluded because of the low temperatures due to the low radiation or strong winds. South exposure was considered preferable. It was considered that urban implementation would only occur below 300 metres and close to the only access to the sea.

The analysis of the attractiveness for *horticulture and fruit production* considered the culture of citruses (Gonçalves and Medeiros, 2005). The climatic conditions analyzed were temperature (accumulated temperature and medium temperature in the coldest months), precipitation (total precipitation and percentage occurring in the summer months) and soil quality. Each of the three criteria was given the same importance in determining the attractiveness. Figure 8-4B shows the resulting attractiveness map, indicating that the eastern coastal zone is moderately suitable for horticulture. Note that the optimal locations are found only in a very small zone in the south.

The attractiveness for *arable farming* (Figure 8-4C) was determined by assessing the territorial capacity for maize and wheat (Gonçalves and Monjardino, 2005), based on temperature, soil capacity and slope conditions. Because of the similarity between the two crops, the definition of attractiveness is the same for both. The evaluated climatic conditions related to temperature accumulation and minimum and maximum temperature limits during certain development stages. The slopes considered suitable for the culture ranged only from 0 to 15%. The areas with no attractiveness do not have either a sufficient soil quality or the right temperature conditions for the crops. The classes of medium and preferable attractiveness are distinguished based on the climatic conditions that allow the cultivation of the latest varieties that are potentially more productive, assuming that there is no water deficit. This assumption can be made in the Azores because of the high precipitation levels during the whole year.

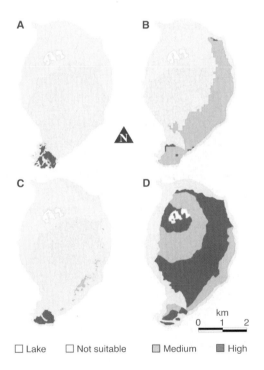

Figure 8-4. Attractiveness for (A) urban use, (B) horticulture, (C) arable farming and (D) pasture/forest.

The territory suitable for *pasture* (Figure 8-4D) was achieved considering both elevation and slope (Gonçalves and Calado, 2005). Regarding elevation three different classes were considered with a decreasing contribution to attractiveness: 0-300 metres, 300-600 metres and more then 600 metres. The slopes were also reclassified in three classes: 0-7%, 7-25% and >25%. Steeper slopes than 25% were considered totally unsuitable for this activity. The attractiveness for *forest* is identical to the attractiveness for pasture.

5.   SIMULATION RESULTS

The first simulation was made for 1590, when there were about 80 inhabitants (Figure 8-5). According to the model in this year, there was already complete occupation of the southern and eastern zones. The rest of the island still had 80% of the area with natural vegetation, which means it was not being very much explored to support the population. The model shows that not all land on the island was needed to keep cattle or grow wheat in order to sustain daily living. Unfortunately no complete records of land use exist from the time of the first settlers.

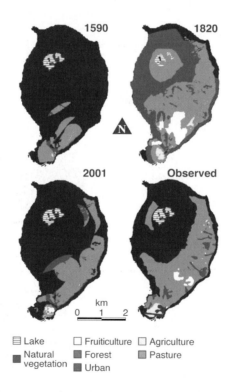

Figure 8-5. Simulated land use for 1590, 1820, 2001 and observed land use (Dentinho *et al.*, 1991). (See also Plate 5 in the Colour Plate Section)

A second simulation is made for 1820, when Corvo's population achieved its historical maximum (910 inhabitants). In this scenario, all the land is completely occupied by the five activities and, similar to other years, the most important sector is pasture. In fact, in 1820, all the usable area is occupied and most of the forest is cut. Historical data show that forest indeed disappeared from Corvo Island in this period and thus underpin the simulation results.

The last simulation represents an estimation of the actual land use. In 2001, Corvo had 440 inhabitants but the important basic activity was no longer the export of cattle but rather the public sector, sponsored by external transferences. This map compares well with actual, observed land use, apart from an underestimate of the area taken by pasture. This can be attributed to a considerable improvement in accessibility after the massive introduction of motorized transportation in the 1980s allowing the use of more distant pastures that was not accounted for in our model. Interestingly enough, a Natural Park was proposed in 2005 for the higher lands which might reduce the area for productive pastures, thus bringing the actual land use closer to our simulation.

6. CONCLUSION

This chapter described the application of a spatial interaction model to simulate past land-use changes based on incomplete information on population, productivities and main exports. From this research we obtained a model that allows a better understanding of the history of man and his environment. Spatial interaction models are interesting tools to analyze the relations between man and the environment since they capture very well the spatial interdependence between the different uses. In combination with an assessment of land attractiveness, this allowed for the reconstruction of historic land-use configurations that, for example, helped understand the major environmental disruptions in the beginning of the 19th century, when the forests were completely cut.

Future research will concentrate on the introduction of dynamic features and the internalization of the area constraints. Furthermore, we want to apply the model to larger islands and explore possible future land-use changes. Especially the current transition from agricultural to urban land, following external financial support to public services in selected villages, is an interesting trend to extrapolate in simulations of future land use. This development is related to a structural change in the economic organisation of the island and calls for a reformulation of the basic set-up of the current spatial interaction model, thus allowing us to further assess the applicability of this type of modelling.

REFERENCES

Azevedo, E.B. (1996) Modelação do Clima Local à Escala Local, Modelo CIELO, *Tese de Doutoramento*, Departamento de Ciências Agrárias, Universidade dos Açores.

Dentinho, T. and Meneses, J.G.A. (1996) Sustainable Development of Urban Areas. Lessons from the Smallest City, Paper presented at the 36th European Congress of ERSA ETH Zurich, Switzerland , 26–30 August.

Dentinho, T., Dias, E., Railey, C., Gomes, H., Rogojo, A., Pinheiro, J., Cardoso, A. and Dentinho, A (1991) *Plano Director Municipal da Ilha do Corvo*, Departamento de Ciências Agrárias, Universidade dos Açores.

Dommen, E. and Hein, P. (1985) Foreign trade in goods and services. The dominant activity of small islands, in *States, Microstates and Islands*, Croom Helm, London.

Gonçalves, J. and Calado, L. (2005) Atractividade para a Agro-pecuária nas Ilhas dos Açores, in Gonçalves, J. and Azevedo, E. (eds) *Estudo das Aptidões do Território dos Açores*, Departamento de Ciências Agrárias, Universidade dos Açores, pp. 19–20.

Gonçalves, J. and Dentinho, T. (2005) Atractividade Urbana nas Ilhas dos Açores, in Gonçalves, J. and Azevedo, E. (eds) *Estudo das Aptidões do Território dos Açores*, Departamento de Ciências Agrárias, Universidade dos Açores, pp. 8–11.

Gonçalves, J. and Medeiros, C. (2005) Atractividade para os Citrinos nas Ilhas dos Açores, in Gonçalves, J. and Azevedo, E. (eds) *Estudo das Aptidões do Território dos Açores*, Departamento de Ciências Agrárias, Universidade dos Açores, p. 7.

Gonçalves, J. and Monjardino, P. (2005) Atractividade para os cereais nas Ilhas dos Açores, in Gonçalves, J. and Azevedo, E. (eds) *Estudo das Aptidões do Território dos Açores*, Departamento de Ciências Agrárias, Universidade dos Açores, pp. 5–7.

Hoyt, H. (1939) *The Structure and Growth of Residential Neighbourhoods in American Cities*, U.S. Printing Office, Washington.

Madeira, A.B. (1999) População e Emigração nos Açores (1766–1820), *Cascais*, Patrimonia.

North, D.C. (1955) Location theory and regional economic growth, *Journal of Political Economy*, June, pp. 243–258.

Tiebout, C.M. (1956) A pure theory of local public expenditures, *Journal of Political Economy*, 64: 416–424.

SREA (2004) Anuário Estatístico da Região Autónoma dos Açores – *Território*, I: 34.

PART III: OPTIMISATION MODELLING

Chapter 9

SPATIAL OPTIMISATION IN LAND-USE ALLOCATION PROBLEMS

W. Loonen[1,2], P. Heuberger[1] and M. Kuijpers-Linde[1,2]
[1]Netherlands Environmental Assessment Agency (MNP), Bilthoven, The Netherlands;
[2]SPINlab, Vrije Universiteit Amsterdam, The Netherlands

Abstract: In densely populated areas, space for development is confined, making spatial planning essential to reconcile the interests of all stakeholders. In the process of policymaking, possible future land-use scenarios are often very valuable as a reference point, but the optimal configuration in terms of costs and effects might provide even more valuable inputs when decisions have to be taken. Tools for exploring optimal land-use configurations are therefore of great interest to policymakers. With these tools, plans can be evaluated and adjusted. Spatial optimisation is a powerful method to explore the potentials of a given area to improve the spatial coherence of land-use functions. In the Netherlands, there are many different planning issues in which multi-objective spatial optimisation can play an important role. This chapter describes two case studies that apply the genetic algorithm approach.

Key words: Spatial optimisation; genetic algorithms; agricultural nitrogen emission; critical load; Kappa statistics; map comparison; spatial coherence; ecological networks; landscape planning.

1.   INTRODUCTION

Within land-use planning, land functions and configurations are constantly re-arranged in order to improve or optimise their performance, while minimising the negative effects on adjacent functions. In densely populated areas, where space is confined, the need for physical planning is particularly great. However, the impacts and effects of changes in land use are often very complex. Therefore, it is very important for policymakers to

E. Koomen et al. (eds.), Modelling Land-Use Change, 147–165.
© 2007 *Springer.*

have tools at their disposal that determine the optimal configuration of land in terms of costs and effects.

In many studies dealing with conflicting goals or competing stakeholders in land-use allocation problems, specific methods are used to get better insights into possible configurations and their implications for the surrounding areas. One of these methods is spatial optimisation: a method designed to minimise or maximise the objectives in spatially explicit studies, given the limited area, finite resources, and spatial relationships between different functions. Spatial optimisation is a powerful method to explore the potentials of a given area to improve the spatial coherence of land-use functions. Other applications include issues of soil erosion (Seixas *et al.*, Chapter 11 in this book), habitat suitability (Seppelt and Voinov, 2002), agricultural sustainability (Zander and Kächele, 1999; Mandal, Chapter 10 in this book) and resource allocation (Aerts, 2002).

A wide variety of optimisation methods is available for this type of problem and each has its specific merits and limitations. The most widely used are mathematical programming methods (linear programming, non-linear programming, mixed-integer programming) and the so-called heuristic methods (genetic algorithms, evolutionary programming, simulated annealing, neural networks). The main differences between these methods include their flexibility in the formulation of the objectives, the scale at which their problems can be solved, and the optimality of their solutions. While linear programming (LP) is a very powerful method, capable of solving problems with very high dimensions (in terms of number of variables, relations and constraints), it has the intrinsic drawback that all relations, constraints and objectives have to be formulated in a linear fashion and it requires the variables to be continuous. This linearity is often not applicable within land-use planning, because of the qualitative character of the relations and the discrete character of (some of) the variables, which have to be optimised. Extensions to LP, such as mixed-integer programming (MIP), where variables are allowed to be discrete, or non-linear programming (NLP), can overcome these limitations at the cost of much longer computing time and the loss of a guaranteed global optimum. Another drawback of these methods is that they often require extensive additional effort in recasting the problem in a feasible framework.

Heuristic methods, like genetic algorithms (GAs), hardly have any restrictions regarding the formulation of the variables and relations, but have the drawback that constraint handling can become cumbersome. These methods generally need a lot of fine-tuning as well as long computation times. An attractive feature of (some of) these methods is that they are flexible in terms of computation environment, requiring limited programming effort and a straightforward use of parallel computing. They are furthermore not geared towards finding a limited number of global optima, but result in sets of near-optimal solutions, offering additional freedom in choosing criteria. That these methods can provide many slightly different configurations with almost the same characteristics is interesting for policymakers. This allows them to evaluate alternative land-use configurations, each with their specific socio-economic impacts, while still achieving optimal results. GAs turned out to be appropriate techniques to optimise land-use configurations in our case studies.

This chapter introduces the basics of GAs and then presents two case studies to show the possible application and importance of these methods for policymakers and land-use planners. In the first case study, we address the problem of atmospheric nitrogen emission caused by intensive agricultural activities resulting in the nearby nitrogen deposition in often sensitive nature areas. The second case focuses on spatial coherence in the National Ecological Network in the Netherlands.

2.   GENETIC ALGORITHMS

A GA is a search technique, originally developed by Holland (1975). GAs have been widely used to solve either single or multi-objective optimisation problems (Sarkar and Modak, 2003; van Dijk *et al.*, 2002; Rauch and Harremoes, 1999). Recently GAs have been applied to spatial optimisation problems (e.g. Aerts, 2002; Vink and Schot, 2002; Matthews, 2001). They are based on the principles of Darwin's evolution theory and search for an optimal solution by simulating processes of evolutionary development of a population of candidate solutions. Feasible candidate solutions are represented as individuals in a population. The values of parameters in an individual solution are considered to be genes, which can be inherited by mating. A fitness function is used to evaluate the suitability of the individuals. The fitness function takes a single individual as input and returns a quantitative measure of the 'goodness' of the solution represented by that individual. The method used in our study is usually initiated by choosing (often at random) an initial population, referred to as generation G_0. From a generation G_t a new generation G_{t+1} is created, by mimicking the evolution process. The most important operations within this procedure are the following (Holland, 1975):

- *Selection* (who survives) of individuals is based on the fitness of the individuals with respect to a fitness function. Individuals with high fitness values (representing better solutions to the problem) will have a higher probability of surviving and entering the 'mating' population, while low-valued individuals will have a high risk of being removed from the population. In this way, individuals with the best genes or characteristics will have better chances of survival and mating.

- *Cross-over* (mixing genetic characteristics) is done by mating and exchanging or recombining characteristics or genes in the offspring.
- *Mutation* (random genetic changes) occurs with a certain chance, often dependent on stagnating objective values. A mutation is a change in some randomly chosen genes. Mutation is especially useful to keep a population from converging to a local optimum too fast.

After *cross-over* and *mutation* have taken place, individuals will be *selected* for the next mating population. This process of selection, cross-over and mutation is repeated until a potential optimal solution is found. This solution is defined here as the best individual in the population when the objective value stagnates and the algorithm is not producing better offspring.

GAs are generally considered to be powerful tools for the solution of unconstrained optimisation problems. However, many real life problems are subject to a large number of constraints. This also holds for the first case study, an emission reallocation problem (see Section 3). Contrary to LP, for example, some constraints cannot easily be implemented into the GA approach. Beasly (2002) discusses four different strategies to deal with constraints:

- *Representation* that assumes generated candidate solutions to automatically satisfy the constraints.
- Use of a *repair operator* that takes care of constraint violations 'afterwards', for instance, by rescaling or projection.
- Distinguishing between *fitness* and *unfitness* by using two measures for each individual. The fitness value is used to evaluate the objectives and the unfitness value represents the degree to which constraints have been violated.
- Use of a *penalty function* that 'punishes' constraint violations by adding a penalty to the fitness value.

The first strategy, *representation*, is, in fact, the most straightforward constraint handler. The algorithm should be formulated such that the candidate solutions always satisfy the given constraints. Unfortunately, this is often a very difficult or even impossible venue. Using a *repair operator* is often useful in cases where the first strategy cannot be applied. If, for example, a variable exceeds a certain maximum value, it can be converted to just that threshold. When using *fitness* and *unfitness* values, every potential solution will obtain fitness and unfitness values, the former depending on how objectives are fulfilled and the latter on the extent to which constraints are violated. Individuals with high unfitness values will stay temporarily within the population, so the new offspring can inherit their potential valuable characteristics. Their unfitness values, however, will cause them to be replaced in time by 'better' individuals according to a decision rule. The last strategy discussed by Beasly is the use of a *penalty* function that is based on the extent to which the constraints are satisfied. The penalty is integrated into the fitness value, so individuals performing well will have a higher fitness and thus more chance of survival. This latter approach is often only considered if other approaches fail.

3. CASE STUDY 1: MINIMISING NITROGEN DEPOSITION FROM AGRICULTURE

In the first case study we explore the capability of GAs to find the optimal solution of a large-scale spatial optimisation problem that was solved in earlier studies using LP. The aim of this approach is to extend the underlying problem with non-linear, discontinuous and qualitative relations. An obvious requirement for an alternative solution method is that it should at least be able to give acceptable solutions to the original problem. Since the LP solutions were globally optimal, the GA-solutions can easily be evaluated by comparing them to the LP-solutions (see Loonen *et al.*, 2006, for more details on this study).

3.1 Introduction

Many plant species and ecosystems have low tolerance levels for available nitrogen. As a result of human activities, the amount of nitrogen available to Earth's ecosystems has more than doubled globally over the past century, causing many plant and animal species to become completely or locally extinct (Vitousek *et al.*, 1997). One of the major threats is the atmospheric input of nitrogen (Galloway *et al.*, 1995). The mean atmospheric deposition of nitrogen in the Netherlands consists of around 70% of ammonium, of which 80% originates in Dutch intensive agricultural activities (RIVM, 2002). For nature areas, this contribution can be even higher because these are often located close to agricultural areas.

Dutch policy aims at the reduction of nitrogen deposition in nature areas to halt the deterioration of ecosystems. Besides measures to reduce total nitrogen emission, reallocation of emissions (in terms of intensification or expansion) can be used to reduce nitrogen deposition in nearby nature areas. In previous studies, LP was used to study the potential of partly reallocating emission sources (Heuberger *et al.*, 1997; Heuberger and Aben, 2002; van Dam *et al.*, 2001). This technique was applied to determine the optimal spatial distribution of agricultural emission sources with minimal atmospheric nitrogen deposition in nature areas, given a fixed level of total agricultural atmospheric nitrogen emission in the Netherlands.

Nature areas within the study area are all classified as a specific nature type, with a corresponding nitrogen critical load value, based on its vulnerability to atmospheric nitrogen deposition (Reynolds and Skeffington, 1999). The critical load for a nature area is then determined by the lowest critical load of occurring nature types.

Combining the ammonium deposition data (resulting from the above mentioned agricultural activities) with other nitrogen deposition data (due to traffic, industrial and foreign sources or background concentrations) results in an evaluation of critical load exceedance in nature areas. For every potential emission raise or decline, but also for a change in the emission pattern, the resulting deposition or critical load exceedance can now be calculated. This property enables the formulation of an optimisation problem – with the goal to minimise critical load exceedances – in terms of an LP problem (Luenberger, 1984). An additional requirement is that the so-called boundary conditions and the goal function (objective functions) are also formulated linearly.

We now present the results delivered by a multivariate optimisation of the spatial distribution of agricultural ammonium emission sources using a GA. The similarity of the GA-solutions is compared to the original LP-outcome by a combination of map comparison techniques. Based on this comparison the applicability of the GA-method, especially for more complex problems, such as non-linear or multiple-goal problems is discussed.

3.2   Model formulation

The multivariate optimisation of the spatial distribution of agricultural ammonium emission sources in the Dutch province of Noord-Brabant is selected as a test case for assessing the potential of GAs in comparison to the previous LP-approach. The optimisation goal is to minimise the effects of atmospheric nitrogen deposition in nature areas. Noord-Brabant comprises approximately 5,000 km², of which some 60% (3,000 km²) is used for agricultural purposes. Although it comprises only 12% of the land surface of the Netherlands, it contains 20% of the Dutch cattle farms, including several areas with high concentrations of intensive bio-industry. Forests and nature reserves in the province occupy a total of 1,300 km².

The research area was converted into a grid of 5,875 cells of 1x1 kilometre. A cell is considered to be a nature cell if more than 25% of the cell actually contains nature, resulting in 1,457 nature cells in this research area. Agricultural activity was assumed possible in the calculations for 3,997 cells. All remaining cells were classified as urban area. Nature cells all have a specific critical load value for nitrogen deposition, depending on the most vulnerable type of ecosystem present (Reynolds and Skeffington, 1999). Deposition values exceeding the critical load are assumed to cause damage to the most vulnerable nature types within the specific grid cell.

Minimum and maximum values for ammonium emissions were determined in cells with agricultural activities. Linear ammonium emission-deposition relations were assumed between cells on the basis of atmospheric properties like temperature, wind speed, sun intensity and precipitation, as well as on surface characteristics and the chemical composition of the atmosphere (van Jaarsveld *et al.*, 2000). Cell-to-cell relations were calculated by the so-called Operational Priority Substances (OPS) model (van Jaarsveld and de Leeuw, 1993; van Jaarsveld, 1995; 2004).

In this study, only agricultural ammonium emission sources were considered for reallocation. All other atmospheric nitrogen inputs to nature areas were assumed to be fixed to the 1999 situation. In the calculations, a number of constraints have to be satisfied. The first constraint is that the total sum of emissions in the research area must stay the same. The rationale behind this is that we want to determine how much the current situation in nature areas can be improved and optimised by re-allocation of emission sources only. Measures to lower total emission were also considered in van Dam *et al.* (2001), but were not considered in this comparative methodological study. A second constraint is used to satisfy the minimum (lower bounds) and maximum (upper bounds) emissions for each cell. The third constraint is that the initial exceedance of the critical loads within nature cells (i.e. the 1999 situation) was not allowed to increase; this is to make sure that the current situation will not deteriorate in these cells. As a fourth constraint, the deposition in other provinces is not allowed to increase. This prevents the GA from moving the deposition problem to neighbouring areas outside the province, by reallocating emissions to the border of the research area. It should be noted that complete protection of nature from nitrogen deposition in the research area is impossible. If all agriculture were to be removed from the province of Noord-Brabant, critical loads for nitrogen would still be exceeded in several nature areas due to deposition from other sources. It is therefore impossible to find a solution under which all nature cells will be protected. For this reason, the objective function of this optimisation problem was set to minimise the sum of exceedances of the critical loads in all nature cells.

3.3   Comparison of maps

The LP and GA optimisation results were compared by analyzing: a) the spatial distribution of optimised emissions and b) the spatial distribution of resulting exceedances of the critical nitrogen loads in nature cells. Kappa Statistics (Sousa *et al.*, 2002) and regression analyses were used to quantify the resemblance of LP and GA output maps for a) and b). Hagen (2002; 2003) developed a useful method by combining the Kappa Statistic tool with the Fuzzy Set theory. The so-called Kfuzzy provides information about the magnitude, nature and spatial distribution of similarity between two maps (where K=1 represents perfect similarity and K=0 no similarity). The main advantage of this method is the use of both the spatial component and the similarity between different categories. Where classical Kappa indices are restricted to cell-by-cell comparisons, the Kfuzzy takes into account both fuzziness of location and fuzziness of category. In practice, a Kfuzzy value of 1 is not likely to occur in large optimisation problems using GAs. To validate the Kfuzzy results of the GA and LP outcomes, the K values were also compared to a 'random distribution', satisfying only the constraints.

Regression analysis was performed to compare the results from LP with GA optimisations. The linear regression has the basic form $y = ax+b$, where y represents the LP-emission value and x the GA-emission value. Near similarity is indicated by regression coefficient values (a) close to 1 and an intercept (b) approaching zero. The regression analysis was applied on the original (1x1 kilometre) results derived from the optimisation, as well as on larger spatial scales (2x2 kilometres and 5x5 kilometres). The regression analysis was performed to evaluate the spatial patterns on an aggregated level. This aggregation was motivated by the fact that the GA does not search in the edges of the search space like LP (Dauer and Liu, 1990). Therefore, in a cell-by-cell comparison, dissimilarities are expected when the LP solution is close to the lower and upper bounds. The dissimilarities will be smoothed by comparison on an aggregate level.

3.4   Results

Because different GA runs do not produce identical solutions, five different GA runs were used for calculating the map similarity. In the actual situation (1999), the critical loads are exceeded in 94% of the existing nature cells, while after optimisation, this is reduced to 82% (for both LP and each individual GA-run). Furthermore, in the actual situation, the exceedance of critical loads is larger than 250 mol per hectare per year in 71% of all nature cells; as opposed to only 27% after optimisation (both for LP and GA). At first glance, the resulting deposition maps after optimisation with LP and GA show almost no differences. The Kfuzzy value for exceedance patterns of the critical loads in nature areas after optimisation with LP and GA varies between 0.94 and 0.96 (compared to a Kfuzzy value of 0.59 in a random situation, satisfying only the constraints). These values indicate almost complete similarity in the maps and thus in the exceedance patterns.

The Kfuzzy value for the similarity of the emission patterns of the LP and GA solutions varies between 0.72 and 0.73 (compared to Kfuzzy value of 0.27 in a random situation, satisfying only the constraints), indicating substantial similarity (Landis and Koch, 1977) between the LP and the GA allocation for emission patterns. The Kfuzzy values comparing emission patterns between pairs of GA-runs show lower similarities than the Kfuzzy values for comparisons between a single GA run and the LP solution. At the same time, the Kfuzzy values comparing deposition patterns show almost perfect similarity. This implies existence of a whole set of emission patterns close to the LP solution, resulting in the same deposition pattern.

Comparison of the two emission maps learns that high emission values are situated at roughly the same locations. However, optimisation performed with LP results in a relatively concentrated distribution of high emissions in the province compared to a smoother distribution derived from the GA optimisation. The concentrated distribution by LP is caused by the fact that the optimal solution has emission values on the extremes, i.e. at the minimum and maximum boundary values set by the constraints. This results in a distribution of emission values situated in the lower or the upper bounds. Due to the aforementioned low sensitivity of local changes in the optimal emission patterns and the intrinsic random character of GAs, the latter will result in a smoother emission pattern.

Another reason for the slight dissimilarities in the LP and GA emission patterns lies in the difference between the objective functions for LP and GA. Local differences in the emission values can cause small differences in deposition patterns for the three components within the objective function, but may result in the same total objective function values.

To quantify the correlation of the LP and GA outcomes across several scales of aggregation, a regression analysis was performed on various resolutions. It was found that the similarity of the emission simulations improved considerably after aggregation of the results to cells of 2x2 kilometres and 5x5 kilometres. For instance, on the 5x5 kilometres aggregation levels, the GA-solution intercepts the y axis at 68 mol per hectare per year (compared to an intercept of 453 mol per hectare per year in a random distribution), with a regression coefficient of 0.95. Therefore, it can be concluded that differences between the LP and GA solutions occur only on a small scale (1x1 kilometre). The emission patterns for both methods are almost similar on aggregated levels (2x2 kilometres and 5x5 kilometres). Figure 9-1 shows a graphical depiction of these results and Loonen *et al.* (2006) provides more details on this specific case study.

We have investigated various concepts and options to improve the efficiency of the applied GA. Constraint handling emerged as the most important issue in our case. For the first constraint (total emission should be kept constant), representation and use of a repair operator were tested. The major problem with the latter was that it was too time consuming, resulting in very long computation times. Representation, especially with respect to the first constraint, performed well. The second and third constraint (i.e. no deterioration of the initial situation, both inside and outside the region) were handled using penalty functions. Although the use of penalties is often rejected in the literature, they proved to be an appropriate method in this application. The major lesson learned here was that the weights used for these penalties should be kept variable throughout the execution of the algorithm so as to prevent premature stagnation; the penalty factors should be kept relatively small in the initial phase of the execution and be gradually increased during the optimisation process.

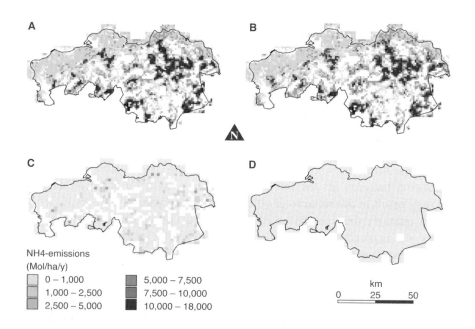

Figure 9-1. Optimised ammonium (NH$_4$) emission following the (A) LP and (B) GA approach and their local differences at aggregated (C) 2x2 km and (D) 5x5 km resolution.

3.5   Discussion

The comparison of LP and GA optimisation for the emission-deposition problem in our case study showed a high degree of similarity for both emission patterns and exceedance patterns. Therefore, the intention is to expand the optimisation problem in the near future with more processes and more constraints or criteria. The model can be a useful tool in future policy making related to land-use problems. Our case study showed the potential of optimisation methods for decreasing ammonium deposition in nature areas.

The optimisation minimised the exceedances of the critical loads to 82% of the area, compared to 94% in the actual situation. The number of cells with an exceedance of less than 250 mol per hectare per year increased substantially: from 29% to 63%. This is a very promising result if we also consider the fact that both national and international policies are causing future emission totals to decline and hence the number of 'clean' cells will increase rapidly.

The calculations also showed that almost identical minimal deposition patterns can be achieved with somewhat different emission patterns. This is an interesting, general feature of GA applications for policymakers, allowing them to evaluate alternative emission distributions on a small scale, each with their specific socio-economic impacts, whilst still achieving optimal results for nature. Another advantage of the method is that large numbers of scenarios with different constraint values and emission totals can be optimised within a relatively short time span.

For optimisation problems where small changes in input values have a large impact on the output, an interesting step in future research will be to initiate a local optimisation after the (aggregated) optimisation of the complete research area. For the problem at hand this will lead to even better results for the similarity in emission patterns between LP and GA. Although differences still occur in emission patterns, the differences in the exceedance of the critical loads are negligible because of the relative simplicity of the selected case.

In our experience, the main disadvantage of GAs is the time-consuming process of constraint handling. This part of the fine-tuning process differs for every constraint and it is impossible to estimate, beforehand, how much effort is needed.

The results of this study are potentially interesting to policymakers, because even small changes in emission patterns can result in higher numbers of clean nature areas in the near future. The proposed procedure results in multiple configurations to achieve this goal, allowing policymakers to take local limitations and considerations into account.

4.   CASE STUDY 2: OPTIMISING ECOLOGICAL NETWORK CONFIGURATION

In the second case study we focus on a problem that, unlike the previous case, cannot be solved with LP. The objective is to optimise the spatial coherence of nature areas in the process of adding new nature areas. We minimise the total sum of the boundary length of nature areas in order to reduce negative external effects like sound and pollution. At the same time, we maximise the number of large connected areas (>1,000 hectares) to reduce extinction risks for several (meta-) populations of species.

4.1   Introduction

Human activities have led to loss and fragmentation of habitat, causing many species to become locally extinct (Ehrlich, 1995; Vermeulen and Opdam, 1995; Vitousek *et al.*, 1997; Gibbs, 2001; Fahrig, 2001; Thomas *et al.*, 2004). Many species and ecosystems depend on suitable areas with sufficient spatial coherence for survival (Berglund and Jonsson, 2003). Unfortunately, these areas are very scarce in highly populated countries. McDonnell *et al.* (2002) used boundary length and total surface area to optimise the spatial cohesion of nature reserves. Minimising the boundary length will not only decrease the vulnerability to pest and weed invasions, but nature reserves will also be less influenced by other effects like noise disturbance and pollution. Large projects are now in progress to create networks of nature areas for improving the spatial coherence of existing nature areas, e.g. Natura 2000 in Europe (EU, 1992) and the National Ecological Network in the Netherlands (LNV, 1990). In the Netherlands, a large area (1390 km²) has been designated by the provincial authorities for transformation into new nature. We will henceforth refer to these proposed new nature areas as the 2003 plans, alluding to the year in which these plans were made (ECLNV, 2004). The following hypotheses will be tested this research:
- Nature area can be allocated more efficiently if spatial optimisation techniques are used with respect to both boundary length and numbers of hectares in the highest surface area class, i.e. more spatial coherence can be achieved by allocating the same surface area to nature reserves.
- The same decline in boundary length as observed in the 2003 plans can be achieved by allocating smaller amounts of new nature.

The 2003 plans do not only focus on the improvement of the spatial coherence, but also use other criteria to determine the best areas for nature development. This research focuses on a limited number of criteria, to allow a straightforward evaluation of the outcomes. Inclusion of all original criteria would obscure the merits of the applied approach. From a computational perspective however, it is possible to include a higher number of criteria and variables.

4.2   Model formulation

The GA used in this research is the same algorithm as used in the previous cases. We use a GA with a very simple criterion: the total boundary length of nature areas. Boundary length can be considered as an inverse measure of spatial coherence. The goal is to minimise the boundary length of all nature areas while increasing the total surface area.

The initial phase of the algorithm application is a random allocation of the new nature cells over the available area. In the search for a better configuration, cells of 100x100 metre are reallocated one by one and the 'new' situation is evaluated in terms of boundary length. A reduction in the total boundary length is considered an improvement and replaces the old configuration. A similar situation in terms of the criteria used will replace the old configuration with a certain probability; when boundary length increases in the new configuration, it is discarded and the old configuration will remain unchanged.

In the Netherlands, 1,390 km² of new nature can be allocated within a potential area of 6,300 km². Because of this large area available for new nature, the number of possible configurations is practically infinite. Heuristic optimisation algorithms will generally not provide one unique solution, but a wide range of configurations with nearly equal scores for the optimisation criterion. For policymakers, this is an interesting feature, allowing them to choose from a set of potential solutions. To estimate the robust part of the optimal configuration, the overlap in cells allocated as new nature in a series of 10 model runs was considered to be the 'most appropriate' area to convert to nature, considering the criteria used.

To test the hypotheses, different scenarios for the allocation of new nature in the Netherlands were developed and evaluated. Increasing amounts of new nature were added and the resulting configurations were compared in terms of the total boundary length and of surface area for the separate nature reserves. The latter was compared by looking at the total surface area in the two classes with the largest areas: 1,000 to 5,000 hectares and over 5,000 hectares. To test the efficiency of the 2003 plans, 25%, 50%, 75% and 100% of the planned 1,390 km² of new nature areas was reallocated, optimising for minimum boundary length.

4.3   Results

As previously mentioned, the calculations do not result in a single solution, but provide a set of potential solutions. Different runs with the same allocation size can have the same boundary length, but slightly different spatial configurations. Minimising the boundary length, however, always results in more spatial coherence by clustering the existing nature areas and filling up gaps. This is exemplified in Figure 9-2 for a small part of the country for one individual simulation carried out for the whole country using the full area of the 2003 plans.

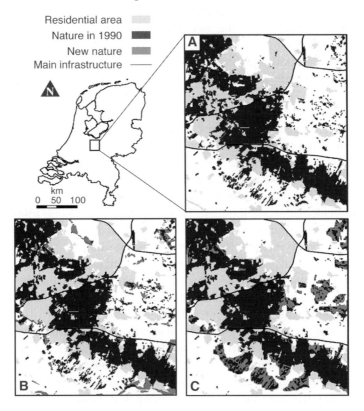

Figure 9-2. Example of (A) 1990 land-use configuration and (B) the 2003 plans compared to (C) optimized nature patterns. (See also Plate 6 in the Colour Plate Section)

A boundary length decline of 23% compared to the 1990 situation is reached, while allocating the same area as in the 2003 plans (1,390 km²). Interestingly enough, the total boundary length in the original plans increases by 21%. An increased coherence can already be achieved with allocating much smaller amounts of new nature. In terms of size of the individual nature areas, an allocation of only 50% of the total area of the 2003 plans already shows more nature within the highest surface area class (> 5,000 hectares) than realized in the actual 2003 plans. An allocation of the full 100% results in an increase of approximately 900 km² in the highest class compared to the original situation and 6,000 km² in the highest class compared to the 2003 plans. There is, however, an optimum with respect to the decline in boundary length following the adding of more nature. To test this, we performed a number of simulations in which an increasing amount of extra nature was added to the originally planned 1,390 km². After adding a total of 1,600 km² the boundary length ceased to decline. This does not imply that it is not useful to add more new nature, but in terms of boundary length loss there is no longer a gain, i.e. the boundary-length criterion loses its significance here.

4.4   Discussion

The optimisation technique applied provides alternative configurations for the allocation of new nature. In this study, we optimised the spatial coherence of nature areas and compared this to the original 2003 plans of the provinces. The application of GA allowed us to allocate new nature more coherently and reach a higher coherence, while allocating less surface area than proposed in the 2003 plans. Classical methods are either not suited to solve this problem or demand very long calculation times (applying Monte Carlo simulations takes years, for example). The method is capable of handling very large numbers of variables and can also handle large numbers of constraints.

In reality, spatial planners have to consider more criteria than spatial coherence alone. Local problems with, for example, the acquisition of land and the following process of conversion into nature also play a role. In practice, this will lead to other solutions that are less optimal from a nature perspective. The optimised configurations can, however, form the starting point for local planners, and can subsequently be adjusted to specific, local circumstances. Alternatively, the relevant local constraints can also be implemented in the optimisation model to calculate a suitable configuration.

An important advantage of the method is that various alternative spatial allocation configurations with almost the same characteristics can be calculated within a relatively short time span, thus offering policymakers a wide range of possible solutions to choose from. Ecological or economic factors, or other factors not explicitly addressed in the model, may then determine which of the proposed configurations is preferable. A difficulty within the optimisation process (but also in the process of policy making in general) is recognizing and evaluating the constraints and criteria of concern for the problem at hand.

Future research may include a distinction in preferred nature types or fixed amounts of different nature types (like swamps, forests or heather fields) in the optimisation. The optimisation techniques can also be used for other land-use allocation problems relating to for example the construction of new residential areas.

5.   CONCLUSION

Based on the experience obtained from the described case studies, we conclude that GAs offer an attractive alternative for the solution of spatial optimisation problems. They proved their potential in optimising a high-dimensional, linear problem implying that they might also perform well on non-linear problems. Of course it is necessary to perform specific tests on this subject. This study showed that problems with a high number of variables can be tackled within the same computation time that is needed in an LP approach. In general, the process of tuning is most time-consuming, and therefore the total procedure takes more time than LP. When non-linearities occur, however, LP is no longer appropriate and GAs become a suitable alternative. They do not impose restrictions on the type of model and can be used in a variety of applications. GAs can, in principle, tackle discontinuous, non-linear or stochastic problems.

Another important advantage of GAs is that they lead to a whole set of solutions, which can be fruitfully exploited by decision makers. For models with long computing times they are especially attractive because they can easily be parallelized or be applied on multiple computers simultaneously. On the other hand, the total number of model runs required is often considerably large and dependent on the specific problem at hand.

There are two major issues to consider in the actual implementation of a GA. First, one has to define functional forms for the procedures of *selection, cross-over and mutation*. This choice will in general have a large influence on the resulting algorithm and should be seen as a necessary part of the fine-tuning phase, which also includes the definition of other degrees of freedom, such as population size or end criterion. A second important subject is *constraint handling*. The most attractive form to deal with constraints is to define the procedures of cross-over and mutation in such a way that the constraints are automatically fulfilled. In general, this cannot be accomplished, and one often has to resort to the use of penalty functions, though this is generally rejected. It was found in our studies that these penalty functions should be adaptive in character, i.e. have weights that can vary through the optimisation, to ensure that the algorithm does not get stuck in local optima. In our experience, both selection procedure and constraint handling are very important with regards to local optima and should therefore receive ample attention.

Although there is both freeware and commercial software available to use GAs, we found that, at least for our case studies, it was necessary to create our own implementation, mainly because most software packages are not intended for dealing with the very large number of variables and/or constraints involved in our spatial optimisation problems.

REFERENCES

Aerts, J. (2002) *Spatial Decision Support for Resource Allocation*, PhD Thesis, University of Amsterdam.

Beasly, J.E. (2002) Population heuristics, in Pardalos, P.M. and Resende, M.G.C. (eds) *Handbook of Applied Optimization* , Oxford University Press, Oxford, pp. 138–157.

Berglund, H. and Jonsson, B.G. (2003) Nested plant and fungal communities; the importance of area and habitat quality in maximizing species capture in boreal old-growth forests, *Biological Conservation*, 112: 319–328.

Dauer, J.P. and Liu, Y.H. (1990) Multiple Objective linear programs in objective space. *European Journal of Operational Research*, 46: 350–357.

ECLNV (2004) *Landelijke Natuurdoelenkaart* [National Nature Purpose Map], ECLNV, Ede.

Ehrlich, P.R. (1995) The scale of the human enterprise and biodiversity loss, in Lawton, J.H. and May, R.M. (eds) *Extinction Rates*, Oxford University Press, Oxford, pp. 214–226.

EU (1992) *Council Directive 92/43/EEC of 21 May 1992*, Council of the European Communities.

Fahrig, L. (2001) How much habitat is enough? *Biological Conservation*, 100: 65–74.

Galloway, J.N., Schlesinger, W.H., Levy II, H., Michaels, A. and Schnoor, J.L. (1995) Nitrogen fixation: anthropogenic enhancement – environmental response, *Global Biogeochemical Cycles*, 9(2): 235–252.

Gibbs, J.P. (2001) Demography versus habitat fragmentation as determinants of genetic variation in wild populations, *Biological Conservation*, 100: 15–20.

Hagen, A. (2002) Multi-method assessment of map similarity, in Ruiz, M., Gould, M. and Ramon, J. (eds) *Proceedings of the 5th AGILE Conference on Geographic Information Science*, Palma, Spain, pp. 171–182.

Hagen, A. (2003) Fuzzy set approach to assessing similarity of categorical maps, *International Journal for Geographical Information Science*, 17(3): 235–249.

Heuberger, P.S.C. and Aben, J.M.M. (2002) Technische Achtergronden bij het rapport "effecten van verplaatsing van agrarische ammoniakemissies: verkenning op provinciaal niveau", (in Dutch), *Report 725501005*, RIVM, Bilthoven, the Netherlands.

Heuberger, P.S.C., Bakema, A.H. and van Adrichem, M.J.A. (1997) Optimalisatie milieurendement; methodiek en berekeningen voor optimale emissiereducties, (in Dutch), *Report 408143003*, RIVM, Bilthoven, the Netherlands.

Holland, J.H. (1975) *Adaptations in Natural and Artificial Systems*, University of Michigan Press, Ann Arbor.

Landis, J.R. and Koch, G.G. (1977) The measurement of observer agreement for categorical data, *Biometrics*, 33: 159–174.

Loonen, W., Heuberger, P.S.C., Bakema, A.H. and Schot, P.P. (2006) Application of a Genetic Algorithm to minimize agricultural nitrogen deposition in nature reserves, *Agricultural Systems*, 88(3): 360–375

LNV (1990) *Natuurbeleidsplan. Regeringsbeslissing.* Ministerie van Landbouw, Natuurbeheer en Visserij, Den Haag.

Luenberger, D. (1984) *Introduction to Linear and Nonlinear Programming*, Addison Wesley, Harlow.

Matthews, K.B. (2001) *Applying Genetic Algorithms to Multi-Objective Land-Use Planning*, PhD Thesis, Robert Gordon University.

McDonnell, M.D., Possingham, H.P., Ball, I.R. and Cousins, E.A. (2002) Mathematical methods for spatially cohesive reserve design, *Environmental Modeling and Assessment*, 7(2): 107–114.

Rauch, W. and Harremoes, P. (1999) Genetic algorithms in real time control applied to minimize transient pollution from urban wastewater systems, *Water Research*, 33(5): 1265–1277.

Reynolds, B. and Skeffington, R.A. (1999) Critical loads, *Progress in Environmental Science*, 1: 371–381.

RIVM (2002) *Milieubalans 2002, het Nederlandse Mmilieu Verklaard*, (In Dutch), Kluwer, Alphen aan den Rijn, The Netherlands.

Sarkar, D. and Modak, M.M. (2003) Optimization of fed-batch bioreactors using genetic algorithms, *Chemical Engineering Science*, 58: 2283–2296.

Seppelt, R. and Voinov, A. (2002) Optimization methodology for land use patterns using spatially explicit landscape models, *Ecological Modelling*, 151: 125–142.

Sousa, S., Caeiro, S. and Painho, M. (2002) Assessment of map similarity of categorical maps using Kappa statistics, ISEGI, Lisbon.

Thomas, C.D, Cameron, A., Green, R.E., Bakkenes, M., Beaumont, L.J., Collingham, Y.C., Erasmus, B.F., De Siqueira, M.F., Grainger, A., Hannah, L, Hughes, L., Huntley, B., van Jaarsveld, A.S., Midgley, G.F., Miles, L., Ortega-Huerta, M.A., Peterson, A., Phillips, O.L. and Williams, S.E. (2004) Extinction risk from climate change, *Nature*, 427(6970): 145–148.

van Dam, J., Heuberger, P.S.C., Aben, J. and van Pul, W.A.J. (2001) Effecten van verplaatsing van agrarische ammoniakemissies: verkenning op provinciaal niveau, (in Dutch), *Report 722501003*, RIVM, Bilthoven, The Netherlands.

van Dijk, S., Thierens, D. and de Berg, M. (2002) Using genetic algorithms for solving hard problems in GIS, *GeoInformatica*, 6(4): 381–413.

van Jaarsveld, J.A. and de Leeuw, F.A.A.M. (1993) OPS: an operational atmospheric transport model for priority substances, *Environmental software*, 8: 91–100.

van Jaarsveld, J.A. (1995) Modelling the long-term atmospherical behaviour of pollutants on various spatial scales, PhD Thesis, University of Utrecht, The Netherlands, also available as *Report 722501005*, RIVM, Bilthoven.

van Jaarsveld, J.A. (2004) The Operational Priority Substances Model: Description and validation of OPS-Pro 4.1, *Report 500045001/2004*, RIVM, Bilthoven.

van Jaarsveld, J.A., Bleeker, A. and Hoogervorst, N.J.P. (2000) Evaluation of ammonia emission reduction on the basis of measurements and model calculations, *Report 722108025*, RIVM, Bilthoven.

Vermeulen, H.J.W. and Opdam, P.F.M. (1995) Effectiveness of roadside verges as dispersal corridors for small ground-dwelling animals: a simulation study, *Landscape and Urban Planning*, 31(1–3): 233–248.

Vink, K. and Schot, P.P. (2002) Multiple-objective optimization of drinking water production strategies using a genetic algorithm, *Water Resources Research*, 38(9):1181, doi:10.1029/2000WR000034.

Vitousek, P.M., Mooney, H.A., Lubchenco, J. and Melillo, J.M. (1997) Human domination of Earth's ecosystems, *Science*, 277(5325): 494–499.

Zander, P. and Kächele, H. (1999) Modelling multiple objectives of and use for sustainable development, *Agricultural Systems*, 59(1999): 311–325.

Chapter 10

SUSTAINABLE LAND-USE AND WATER MANAGEMENT IN MOUNTAIN ECOSYSTEMS
A case study of a watershed in the Indian Himalaya

S.K. Mandal
National Institute of Public Finance and Policy, New Delhi, India

Abstract: The chapter examines the problem of choice of land use, energy input and technology that ensures economic viability and ecological sustainability in the Himalayan mountains. As the nature of the problem depends upon the local characteristics of the ecosystem, we propose to develop a model at the watershed level. This work develops a quantitative optimisation framework of net revenue maximisation using linear programming for structuring and articulating the problem of choice. The constraint of water availability has been taken into account to show how it drives the choice of technology and land use. The aims are to identify cost effective technologies, optimal land-use patterns and input combinations and to prescribe policies for adopting these technologies for rural situations with similar eco-regional and agro-climatic conditions.

Key words: Land use; sustainable development; mountain ecosystem.

1. INTRODUCTION

Economic development, whether in uplands or on plains, has involved continuous interaction between the processes of nature and the efforts of human beings to improve their material well-being. While the environmental challenges of development have induced many scientific discoveries and innovations in technology and social organization that have significantly contributed to relaxing the constraints of the carrying capacity of nature, the efforts to develop have sometimes resulted in environmental degradation, economic and social stagnation, and human suffering. The latter has been

E. Koomen et al. (eds.), Modelling Land-Use Change, 167–180.
© 2007 *Springer.*

caused partly by inefficient resource allocation and management and partly by explicit neglect of environmental concerns in the development process.

The character of this ecological degradation, however, differs with the state of development of an economy, and its ecological and socio-cultural setting. The overpopulated, poverty-stricken, bio-mass-based developing economies face environmental problems due to both the pressure of population and the unsustainable use of resources. The latter arises from the failure of social and economic institutions to resolve the problems of property rights, externalities, and those of income and asset distribution. The mountain areas in developing countries, like India, have specific environmental problems of degradation or conservation due to the features of verticality in physiography, resource richness, biodiversity and ecological fragility. The economy of mountain society revolves around the primary economic activities like agriculture, animal husbandry and forestry. Industrial development often becomes infeasible as a development strategy in the mountains due to the high cost of energy and transport for carrying raw materials to production sites and products to markets. The higher costs of these accounts are likely to offset the higher labour productivity of industry *vis-a-vis* the primary sector. As all primary activity involves land use, the issue of land use becomes crucial in any discussion of environmental sustainability of the development process in the mountains.

This chapter discusses the problem of sustainable development in the mountains of India, particularly in the Himalayan region, and presents a case study of land-use and energy planning in a micro-watershed to illustrate how ecological conservation can be enhanced by rational use of land with choice of appropriate technology, even if we ignore the monetary value of ecological services rendered due to the use of land as forest. The case study uses optimisation (linear programming) for the alternative static allocation of land resources and available technologies. The chapter also discusses the policy implications of such a case study in respect of the strategy of mountain development in India.

2.   PROBLEMS OF SUSTAINABLE DEVELOPMENT IN MOUNTAINOUS AREAS

Mountain heights play a crucial role in the climatic conditions of the tropical region of India and determine the regional hydrological conditions. The glaciers and watersheds of the mountain regions have been the sources of innumerable water streams which form into rivers and major flows of surface water in the lower plains. The slopes of the hills have mostly been covered by forests which often contain a rich biodiversity of plants and animal organisms. Today, such forest eco-systems are very often fragile in many parts of the mountainous regions of India. The fragility has been due to the extremely leached and poorly developed soil conditions of the forest ecosystem (Rao and Saxena, 1994).

In India, the mountain regions with forest resources have been inhabited by human population in scattered settlements. In the early stages of development of the economy, the traditional societies in these settlements had evolved their livelihoods in tune with the ecological processes of the region. The pattern of land use, the cropping pattern in agriculture and other practices relating to primary activities in most of the tribal societies of the hills took account of nutrient recycling and soil conservation, e.g. by practicing crop rotation and leaving the plot fallow for some time (Ramakrishnan *et al.*, 2002).

With the growth of human population and livestock, there has been a change in the land-use pattern, forest land being converted into agricultural land or forest land being overused for animal grazing – both causing land degradation. The increasing dependence on rain-fed agriculture on slopes or terrace cultivation by removing forests to meet the increasing food requirement of a growing population has caused soil erosion in the hills. About 90% of the total cultivated area in the mid hills of the western Himalayas is under rain-fed agriculture (Bhatnagar and Kundu, 1992). The model based exercises try to identify technology that will reduce land under rain-fed agriculture and allow its conversion to a more sustainable use which can halt the process of soil erosion.

Cattle, goats and sheep in the western and central Himalayas and pigs and poultry in the eastern Himalayas constitute important livestock wealth, while yaks are reared in Alpine areas. Due to land holdings being very small, livestock supplement the income of poor households and are considered as constituting a capital asset. Animal dung is used as fertilizer. The energy requirement for land preparation and transportation in agriculture is entirely met from bullock power. Whilst the high utility of livestock has led to an increase in its population, this has led to overgrazing due to lack of exclusive fodder crop farming in the mountains. Mismanagement of forests contributes to overgrazing and forest degradation (Singh *et al.*, 1994). The reserved forests are managed by the forest department mainly to earn revenue. Gradually these forests have been converted into *chir* pine and *deodar* forests which have high commercial value; broad-leafed species like *oak, kafal, sandan, bauhinia, ficus and hatab,* which supplied fodder and fuel-wood, have gradually dwindled. The pressure for fuel-wood and fodder consequently fell on the civil and community forests, which started shrinking. Efforts to diffuse grazing pressure on land in local animal husbandry systems do exist. The animals are sent to high altitudes for grazing in the summer months and a significant portion of fodder is obtained from crop residue (Rao and Saxena, 1994). Nevertheless, a trend of increasing pressure of livestock on forests is obvious. The optimisation model of land use that is outlined in the following section takes into account the fodder requirement and the possibility of animal energy utilization for economic activities.

Energy is demanded in the hills for cooking, lighting and heating in the household. The possibility of irrigated cultivation of fruit trees, herbs and medicinal plants and vegetables may be considered in small areas of land if electricity is available for such activities. It is estimated that about 1.1 trillion cubic metres of water flow every year down the Himalayas. The use of hydro-electricity for household purposes would reduce the pressure on forests for fuel-wood. Decentralized and small-scale management of micro-hydro-electric power systems, involving people participation and adapted to mountain constraints, appear more suitable, particularly for meeting the energy needs of marginal areas. The Himalayas offer a potential of generating 28,000 megawatt of electricity (Saxena *et al.*, 1994).

Technological options for the supply of energy for agriculture and household needs have to be assessed and explored. Fuel options which are economically efficient and ecologically sound need to be identified. Energy options with least intensity of carbon-dioxide emission would be suitable for fragile mountain ecosystems. In the model, several fuel options for energy use in irrigation, cooking, heating and lighting have been considered. In particular, the impact of a micro-hydro-electric power system on land use has been estimated.

With deforestation, unsustainable agriculture and their consequent impact on soil, water, vegetation cover and biodiversity, the carrying capacity of the mountain ecosystems for the human and livestock populations has declined over time. With a growing population, the development of ecological degradation has often led to out-migration of able-bodied males of working age to the plains for earning a livelihood and sending remittances back to the hills. In many places, as indicated by some of the primary surveys, this creates populations dominated by dependents consisting of the old and the children who are being looked after by the adult women staying back in the villages. Most agricultural work related to growing food and collecting fuel wood and water, which is all quite physically strenuous, is being carried out by the women. This adds a gender dimension to the pattern of livelihood and quality of life in the hills and raises concern for the well-being of the women due to stress caused by dwindling life support as provided by the ecosystem.

3.   A CASE STUDY OF OPTIMAL LAND USE

The discussion in the preceding section points to the importance of land-use patterns in determining both ecological sustainability of an ecosystem and the economic well-being of the people inhabiting the region. We submit below an optimisation model of land use for the Hawalbag watershed region and summarize the results in the next section in order to illustrate the real extent of conflict between developmental needs and environmental concerns. The analysis based on the model essentially focuses on the existing patterns of land use and its connectivity with various economic activities in the watershed and compares it with the optimal pattern of land use for the region. The comparison points to the potential of combining efficient choice of technology and land use with environmental conservation in similar watershed regions in the mountains. It illustrates how sometimes inefficient land-use and technological choice in the hills cause both loss of conventional economic value as well as ecological resources like top soil, air quality *et cetera.*

The optimisation model articulates the choices in the use of land, technology and natural resources for alternative purposes with the objective of net revenue maximisation from the major primary activities of the watershed economy, subject to meeting the basic need for food and energy by human beings as well as fodder for livestock, the latter being an important resource providing support to the mountain economic system. A linear programming framework has been chosen for the model primarily for its capacity to elegantly map intricate interrelationships among a large number of variables in a systematic manner. Such a framework provides the scope for optimising an objective function subject to a large number of constraints expressed as demand-supply equations.

The objective function estimates the value of returns from all land-use activities which include agricultural output (foodgrains and vegetables), fodder from pastures and forest products. The cost items include cost of cultivation of foodgrains and vegetables in agricultural land, fodder in pasture land, forest products in forest land and cost of conversion of land from one use to the other. It also includes the cost of household energy use for cooking, heating and lighting. The revenue from output of each of these land-use activities is valued at current market price. The cost computations of the inputs are discussed below in detail.

The total requirement of each input and its aggregate supply from various sources is specified in the form of demand-supply equations. The coefficients express input demand for a unit of output and the unit cost of supply from available sources respectively. Additionally, there are upper bounds to input availability in the case of limited supply.

The lower bound for food, fodder and energy requirement is specified for the entire human and livestock population of the watershed. These requirements are based on the primary survey and take into account the food habit of the population. Agricultural land allocation has to be done in a manner such that the organization of crop production activities in different seasons and in different types of land can meet the basic food requirement determined by the consumption pattern of a given population. The allocation of pasture land has to ensure fodder for a given size of livestock. The possibility of obtaining fodder from agricultural land in the form of agricultural residues and forest land in the form of leaf and grass has been allowed.

The surplus land that remains after meeting the basic needs of the watershed is devoted to the most market value adding use among the various options. The model considers the bounds of the availability of total land and water resources as given. To be more precise, the range of options that the model attempts to articulate in the case study covers: (1) the *use of land*, (2) the *choice of technology* and (3) *choice of fuel source*. This is discussed below in more detail.

The *use of land* can be either: agriculture with or without irrigation, pasture and forestry after allowing for conversion from one use to the other. Choice of cropping pattern along with seasonality has been explicitly considered in the model. In the current situation, only 40 acres can be irrigated from natural streams. The cost of an irrigation network has been simulated into the modelling structure that would allow for irrigated cultivation of around 400 to 425 acres depending on water availability and cropping pattern. The irrigation network would consist of pump sets and engines that would lift the water from the river bed to a height from where water can be canalized to agricultural land. Electricity required for pumping water would be supplied from the micro-hydro-electric plant which can be located on the main river of the watershed. The other options considered for the supply of energy for irrigation are biogas, diesel and animal power (lift irrigation). The cost coefficients of irrigation per acre of different crops have been computed by taking into account annualized capital cost, operational cost, end use cost and the amount of water required per acre of cultivation of different crops. The amount of area cultivated in the two seasons would vary due to seasonal variation in water availability and water requirements of cultivated crops. The total requirement and options for supply of other inputs in rain-fed and irrigated agriculture are discussed below. The cost coefficient per unit (acre) of pasture land takes into account the cost of seeds, fodder plants and labour required for growing and maintaining the pasture. The cost coefficient of forest land takes into account the conventional cost items of raising forests. The input requirement for a given productivity has been expressed as fixed coefficients for different categories of land use. The model has been designed to take into account cost of land conversion from present land-use pattern to an alternate use. This would indicate the average investment requirement for an alternate pattern of land use.

The *choice of technology* is determined by the use of (a) seed, (b) water, (c) organic and chemical fertilizer, (d) animal energy for land preparation and grain transportation, and (e) human labour. The requirement of each of these inputs is expressed as coefficients associated with the land type (rain-fed and irrigated) allocated for cultivation of foodgrains and vegetables in the monsoon (*kharif*) and winter (*rabi*) seasons. The requirement of nitrogen (N), potassium (K) and phosphorus (P) has to be met from chemical and organic fertilizers. The availability of organic fertilizer is associated with the dung produced by the livestock population and slurry from biogas digester. The seasonal nature of demand for animal labour in different crop producing activities in overlapping time has been synchronized such that the given draught animal population can be engaged optimally. The economic cost of all the agricultural inputs has been calculated. The cost of fertilizer has been calculated net of taxes and subsidies. The cost of animal and human labour, owned and hired, is valued at the prevailing market wage rate. The limit to supply of human labour has been determined by the available working population in the watershed.

The *choice of fuel sources*, as being either commercial or non-commercial, and their relation with food, fodder, fertilizer and land-use pattern is included in the model. For example, crop waste from agriculture can be used either as fodder or biomass fuel or compost fertilizer. Biomass crop waste fuel is a substitute of commercial and non-commercial energy forms. Dung from livestock can be used alternatively for fertilizer or energy used for household activities; livestock also provides energy for agriculture and rural transportation. The range of choices for household cooking energy includes animal dung, crop residue, fuel-wood, biogas, kerosene, liquid petroleum gas (LPG), electricity and coal; for lighting energy, it includes kerosene, electricity and biogas; and for space heating, it includes animal dung, fuel-wood, electricity and coal. Energy inputs like biogas and micro-hydro-electric power have been included among the options to understand their economic viability as a locally available non-conventional and renewable energy option in the context of development in the remote mountain regions. The total availability of human excreta has been considered as an input to biogas plants; the choice of use of animal dung as input in the digester has also been allowed. For cooking and lighting energy options, economic cost per unit (joules) of useful energy has been considered by taking into account the efficiency of the end-use device, including end-use cost (NCAER, 1978; TERI, 1989). For non-traded inputs like animal dung, crop waste and fuel-wood, the labour cost of time spent in collecting these inputs has been considered. The demand for cooking, lighting and space heating has been obtained from primary survey of household energy demand in the watershed.

The analysis also addresses some ecological implications of the optimal land-use choice: (a) emissions in the form of carbon dioxide from agricultural processes and household energy use, and (b) soil erosion. Total emission of carbon is determined by the choice of fuel in cooking, heating and energy use in agriculture. The emission coefficients for each of the fuel sources used in the energy using activities have been obtained from secondary sources (NCAER, 1978; TERI, 1991). Electricity from the micro-hydro-electric plant does not cause any emission. The net emission takes into account carbon sequestered in the process of growing forests. The rate of carbon sequestered per acre of forest in the Himalayan mountains has been obtained from secondary sources (Houghton, 1996; Lea *et al.*, 1996). The rate of soil erosion has been calculated by using the universal soil loss equation (Wischmeier and Smith, 1978). It takes into account coefficients of erosion due to rainfall and soil quality, length and slope of land in different land-use activities, and management and erosion control practices like erecting bunds and terraced farming. Broad categories of land-use activities, i.e. rain-fed and irrigated agriculture, pasture and forests have been considered to calculate the total amount of annual soil loss. Erosion coefficients of four ranges of slopes, i.e. 0 to 5 degrees, 5 to 10 degrees, 10 to 45 degrees and above 45 degrees have been obtained for each of the above categories of land use. All coefficients used in the calculation of soil loss have been obtained from the Watershed Management Directorate of the province located at Irgad.

The chosen area for the study is the Hawalbag watershed, located on the bank of river Kosi, in the Almora district of Uttaranchal province – in the central Himalayas between altitude 1,000-2,000 metres, where human activities have been widespread in terms of population growth, deforestation, extension of agriculture on mountain slopes, growth of livestock and demand for energy resources. The area spreads over 6,088 acres and contains a human population of 4,780 and a livestock population of 3,729 distributed in 15 villages. Details of the land-use pattern are given in Table 10-1.

The model has been estimated on the basis of data obtained from primary survey sources conducted in the Hawalbag area and supplemented by the geographical information system database on land use recorded on a scale of

1:50,000 for Almora district of the Forest Research Institute, Dehra Dun. The exogenous variables of the model describe the basic needs of the people in the concerned watershed region for a given year. The estimated model considers the economy of the watershed region to be representative for illustrative purposes.

4. RESULTS

The major feature of the results of the model has been that out of 6,088 acres of land use, the area for agriculture with irrigation and rain-fed agriculture should be 410 acres and 21 acres respectively, as against 40 acres of existing net sown irrigated area and 2,740 acres of net sown rain-fed agriculture (Tables 10-1). The land under pasture should also decrease from the existing 1,371 acres of use to 927 acres. The forest land under use should increase from the existing 1,533 acres to 4,451 acres of land. Cultivable wasteland and land under trees and shrubs would also be converted to forest land. Uncultivable wasteland and land under non-agricultural use, which is used for residential buildings, schools, hospitals, post offices *et cetera* has not been considered for land conversion. The implication of land conversion from present land use to the simulated optimal land use is given in Table 10-2.

Table 10-1. Current and simulated, optimal land-use areas [acres] in the Hawalbag micro watershed

Land-use type	Current area	Optimal area
Net sown area (irrigated)	40	410
Net sown area (rain-fed)	2,740	21
Pasture land	1,371	927
Forest land	1,533	4,451
Non-agriculture uses	273	273
Cultivable waste land	18	-
Uncultivable waste land	6	6
Fallow land	-	-
Shrubs and trees	107	-
Total land	6,088	6,088

Table 10-2. Land conversions following the optimisation analysis

Land conversion	Area [acres]
Rain-fed to irrigated agricultural land	370
Rain-fed agricultural land to forest land	2,349
Pasture land to forest land	444
Cultivable waste land to forest land	18
Land under trees and shrubs to forest land	107

The results of crop production activities associated with the agricultural land type (rain-fed and irrigated) allocated for cultivation of foodgrains and vegetables in the *kharif* and *rabi* seasons in the simulated optimal land use which would meet the basic requirement of foodgrains and vegetables of the population of the watershed are summarized in Table 10-3 .

Table 10-3. Optimal land allocation for agriculture

Land type	Seasons	Crops	Land requirement (acres)
Irrigated	Kharif	Rice	253
Irrigated	Kharif	Vegetable	61
Irrigated	Rabi	Wheat	318
Irrigated	Rabi	Vegetable	26
Irrigated	Rabi	Mustard	12
Irrigated	Rabi	Potato	52
Rain-fed	Kharif	Rice	10
Rain-fed	Kharif	Madua	10
Rain-fed	Rabi	Wheat	10
Rain-fed	Rabi	Potato	10
Rain-fed	Rabi	Mustard	-

The optimisation results emphasize the economisation of land use under agriculture by shifting acreage from inefficient rain-fed agriculture on slopes to irrigated agriculture with the use of inorganic fertilizer or organic manure in valleys as far as possible. The use of pasture land should also be kept at the minimum by efficient resource use and all surplus land, after meeting the need of food and fodder, should be transferred for use in forestry. It is the net added value of forestry products which makes forestry an attractive option purely on economic grounds.

Even without taking into account the ecological value of forests, the revenue maximisation objective would warrant the transfer of land from agricultural and pastoral use to forest, subject to the constraints of meeting the basic needs of food and fodder within the watershed. The simulated decline in the agricultural area essentially reflects the assumed economic inefficiency of rain-fed agriculture.

The implication of further growth of population with respect to food demand has been worked out. An increase in food demand by 1% would require agricultural land to grow by 1.85%. Given the capacity of irrigation infrastructure, the required increase for agricultural land would have to be met from extension of agriculture to rain-fed land. Thus, increase in demand for food would put pressure on pasture land and forest land which would be set to decline gradually.

With respect to water, the results of the model indicate that its availability for irrigation is a key factor that influences allocation of land for different uses. Due to the lack of irrigation facilities, the potential of water is not fully realized now. The maximum potential of sustainable water use permits 410 acres of irrigation. The optimal land-use pattern makes full use of the irrigation potential for production of foodgrains and vegetables, making the water constraint binding and requiring small acreage of land use under rain-fed conditions. An installed capacity of 20 kilowatt hydro-electric plant and six engines of 20 horsepower will be required to pump water to a height of 165 feet from where the water can be canalized to irrigate the required amount of land. The adoption of this technology would considerably reduce land use under rain-fed conditions.

The results of the model indicate that there is sufficient scope for increasing productivity in agriculture by better management of local resources like dung and other biomass-based manure. A technological intervention for anaerobic digestion of dung would greatly increase the fertility potential of locally available organic manure. About 43% of the dung generated will optimally flow to the anaerobic digester to meet the entire requirement for N, P and K of the watershed. Further, the anaerobic technology will initiate substitution away from chemical fertilizer which would make agriculture better environmentally sustainable.

On the livestock management, the optimisation of net revenue points to crop residue, fodder grown in agricultural lands (when it is left fallow between seasons) and grazing in pasture and old forests as the major sources of fodder. Crop residue can contribute 2,059 tons (22% of total fodder requirement). Fodder from agricultural land contributes marginally, i.e. 0.11% of total fodder requirement. The major share of fodder comes from grazing pasture. In this particular exercise, it has been assumed that the livestock does not graze in the new forest area since allowing grazing while regenerating of forests will decrease the chances of survival of the plants. Grazing may be allowed in a full-grown forest. About 2,556 acres of forest will be required to sustain the livestock population if grazing in full-grown forests is allowed; in that case, no pasture land would be required. So, as forests start regenerating, pasture land may be gradually transformed into forest. The implication of increase in livestock population has been worked out, indicating that a 1% increase in fodder demand would increase allocations to pasture land by 1.28%. The results show that it is better to use crop residue for fodder than for other uses like compost fertilizer or for cooking fuel.

The energy requirement for land preparation and local transportation of foodgrains and vegetables can be met from the available animal energy in both seasons of cultivation. According to the results of the model, 80 and 5% of animal energy available from draught animals will be utilized in irrigated and rainfed cultivation respectively in the *kharif* season. In the *rabi* season 87 and 5% of the available animal energy in irrigated and rainfed cultivation will be utilized. For local transportation 35% of the available animal energy will be utilized in irrigated cultivation in the *kharif* season and less than one percent will be used in rain-fed cultivation. In the *rabi* season 70% of the available energy will be utilised in irrigated and 2% will be utilised in the rain-fed cultivation. Thus animal energy availability will be sufficient for land preparation and agricultural transportation in the watershed.

The requirement for inanimate energy for cooking, lighting and space heating should be ideally supplied, as shown by the results of the model, by electricity from micro-hydro-electric units which can be set up to tap the hydro-energy potential of the region. An installed capacity of 385 kilowatt will be sufficient to supply the required amount of cooking energy for the entire watershed. For the purpose of light energy, an installed capacity of 380 kilowatt will be sufficient and for space heating in the cold mountain region an installed capacity of 360 kilowatt will be required. In any situation of scarcity of electric power because of inadequate investment to utilize such potential, LPG and soft coke would be the next best options for cost economisation for cooking and space heating respectively. Dung is to be mainly used for organic fertilizer. The optimisation model warrants a part of it to be used in an anaerobic digester to produce slurry for fertilizer. Biogas turns out to be uneconomical due to low productivity in cold mountain regions.

The environmental impact of the change in land use and related activity pattern as per the net revenue optimisation model would be favorable in respect of top soil loss, carbon emission and carbon sequestration. In the present land-use pattern, the extent of soil erosion works out to 8,709 tons; according to the simulated land-use pattern, the soil loss works out to 2,003 tons, thus, total soil erosion will be reduced by 77%. A large amount of agricultural land located on the slopes can be released for afforestation by increasing the area under cultivation in the valley land. The carbon emission due to energy consumption would also be drastically reduced by the utilization the hydro-potential of the region. This would require, of course, the mobilization of a capital fund and institutional arrangements for the implementation of small power projects.

Finally, transfer of land to forest use will facilitate substantial carbon sequestration in the region by substantive amount, the net sequestered amount being 5,466 tons of carbon as per the optimal solution. Afforestation would also have favorable impact on the employment situation due to expansion of forestry-based activities. However, such land-use change in favour of forests would also demand appropriate institutional arrangements to be in place.

5.   CONCLUSION

In the light of the case study referred to above, it is important to note that environmental conservation of resources is intimately linked with the pattern of land use and technology in the mountains. It is also interesting to note that profit or net revenue maximising allocation of resources in terms of land use goes often along with environmental conservation of resources like top soil, water resources *et cetera*. It is, in fact, the choice of land use along with associated economic activities in the hills which is the crucial factor in characterizing the developmental process in the mountains.

The optimisation results show that an economically viable and ecologically sustainable pattern of resource use in the mountains is possible that can provide the required food, energy and fodder and can, simultaneously, reduce the process of degradation by controlling soil erosion to a large extent. This pattern will also improve air quality and reduce the possibility of global warming through carbon sequestration by increasing the area under forests. The optimal land and energy-use pattern is not driven by self-perpetuating economic considerations due to the perception of food security of poor people living in far flung mountain areas. Food shortages in such areas can be conspicuous during times of landslides and natural disaster that keep the area and supply lines cut off for long periods. In general, poor people in many regions are driven by concerns of food security. Hence, they would ensure that the land they possess should guarantee availability of foodgrains and fodder to the extent possible. The perception about food security can change with the process of development of the region and economic well-being of the population. Lack of capital and technological know-how also prevent communities of such regions to adopt a land and energy-use pattern as outlined in the case study above. Such problems can be addressed by transferring funds to the village local governments from the federal governments in the course of democratizing the development process. Assigning carbon credit, as envisaged in the Kyoto Protocol may also contribute to solving the problem of capital crunch.

In terms of a sectoral strategy of development, the case study suggests that forestry and livestock raising mainly for livelihood will be economically beneficial and ecologically sustainable in the hills. Agriculture should be confined mainly to the valleys, except for such plantations which can be grown on gradients without degradation of the soil-water system. This would not necessarily cause a deficit of foodgrains. However, if there is a deficit in some cereals, pulses *et cetera,* in the mountains, it needs to be imported from the plains. While the mountains provide ecological support to the plains through the major flows of surface water and forest resources, the plains need to supply, in return, foodgrains in which there is a deficit and other industrial goods to ensure life support on the hills.

What is important is both efficient choice of technology from the overall point of view of resource economisation as well as careful consideration of environmental or ecological costs and benefits in addition to the conventional ones.

The analysis of environment related problems in any region should take account not only of the limits of nature in decisions of economic choice, but should also analyse the impact of economic choices on ecology and take account of the feedback effect of ecological changes on the economic system to understand the dynamics of long-run processes of economy, society and nature.

REFERENCES

Bhatnagar, V.K. and Kundu, S. (1992) Climatological Analysis and Crop Water Use Studies for Sustainable Improved Rainfed Agriculture in Mid Hills of U.P., Vivekananda Parvatya Krishi Anusandhan Shala, Almora.

Lea, H., Zhou, D., Jung, Y. and Sathaye, J. (1996) Greenhouse Gas Emissions Inventory and Mitigation Strategies for Asia and Pacific Countries, *Ambio*, 25; 4.

Houghton, R.A. (1996) Converting Terrestrial Ecosystems from Sources to Sink of Carbon, *Ambio*, 25; 4.

NCAER (1978) Domestic fuel survey with special reference to kerosene –Volume II. National Council of Applied Economic Research (NCAER), New Delhi.

Ramakrishnan, P.S., Rai, R.K., Katwal, R.P.S. and Mehndiratta, S. (2002) Traditional Ecological Knowledge for Managing Biosphere Reserves in South and Central Asia, UNESCO and Oxford & IBH, New Delhi.

Rao, K.S. and Saxena, K.G. (1994) Sustainable Development and Rehabilitation of Degraded Village Lands in Himalaya, G.B. Pant Institute of Himalayan Environment and Development, Almora.

Saxena, K.G., Kumar, Y., Rao, K.S., Sen, K.K., Rana, U., Majila, B.S., Pharrsawan, A., Singh, G. and Nehal, S. (1994) Hydropower Management for Sustainable Rural Development in Remote Un-electrified Zones of Himalayas, Himvikas Publication No. 7, G.B. Pant Institute of Himalayan Environment and Development, Kosi, Almora.

Singh, J.S., Pandey, U. and Tiwari, A.K. (1994) Man and forests: a central Himalayan case study, *Ambio*, 13: 80–87.

TERI (1989) Application of a Rural Energy Model in Energy Planning Parts A & B, Tata Energy Research Institute (TERI) New Delhi.

TERI (1991) Strategies for Mitigation of Environmental Stress in Garhwal Himalayas, Report submitted to MOEF, Tata Energy Research Institute (TERI) New Delhi.

Wischmeier, W.H and Smith, D.D. (1978) Predicting rainfall erosion losses; A guide to conservation planning. Agricultural Handbook No. 537, USDA.

Chapter 11

GENETICLAND: MODELLING LAND-USE CHANGE USING EVOLUTIONARY ALGORITHMS

J. Seixas[1], J.P. Nunes[1], P. Lourenço[1] and J. Corte-Real[2]

[1]*Departmento de Ciências e Engenharia do Ambiente, Faculdade de Ciências e Tecnologia, Universidade Nova de Lisboa, Caparica, Portugal;* [2]*Centro de Geofísica de Évora e Departamento de Física da Universidade de Évora, Portugal*

Abstract: Future land-use configurations provide valuable knowledge for policy makers and economic agents, especially under expected environmental changes such as decreasing rainfall or increasing temperatures. This chapter proposes an optimisation approach for modelling land-use change in which landscapes (land uses) are generated through the use of an evolutionary algorithm called *GeneticLand*. It is designed for a multiobjective function that aims at the minimisation of soil erosion and the maximisation of carbon sequestration, under a set of local restrictions. *GeneticLand* has been applied to a Mediterranean landscape, located in southern Portugal. The algorithm design and the results obtained show the feasibility of the generated landscapes, the appropriateness of the evolutionary methods to model land-use changes and the spatial characteristics of the landscape solutions that emerge when physical drivers have a major influence on their evolution.

Key words: Spatial planning; long-term; land use; climate change; optimisation, evolutionary computing; Mediterranean landscape.

1.   INTRODUCTION

Land-use models have been developed during the last decades to answer land-use planning issues. Several methodological approaches have been adopted to produce land-use scenarios for a wide range of time scales, with emphasis on the next cycle of decision-making policy (i.e. the next years).

E. Koomen et al. (eds.), Modelling Land-Use Change, 181–196.
© 2007 *Springer.*

The need to reason in a longer time scale has appeared recently, in part motivated by the climate change concern. Rounsevell *et al.* (2005), for example, have presented scenarios of future agricultural land use in Europe to 2080. These authors approach the issue of land-use planning as a solution-based process derived from scenario writing: they consider, from the beginning, different policy scenarios to which a landscape should be adapted, in most cases with the help of technology. However, several examples have shown that physical drivers, such as climate and morphology, have conditioned the evolution of landscape productivity and configuration, especially where they are close to some vulnerability threshold. This is the case with, for example, agriculture and forestry productivity in European southern regions that are characterised by a decrease of water availability, resulting from low soil storage capacity and decreasing precipitation trends. The importance of physical drivers in land-use planning is exacerbated within the context of long-term planning (e.g. more than 30 years).

Considering this time framework, an approach to land-use planning as a goal-oriented process is proposed. The goal relies on the discovering of expected behaviours of a landscape to key drivers of change (mostly environmental and morphological drivers), identifying concerns, assessing trends and spatial configurations and producing knowledge after which policy scenarios should be designed. For example, discovering long-term future landscapes (understood as a spatial configuration of land uses) under a changed climate scenario, in a goal-oriented process, should result in different expected behaviours to which technology, social and economic scenarios should be designed in order to prevent or accommodate them. On the contrary, discovering long-term future landscapes in a solution-based process requires, firstly, several scenario assumptions, as for example described in the special report on emissions scenarios (or SRES by Nakicenovic *et al.*, 2000), to which adapted landscapes can be determined in response to a set of allocation rules.

This work proposes to reason about very long-term land-use change following an optimisation approach, where several applications of the objective function result in different landscape solutions in order to facilitate the identification of emergent spatial patterns. Landscapes can be understood as complex systems, considering their characteristics of spatial self-organization, and non-linear behaviour to long-term drivers. However, one may assume that a landscape under specific climate and morphological conditions, for example a Mediterranean region, will evolve in a way, if: 1) one considers no policy assumptions, and 2) a set of constraints is respected in order to keep spatial coherence and physical feasibility in the future.

The formulation of future landscape generation as an optimisation problem relies on more than one objective function and, at least, two sets of constraints. There is a wide application of techniques to land-use optimisation problems (Pongthanapanich, 2003). However, most of these techniques were not developed to allow objectives that require spatial data and are unable to handle spatial objectives. Ducheyne (2003) presents an overview of the disadvantages of classic optimising procedures to handle spatial data and spatial tailored objective functions. Linear programming, for example, uses continuous variables which are not suitable when spatial integrity is of concern. Integer and mixed integer programming overcome this problem, but in order to explicitly formulate the spatial requirements, they have computing power problems. Heuristics have also been proposed to handle complex optimising problems, but they are essentially single objective optimisers. Usually, they require multiple objectives to be reformulated into a single objective function and this hampers the search for the trade-off front between these objectives.

The approach of evolutionary computation to deal with land-use generation as an optimisation task, accommodating its explicit spatial dimension and multiple objective functions, was adopted. *GeneticLand*, the name for the evolutionary algorithm designed to generate landscapes, accommodates two objective functions – minimisation of soil erosion and maximisation of carbon sequestration – and a set of physical feasibility and suitability constraints to land uses, as well as spatial coherence constraints. In order to illustrate the impact of physical properties in long-term planning, optimal landscape configurations will be produced considering the actual climate as well as future climate scenarios.

2. METHODOLOGY

Modelling land-use change is addressed as a multi-objective optimisation problem in which landscapes, defined as a set of spatially organized land uses, are generated by means of evolutionary computation (EC). This field of mathematics contains a number of techniques, such as the genetic algorithms discussed by Loonen *et al.* (see Chapter 9), that have been applied successfully in search and optimisation problems across a variety of domains.

Evolutionary algorithms are based on the principles of natural selection. The rationale is that fit individuals survive and propagate their traits to future generations and unfit individuals have a tendency to die. In this type of algorithm, an individual corresponds to a possible landscape solution. The task of the EC is to search for a good land-use assignment under certain constraints.

As opposed to classical optimisation methods, which often combine multiple objectives into a single objective by assigning different weights to each of them, the field of EC has developed methods allowing a diverse set of solutions which incorporate the tradeoffs that are intrinsic to the problem at hand. It is left up to the decision makers which of the different alternative solutions should be chosen, usually based on higher level information (Deb, 2001).

2.1   Algorithm formulation

The evolutionary algorithm for land-use generation applied here, *GeneticLand*, operates on a region represented by a two-dimensional array of cells. For each cell, there is a finite number of possible land uses. The task of the algorithm is to search for an optimal assignment of these land uses to the cells, evolving the landscape patterns that are most suitable for the various objectives satisfying the set of restrictions.

2.1.1   The objective functions

The *GeneticLand* algorithm considers two goals: minimisation of soil erosion and maximisation of carbon sequestration. Soil erosion was selected due to its importance in the Mediterranean landscapes and the perspective of its evolution under climate scenarios characterized by annual reduced rainfall patterns and increased flash floods (McCarthy *et al.*, 2001). Each landscape solution, provided by the *GeneticLand* algorithm, is validated by applying the USLE (Universal Soil Loss Equation), with the best solution being the one that minimises both the global landscape soil erosion value and lowers the local erosion values below a manageable threshold (10 ton.ha^{-1}.y^{-1}). The USLE predicts the long-term average annual rate of erosion with the following expression:

$$Global\ soil\ loss = \Sigma_i\ [R * LS * K * C * Landscape_i] \qquad (1)$$

where:
- i is a pixel of the landscape;
- R is the rainfall factor: the greater the intensity and duration of the rain storm, the higher the erosion potential;
- LS is the slope length factor: the steeper and longer the slope, the higher the risk for soil erosion;
- K is the soil erodibility factor: a measure of the susceptibility of soil particles to detachment and transport by rainfall and runoff;

C   is the crop/vegetation and management factor used to determine the
       relative effectiveness of crop management systems in terms of
       preventing soil loss.

Each landscape (*Landscape$_i$*) solution, generated by *GeneticLand*, is
multiplied by the appropriate USLE factors resulting in a long-term average
annual soil loss in tons per hectare per year. The global landscape soil
erosion value is the sum of average annual soil loss calculated for all pixels,
except for those with values less than 10 ton.ha^{-1}.y^{-1}, which are considered
negligible. The optimal landscape is the one that, selected from a set of runs,
minimises the global soil erosion value. The *GeneticLand* algorithm is
implemented in a way that all factors may be changed in order to consider
different data sets. Therefore, and since the impact of physical properties in
long-term planning is of interest, two optimal landscape configurations were
generated by the *GeneticLand* algorithm, one considering actual climate
drivers and another one taking into account future climate scenarios. These
climate scenarios were accommodated through factor *R* in Eq. (1).

Maximisation of carbon sequestration was considered due to its
importance to the carbon cycle and under climate change scenarios. Each
landscape solution is evaluated in terms of its carbon uptake according to the
following expression:

$$Global\ carbon\ uptake = \Sigma_i\ C\ uptake * Landscape_i \qquad (2)$$

where *i* represents a pixel of the landscape. Each landscape (*Landscape$_i$*)
solution is multiplied by the average carbon uptake indicators (Instituto do
Ambiente, 2003) for each land use, with the best solution being the one that
maximises the global landscape carbon uptake. The global landscape carbon
uptake value is the sum of carbon uptake calculated for all pixels. The *C*
uptake coefficient in Eq. (2) varies according to actual and future CO_2
atmospheric concentrations, and the *GeneticLand* algorithm generates
different landscape solutions.

2.1.2   The set of restrictions

The assignment of land-cover classes in future landscapes by the
GeneticLand algorithm is conditioned by two types of constraints: physical
constraints, concerning geomorphological variables, and landscape ecology
indices at the patch, land-use and landscape levels. The physical constraints
were stated from the analysis of land-cover maps regarding the distribution
of land-cover classes against four different variables: 1) soil type; 2) slope;
3) aridity index, derived from the ratio between precipitation and

Thornthwaite's potential evapotranspiration (P/PET); and 4) topographic soil wetness index (TSWI) (Beven, 2000). Therefore, based on historical data, the constraints state the allowances to assign a specific land use to a specific pixel, considering the physical and climate properties that occur in that pixel. The inclusion of climate change scenarios on land-use modelling was accommodated in the restrictions instantiations, namely in the aridity index, since all the other factors can be assumed constant over time.

Landscape ecology restrictions were used to assure the spatial coherence of the landscape. Landscape ecology provides indices that help the characterization and quantification of landscape structure, function and change (McGarigal and Marks, 1995). Two indices were selected: the patch size for each land use, and the adjacency index, named contagion.

A patch is defined as a non-linear surface that differs in the appearance, shape and complexity, and includes a single pixel or a set of adjacent pixels of the same land-use class. The patches vary in size, form, type, heterogeneity and characteristics of edge. The size is an important aspect of a patch, since it governs the circulation of nutrients through the landscape and the amount of species in a region. *GeneticLand* considers a range for the patch size (min and max) for each land-use class. Contagion is a landscape ecology index that measures the probability of 'adjacency' of cells (pixels) of the same land-use class. This index measures the degree of dispersion or aggregation of the landscape elements where high values (maximum 100) are from landscapes with few patches of great dimension, while low values (minimum 0) show landscapes with many dispersed units.

2.2   Algorithm implementation

Multi-objective evolutionary algorithms (MOEAs) are applied where there is no single optimal solution, but rather a set of alternative solutions. These solutions are optimal in the wider sense that no other solutions in the search space are superior to them when all objectives are considered. This idea has been incorporated successfully in a number of MOEAs, and the main reason for that is due to the fact these algorithms work with a population of solutions rather than with single solutions.

For the evolutionary algorithm, the Pareto archived evolution strategy (PAES) developed by Knowles and Corne (1999) was used. This algorithm is perhaps the simplest evolutive scheme for multi-objective optimisation and it is based on an extension of the (1+1) evolution strategy (Schefel, 1981). It starts with a random solution (one parent) and then it generates one offspring by means of a mutation operator. If the offspring is a better solution than the parent, it replaces the parent for the next generation (iteration). In the opposite case, the offspring is discarded. This process is repeated a number of times until a specified stopping criterion is satisfied. The PAES algorithm maintains an archive of non-dominated solutions, each of which cannot be said to be better that the other. The PAES algorithm was selected, rather than a more sophisticated evolutionary algorithm, due to the very large problem dimension as a landscape extension of several km^2. The mutation operator implemented in *GeneticLand* changes a cell to a different land use. In addition, it also changes a number of surrounding cells to the new land use, because otherwise the mutation operator would generate invalid solutions with a very high frequency. When constraints are violated by means of the mutation operator, the fitness of the solution is penalized by a certain amount. The more a constraint is violated, the more the solution is penalized.

The algorithm was run for a total of slightly over one million iterations. Theoretical results (Mühlenbein, 1992) refer that, for a simple unimodal function, the average number of iterations that the (1+1) EA needs to find the optimal solution is exp(1)*N*log(N/2), where N denotes the number of decision variables of the problem. That result was derived for solutions coded as binary decision variables, which is not the case with landscapes, with more than two land-use classes. Considering that the most difficult step is to optimise the last gene (which means not mutating the correct genes and mutating the incorrect one to the right value), *GeneticLand* approximates the number of iterations multiplied by a factor of α, equivalent to the number of land-use classes. The simulations of *GeneticLand* were run on a Linux cluster.

3.   APPLICATION

The *GeneticLand* algorithm was designed and tested in an area located in southern Portugal, within the Guadiana watershed, characterized by a semi-arid climate, poor soils and low productivity, within the region presented in Figure 11-1. Main land uses include annual agriculture, mostly wheat, mixed agriculture, usually spots of agriculture within natural vegetation, as well as vast areas of shrubs, usually species adapted to water stress conditions, like rock-roses (*Cistus ladeniferus*), and small spots of forests. This area was chosen due to its recognized ecosystem fragility and high vulnerability to climate change (European Environment Agency, 2005), which fulfils the motivation for this work, relating to long-term land-use planning under climate change scenarios.

The minimisation of the soil erosion objective function was supported by climate and geomorphological data sets, as well as a susceptibility factor of land uses to soil loss, as presented in Table 11-1 for the Mediterranean landscape (Pimenta, 1998) under study. The maximisation of carbon uptake on carbon uptake coefficients associated with each land cover, is shown in Table 11-2, for both the actual and climate change scenarios of CO_2 atmospheric concentrations. For the current study, the climate scenario referring to a 660 PPMV (parts per million volume) CO_2 concentration was used.

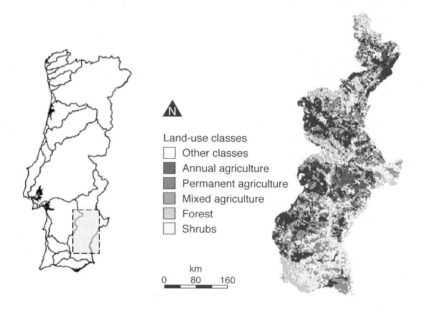

Land-use classes
- ☐ Other classes
- ■ Annual agriculture
- ■ Permanent agriculture
- ■ Mixed agriculture
- ☐ Forest
- ☐ Shrubs

km
0   80   160

Figure 11-1. Study region and land-use map, from CORINE 1987.

Table 11-1. Crop/vegetation management (C) factor of the distinguished land-use classes

Land-use classes	Susceptibility factor of land uses to soil loss
Forest	0.1
Shrubs	0.02
Permanent agriculture	0.1
Annual agriculture	0.3
Mixed agriculture	0.3

Climate variables include: 1) a rainfall data set: long-term monthly climate measurements from the Portuguese Water Institute, downscaled using geostatistical methods (Nicolau, 2002); and 2) a temperature data set: long-term monthly climate measurements from the Portuguese Meteorological Institute, downscaled using conventional interpolation methods. Potential evapotranspiration was derived from the Thornthwaite's equation (Lencastre and Franco, 1992).

Table 11-2. Carbon uptake coefficients for specific land uses

Land-use classes	Carbon uptake indicators(t C/ha/year)*		
	Current carbon concentration: 360 PPMV CO_2	Scenario: 495 PPMV CO_2	Scenario: 660 PPMV CO_2
Forest	1.6	1.84**	2.08**
Shrubs	0.4	0.44***	0.47***
Permanent agriculture	0.5	0.60****	0.69****
Annual agriculture	0	0	0
Mixed agriculture	0.1	0.12****	0.14****

*Climate change estimates based on the response of plant radiation-use efficiency to increased CO_2 concentrations; **Calculations for *Quercus suber* and *Quercus ilex (*Maroco *et al., 2002, and* Tognetti *et al., 1998),* ***Calculations for *Cistus sp (*Tognetti *et al., 2000);* ****Calculations for *Olea europaea (*Tognetti *et al., 2001)*

The climate variables from climate change scenarios, used as the physical drivers of long-term land-use planning, consider rainfall and temperature changes. These are derived from the Had-AM3 Global Circulation model for the SRES emission scenario A2, in which rainfall is downscaled through a statistical method based on weather circulation patterns (Qian *et al.*, 2002) . Figure 11-2 shows expected changes for the A2 emission scenario predicted by the Had-AM3 model, compared with the current normal climate, for the study region. Erosivity is expected to decrease, due to a decrease in annual precipitation, while aridity will be exacerbated, due to a temperature increase. It is assumed that no changes in the geophysical structure of the landscape (slope, soil type) and the vegetation types and patterns will occur under climate change.

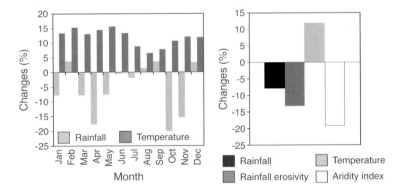

Figure 11-2. A2 scenario changes compared to current climate as predicted by the Had-AM3 model: monthly rainfall and temperature (left) and some genetic algorithm parameters (right).

Physical characteristics of the landscape, selected for the design of the *GeneticLand* constraints, as presented previously, are illustrated in Figure 11-3: the aridity index includes seven classes ranging from very dry (1) to very humid (7); slope is considered in fifteen classes from flat (1) to very steep (15); and topographic soil wetness index consist of thirteen classes varying from very low (1) to very high soil humidity (13).

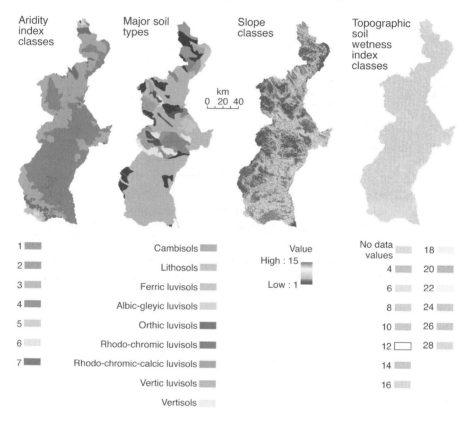

Figure 11-3. Selected physical characteristics of the landscape under study, used to design the constraints in the *GeneticLand* algorithm. (See also Plate 7 in the Colour Plate Section)

Spatial analyses, such as overlaying and cross-classification, were performed on datasets related to land use, geomorphology and landscape ecology to derive an appropriate set of restrictions. Table 11-3 presents the constraints stated for each land cover in soil type 1. For example, annual agriculture only occurs in areas with slope classes ranging from 1 to 4, evapotranspiration classes from 2 to 3 and soil wetness values from 6 to 28, while forests occur in every slope class, in areas where evapotranspiration ranges from 2 to 7, and soil wetness classes from 4 to 16. This procedure was performed for every land cover in every soil type, providing a complete set of geomorphological constraints to feed the *GeneticLand* algorithm.

Table 11-3. Example of the constraints feeding the *GeneticLand* algorithm, concerning the land covers occurring in soil type 1 (Cambisols) for the case study

	Slope		Evapotranspiration		Soil wetness	
	Min.	Max.	Min.	Max.	Min.	Max.
Annual agriculture	1	4	2	3	6	28
Permanent agriculture	1	7	2	5	4	16
Mixed agriculture	1	6	2	7	4	26
Forest	1	15	2	7	4	16
Shrubs	5	8	4	6	4	4

4. RESULTS AND DISCUSSION

A small subsection of the study region, representing an area of about 9 km^2 (100 columns and 100 rows), was selected to test the implementation and performance of the *GeneticLand* algorithm. As previously explained, and according to the evolutionary methods in general, each run of the *GeneticLand* algorithm generates a set of solutions, which evolve from an initial random image, generated with different random seeds. In this experiment, two sets of 15 landscape solutions were generated, for each climate scenario (current and climate change). The solutions' tradeoff between the two objective functions, more soil erosion corresponds to more carbon uptake, for one experiment can be evaluated in Figure 11-4.

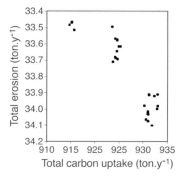

Figure 11-4. Tradeoffs between the two objective functions results, from 30 landscapes generated by the *GeneticLand* algorithm.

Although there are differences among the 30 landscape solutions, all the solutions comply with the stated restrictions (more than 91% in area). An in-depth analysis of the solutions reveals that carbon uptake increase is preferred to soil erosion decrease, when the *GeneticLand* algorithm assigns land-use classes. This is a consequence of the imposed constraints on this landscape, since the choice of shrubs, the land use more appropriate to prevent soil erosion, is restricted in part of this area. Figure 11-5 presents one of the landscape solutions generated by the *GeneticLand* algorithm both for the current climate and the climate change scenario. The landscape mosaic is due to the patchy nature of some restrictions, like slope and wetness index. Furthermore, there is not a clearly preferable land-cover class. If such a class would exist, a more homogeneous landscape would be generated.

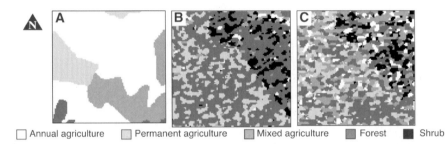

□ Annual agriculture   ▨ Permanent agriculture   ▨ Mixed agriculture   ■ Forest   ■ Shrub

Figure 11-5. Land-use map of (A) the test area, (B) landscape solutions generated by the *GeneticLand* algorithm for current climate and (C) climate change scenarios.

Soil erosion patterns are shown on Figure 11-6. Comparing those associated with current land use to the solutions generated by the *GeneticLand* algorithm for both the current and climate change scenarios, it can be concluded that the algorithm succeeds in reducing the overall soil erosion rates. However, soil erosion patterns derived from landscapes under the climate change scenario reveal slight higher soil erosion rates when compared to landscapes under current climate scenario, as expected.

Finally, Figure 11-7 shows the comparison of global results concerning land-use classes, soil erosion and carbon uptake, between current landscape and those generated by the multi-objective *GeneticLand* algorithm, for the current climate and climate change scenarios. On average, the *GeneticLand* algorithm was able to generate landscapes with increases of carbon uptake up to more than 840% and 660%, for current and climate change scenarios respectively, when compared with actual land uses. At the same time, the algorithm was able to reduce soil erosion by about 65% and 46%, accordingly. Also, it should be noticed that areas of serious soil erosion (>100 ton.ha^{-1}.y^{-1}) were reduced more than 60% in both cases, when compared with actual land uses. This is a very interesting result, because the reduction of serious soil erosion areas was not a requested goal, but the algorithm chose to improve its solution in those pixels.

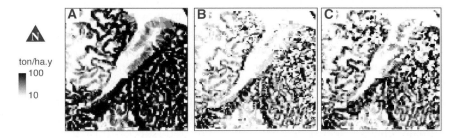

Figure 11-6. Soil erosion patterns derived from (A) current land cover and (B) landscape solutions for current climate and (C) climate change scenarios.

Concerning the spatial distribution of land uses, it should be noted that the landscape solutions generated by the *GeneticLand* algorithm are characterized by a mosaic of land uses, with a high degree of spatial heterogeneity, when compared with the actual landscape. Although the impact of resulted fragmentation in the landscape was not evaluated, the mosaic model of land-use spatial distribution has been proposed as an adapted strategy of the landscapes to promote ecosystem sustainability in general, and preventing soil erosion, in particular.

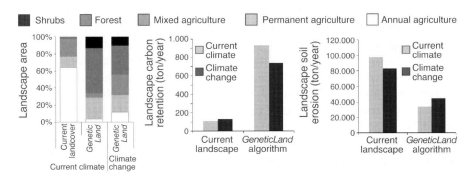

Figure 11-7. Aggregated results concerning land-use change, carbon uptake and soil erosion for current and changed climate conditions, compared for the current and simulated landscape of the *GeneticLand* algorithm.

5.   CONCLUSION

This chapter proposes to reason on the very long term land-use change as a goal-oriented process, aiming to discover expected behaviours of a landscape to key drivers, like climate, after which policy scenarios should be designed. Land-use modelling was approached as an optimisation problem, formulated for two objective functions (soil erosion minimisation and carbon uptake maximisation), and subject to a set of physical and spatial constraints. The evolutionary algorithm, named *GeneticLand*, was designed and tested in a small area located in southern Portugal, within the Guadiana watershed.

GeneticLand generated landscape solutions for current climate and climate change scenarios that accommodated successfully the objective functions, while complying in more than 90% in the area with the restrictions. For both climate scenarios, *GeneticLand* algorithm was able to produce landscape solutions that increase carbon uptake very significantly when compared with current land uses, while at the same time reducing soil erosion. Landscape solutions are characterized by spatial heterogeneity of the land uses, which is appointed as an appropriate strategy to promote ecosystem sustainability in general, and to prevent soil erosion, in particular. Comparing the results for the two climate scenarios, there are no significant differences, although mixed agriculture was slightly preferred. Major conclusions are the appropriateness of the evolutionary methods to model land-use changes, and the spatial characteristics of the landscape solutions that emerge when physical drivers have a major influence on their evolution.

Some limitations should be considered for further development, including: (1) the physical constraints were derived from CORINE 1987 spatial analysis but should rather be derived from a time series analysis in order to accommodate land-use change dynamics; (2) the algorithm is only allowed to choose the same land-use classes as the current land-use map, but other classes should also be allowed; (3) physical suitability of land uses, stated in the set of constraints, is kept constant in the future, but some aspects should be adjusted as is the case with forests that can adapt to other evapotranspiration classes than 2 to 7; (4) improvement of the implementation strategy to run images larger than 100 by 100, in order to handle more complex landscapes.

ACKNOWLEDGEMENTS

This work was funded by the Portuguese Foundation for Science and Technology, under Contract POCTI/MGS/37970/2001. The authors would like to thank F. Lobo and P. Condado from Universidade of Algarve for the discussion of the concept and the implementation of the *GeneticLand* algorithm.

REFERENCES

Beven, K. (2000) *Rainfall-Runoff Modelling, The Primer*, John Wiley & Sons, Chichester.
Deb, K. (2001) *Multi-Objective Optimization using Evolutionary Algorithms*, John Wiley and Sons, Ltd, Chichester.
Ducheyne, E. (2003) *Multiple Objective Forest Management using GIS and Genetic Optimization Techniques*, PhD Thesis, University of Gent, Gent.
European Environment Agency (2005) *Vulnerability and Adaptation to Climate Change in Europe*, EEA Technical Report no. 7/2005, July. http://reports.eea.eu.int/ technical_report_2005_1207_144937/en.
Instituto do Ambiente (2003) *Programa Nacional para as Alterações Climáticas, Vol. 8: Florestas e Produtos Florestais* Instituto do Ambiente, Faculdade de Ciências e Tecnologia, CEEETA, Lisboa.
Knowles, J.D. and Corne, D.W. (1999) The Pareto Archived Evolution Strategy: A New Baseline Algorithm for Pareto Multiobjective Optimisation, in *Proceedings of the 1999 Congress on Evolutionary Computation (CEC'99), Volume 1*, pp. 98–105.
Lencastre, A. and Franco, F.M. (1992) *Lições de hidrologia*, New University of Lisbon Editorial Services, Lisbon.
Maroco, J.P., Breia, E., Faria, T., Pereira, J.S. and Chaves, M.M. (2002) Effects of long-term exposure to elevated CO2 and N fertilization on the development of photosynthetic capacity and biomass accumulation in Quercus suber L, *Plant Cell and Environment*, 25 (1): 105–113.
McCarthy, J., Canziani, O., Leary, N., Dokken, D., and White, K. (2001) *Climate Change 2001: Impacts, Adaptation, and Vulnerability, Contribution of Working Group II to the Third Assessment Report of the Intergovernmental Panel on Climate Change*, Cambridge University Press.
McGarigal, K., and Marks, B.J. (1995) FRAGSTATS: spatial pattern analysis program for quantifying landscape structure, USDA For. Serv. Gen. Tech. Rep. PNW-351.
Mühlenbein, H. (1992) How Genetic Algorithms Really Work: Mutation and Hillclimbing. in Männer, R. and Manderick, B. (eds) *Parallel Problem Solving from Nature*, Elsevier Science, Amsterdam, pp. 15–25.
Nakicenovic, N., Alcamo, J., Davis, G., de Vries, B., Fehann, J., Gaffin, S., Gregory, K., Grubber, A., Jung, T.Y., Kram, T., Emilio, la Rovere, E., Michaelis, L., Mori, S., Morita, T., Pepper, W., Pitcher, H., Price, L., Riahi, K., Roehrl, A., Rogner, H., Sankovski, A., Schelesinger, M.E., Shukla, P.R., Smith, S., Swart, R.J., van Rooyen, S. Victor, N. and Dadi, Z. (2000) *Special Report, on Emissions Scenarios*, Cambridge University Press, Cambridge.
Nicolau, M.R.R.C. (2002) *Modelação e mapeamento da distribuição espacial de precipitação – uma aplicação a Portugal continental*, PhD Thesis, Faculty of Sciences and Technology, New University of Lisbon.
Pimenta, M.T. (1998) Directrizes para a aplicação da Equação Universal de Perda dos Solos em SIG: factor de cultura C e factor de erodibilidade do solo K, INAG, Lisbon.

Pongthanapanich, T. (2003) *Review of Mathematical Programming for Coastal Land Use Optimization*. University of Southern Denmark, Department of Environmental and Business Economics IME Working Paper 52/03.

Qian, B., Corte-Real, J. and Xu, H. (2002) Multisite stochastic weather models for impact studies, *International Journal of climatology*, 22: 1377–1397.

Rounsevell, M., Ewert, D.A.F., Reginster, I., Leemans, R. and Carter, T.R. (2005) Future Scenarios of European agricultural land use. II. Projecting changes in cropland and grassland, *Agriculture Ecosystems & Environment*, 107(2–3): 117–135.

Schwefel, H.P. (1981) *Numerical Optimization of Computer Models*, John Wiley & Sons, Chichester.

Tognetti, R., Johnson, J.D., Michelozzi, M. and Raschi, A. (1998) Response of foliar metabolism in mature trees of Quercus pubescens and Quercus ilex to long-term elevated CO2, *Environmental and Experimental Botany*, 39: 233–245.

Tognetti, R., Minnocci, A., Peñuelas, J., Rasci, A. and Jones, M.B. (2000) Comparative field water relations of three Mediterranean shrub species co-occuring at a natural CO2 vent, *Journal of Experimental Botany*, 51(347): 1135–1146.

Tognetti, R., Sebastiani, L., Vitagliano, C., Raschi, A. and Minnocci, A. (2001) Responses of two olive tree (Olea europaea L.) cultivars to elevated CO2 concentration in the field, *Photosynthetica*, 39(3): 403–410.

PART IV: INCORPORATION OF NEW MODELLING APPROACHES

Chapter 12

MICROSIMULATION OF METROPOLITAN EMPLOYMENT DECONCENTRATION
Application of the UrbanSim model in the Tel Aviv region

D. Felsenstein, E. Ashbel and A. Ben-Nun
Department of Geography, Hebrew University of Jerusalem, Israel

Abstract: Employment deconcentration has become a major issue on the policy and planning agenda in many metropolitan areas throughout the western world. In recent years, growing evidence indicates that in many developed countries, the deconcentration of employment - particularly of retail centres and offices - has become a key planning issue. This chapter uses the *UrbanSim* forecasting and simulation model in order to investigate some of the projected changes in land use, land value and sociodemographic characteristics of metropolitan areas undergoing employment deconcentration. The process of model application in the Tel Aviv metropolitan context is described. Two land-use scenarios of very different scales are simulated: a macro-level scenario relating to the imposition of an 'urban growth boundary' and a micro-level scenario simulating the effects of a shopping mall construction in different parts of the metropolitan area. The results are discussed in terms of the potential and constraints of microsimulation for analyzing metropolitan growth processes.

Key words: Employment deconcentration; land use; microsimulation; *UrbanSim.*

1. INTRODUCTION

The spatial deconcentration of retail, manufacturing and office activities is a very visible phenomenon in large metropolitan areas. 'Edge city' development is an entrenched metropolitan phenomenon in the US (Lang, 2003). In Europe, incipient edge city development is becoming an increasingly familiar sight on the edges of many large metropolitan areas. Whether on the outskirts of Amsterdam, Paris, Madrid, Bristol or Prague, the picture of out-of-town office developments, shopping centres or industrial

E. Koomen et al. (eds.), Modelling Land-Use Change, 199–217.

parks seems to repeat itself. With nearly 80% of Europe's population living in cities, recent years have witnessed a steady shift in population between city centres and suburban areas. Consequently, urban densities in the centres of major European metropolitan areas have been constantly declining in many cities. While this change is not uniform across countries or even within countries themselves, there is no doubt that a more polycentric European metropolitan area is emerging (Bontje, 2001; Kratke, 2001).

Very little, however, is known about the prospective effects of employment deconcentration (or non-residential sprawl) in a non-US context. Employment deconcentration is taken to mean here the movement of economic activities (industry, retail, services) from the centre to the urban fringe or the relative decline of employment in the city centre versus the urban periphery. The latter can result not just from movement from the centre to the fringe but from in-situ growth in the urban perimeter or in-movement to the fringe area from outside the region. Deconcentration can be measured by relative employment densities, land consumption, floorspace and similar metrics. We can hypothesize that in developed countries outside the US (and especially in European-type countries), the effects of employment deconcentration will be very different to those arising from residential deconcentration. While employment sprawl has received little systematic attention, a wealth of anecdotal evidence points to it raising questions of efficiency and equity as contentious as those raised by residential sprawl.

On the one hand, it can be viewed as a response to needs and free choice in the market. Allowing firms and offices to move to suburban locations will encourage the creation of jobs that would not be produced in the dense and expensive inner parts of metropolitan areas. This leads to a rational and efficient allocation of resources (land, jobs *et cetera*) and a higher quality of life. Producers are expected to make higher profits in suburban locations and will also create employment that would not have been produced in the dense and expensive inner parts of metropolitan areas. Consumers and workers gain as deconcentration of offices, 'big box' retailers and factories to the urban fringe provide more employment choice and greater services at reduced prices.

On the other hand, many contest this benign view of employment deconcentration and present a string of equity and welfare issues that are affected by this process (Persky and Wiewel, 2000). These can be classified as socioeconomic issues pertaining to the spatial mismatch of employment, job opportunities, community cohesion, costs of infrastructure provision and accessibility, environmental effects such as noise, congestion, pollution, groundwater quality and resource effects relating to the loss of open space, the consumption of agricultural land *et cetera*. When negative deconcentration effects predominate, these patterns of development can undermine the viability of inner/central cities, and the decline of central cities is likely to harm the quality of life of residents in suburban locations as well. If the positive aspects of deconcentration predominate, the reverse will be the case. These issues have been debated extensively in the North American context (Ding and Bingham, 2000; Felsenstein, 2002; McMillen and McDonald, 1998) and increasingly appear on the European urban agenda (Urban Audit, 2000).

This chapter explores the effects of employment deconcentration going beyond the measurement and morphology of metropolitan change that has attracted much attention in the literature (Ewing *et al.*, 2002; Galster *et al.*, 2001; Torrens and Alberti, 2000). We attempt to address the broader issues of deconcentration by simulating the wider socioeconomic impacts associated with this process. To capture these effects, we use the *UrbanSim* land-use simulation model (Waddell, 2002; Waddell *et al.*, 2003) to forecast land-use change in two metropolitan areas and to explore the resultant socioeconomic and demographic changes that they imply.

UrbanSim continues a microsimulation tradition in land-use modelling and extends the earlier work of Wegener (1982), Mackett (1990) and Simmonds (2001). Contemporary efforts in this area are centered on both improving the economic modelling mechanisms at the base of the allocation procedures that drive the models and on extending the level of disaggregation at which the model operates. The PECAS modelling system (Hunt and Abrahams, 2005), for example, uses a spatial input-output approach to capture the exchanges between producers and consumers in an equilibrium framework. The *UrbanSim* system is grounded in a random utility approach in which the main agents in the land market (workers, households, firms, institutions and developers) attempt to maximise their utility and make choices accordingly. In this way, a price structure emerges (capitalized via land prices) and markets clear. All agents are modelled at a very fine level of analysis (the grid cell) which allows for dynamic microsimulation at a level of disaggregation that captures the full behaviour of the individual agent.

The spatial context for this study is the Tel Aviv metropolitan area which has experienced accelerated patterns of employment deconcentration since the mid 1980s. This chapter proceeds as follows. Section 2 describes the *UrbanSim* applications and the process of data preparation undertaken for the two case study areas. In Section 3, we present results arising from two types of simulations relating to employment deconcentration. The first relates to simulating an attempt to deal with employment deconcentration through regulatory instruments such as imposing an urban growth boundary (UGB) or restrictive zoning. The second relates to the micro-based implications of deconcentration arising from particular development events such as the building of new industrial parks, shopping malls and the like. In all cases we compare results of the 'with event' scenario case with results for the baseline 'without event' (business as usual) case. Finally, some tentative ideas about the microsimulation of urban growth processes and their planning implications are presented in Section 4.

2.   MODEL AND APPLICATION DESCRIPTION

2.1   Modelling system

UrbanSim is a land-use modelling system for scenario simulation and policy analysis developed by the Center for Urban Simulation and Policy Analysis at Washington University, Seattle, USA. It comprises a series of interlinked models that together form a dynamic activity-based system that simulates the activities of three major urban 'actors' that interact with each other in the land market; a) grid cells that represent the land parcels of the study area and their physical traits; b) households and their characteristics and c) jobs represented by workers. The unique attributes of the model include its high resolution of prediction (150 metre by 150 metre grid cells), full integration with GIS systems and a modelling approach based on microeconomic and behavioural foundations. The full workings of this system have been outlined elsewhere (Waddell, 2002; Waddell *et al.*, 2003). Here, we limit ourselves to a short description of the integrated models, depicted in Figure 12-1, that make up the system.

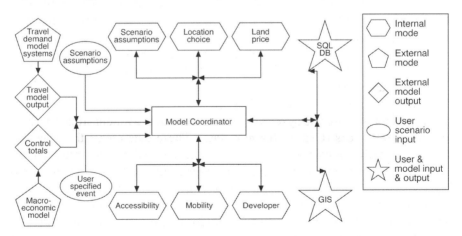

Figure 12-1. The *UrbanSim* land-use modelling system.

Two *external models* serve the system. The first is the '*macro-economic model*' which is used to predict the changes in annual household and employment totals within the study area. The data created by this model is imported into the different model parts and used as a guideline (control total) for the different model components. The macro-economic model creates predictions for the changes in the number of households, by size and race, as well as the change in employment by sectors. The second external model is the '*travel model*'. The travel model is used to create the composite utility of getting from one travel analysis zone (TAZ) to another, given the available travel modes. This data is created externally and then imported into the model to create accessibility measures between different grid cells.

Six separate *internal models* simulate the different actions of the three urban 'actors'. The '*economic and demographic transition model*' uses the control totals created by the exogenous macro-economic model to create new households and jobs which will be added into the study area. In cases where the number of households (in a specific group) or jobs (in a specific sector) has declined, the transition model removes those households and or jobs from the study area. The '*employment and household mobility model*' simulates the decision of households and jobs to change location within the study area during each year of simulation. The model creates a list of households and jobs which have decided to move from their current location within a specific year and extracts them for relocation. The '*household and employment location choice model*' simulates the location decisions taken by the households and jobs in the study area. This includes all households and jobs created by the transition model as well as the households and jobs which have decided to change location in the mobility model. The '*real estate development model*' simulates the actions of real estate developers within the study area. The model predicts the grid cells that will encounter a development event and the type of development that will result. The '*land price model*' simulates the changes occurring within the real estate market using a hedonic regression of the land value on the attributes of the land parcel and its surroundings. Finally, the '*accessibility model*' combines the data created by the external travel model and the land-use data in order to create an accessibility matrix between different grid cells.

The input data for the system are imported into the model from a number of different sources (GIS, tables *et cetera*). These data create the base year from which the model runs as well as the coefficients for the different internal models and the scenarios. None of the different models listed above connect directly. The interaction between the models is done within a 'Model Coordinator' module and is then exported back in to the different models parts. The exported data are the result of the model prediction. The data can be exported for each simulated year as well as for specific years only. These data can be fully integrated with GIS layers allowing for further examination as well as improved visualization.

2.2   Study area

The Tel Aviv metropolitan area lies on the western shoreline of Israel (Figure 12-2). This region has 2.98 million inhabitants and a million employees in an area of approximately 1,683 km^2 and is the largest metropolitan area in Israel. The Tel Aviv metropolitan area is the economic heart of Israel and produces approximately 49% of the country's GNP. The residential and employment deconcentration processes in the Tel Aviv metropolitan area began during the 1980s. The rising levels of car ownership, the improvement in living standards and the mass immigration from the former communist countries of Eastern Europe created growing pressure for suburban residential, commercial and industrial development. These pressures created a metropolitan region which, today, is increasingly suffering from congestion, lack of open space, air and noise pollution *et cetera*. Although the new Israel National Outline Plan #35 (TAMA 35) tries to confront these sprawl-related problems, no real spatial modelling has been done to try and predict the effects of these processes on the quality of life in the future.

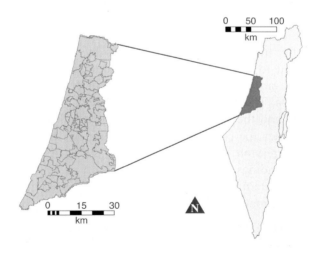

Figure 12-2. The Tel Aviv metropolitan area.

2.3 Data description and preparation

The process of constructing the Tel Aviv *UrbanSim* databases, required the use of a number of data sources and software products such as ArcInfo 9.0, Excel, and Access. The data collected was used to create the grid cells, households, and jobs databases as well as the data needed for the control totals, relocation rates and the different model coefficients (Table 12-1).

Table 12-1. Sources and formats of input data

Theme	Data source	Format
Households	National Census, 1995	Grid cells
Jobs	Travel Survey, 1996	Tables
Relocation rates	Labor Force Surveys, 1995-2003	Tables
Control totals	Israel National Plan for the year, 2020	Tables
Land use	Hebrew University (HUJI) GIS Database	Grid cells
Historical events, land prices	The Israel Lands Administration.	Grid cells
Accessibility	Tel Aviv Metropolitan Area Travel Model (NTA Corp.)	Grid cells

The most extensive data available on the metropolitan area are available in the national census of 1995 and the travel census of 1996. These two sources determined the way in which the data were collected and implemented. In order to keep the information as exact as possible, the grid cell size selected for the Tel Aviv application of the *UrbanSim* model was 250 metre by 250 metre. This size allowed us to also include the data available in the smallest census tracts, covering 500 metre by 500 metre, without losing any information. Using GIS, a fishnet of grid cells of 250 metre by 250 metre was created covering the whole metropolitan area creating the base-year grid cell database.

The division and insertion of the census data into the base year database was done using standard GIS and database management software. Each grid cell was allocated a census tract to which it corresponded. The data from each census tract were transferred into the grid cell using a GIS join command. In cases where there was more than one grid cell per census tract, the data were divided in an equal fashion between the different grid cells. This was based on the definition of the census tracts as homogenous units. The households in each census tract were divided into separate entities with their spatial location based on the census tract from which they came, creating the households database.

In order to complete the grid cells database, with information that was not available through census tracts, such as percentage road or percentage water, additional data were imported into the database using GIS layers from

the Hebrew University (HUJI) GIS database. All data were converted to a
raster format, following the import process shown in Figure 12-3.

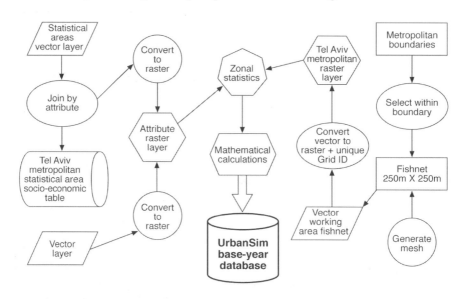

Figure 12-3. Base-year database construction: the raster and vector import process.

The jobs database was created using the national travel survey (1996).
This includes data about the movement of workers, from different
employment sectors, from and to work. These data were used to allocate
each job within the Tel Aviv metropolitan area to a grid cell according to the
census tract the job belonged to. Having created the base-year database (grid
cells, households and jobs), the other tables had to be updated according to
the Tel Aviv guidelines. The control totals for the employment and
households were taken from the Israel National Plan for 2020. The
accessibility and travel analysis zones data were collected from the Tel Aviv
metropolitan area travel model developed by the Tel Aviv transit authority
(NTA) and the relocation rates for jobs and households were taken from the
Israel Labour Force survey between the years 1995 and 2003. Land price
data both current and historical were made available by the Israel Land
Administration (ILA).

The Tel Aviv application of *UrbanSim* was accomplished mainly under
textbook conditions. Relatively large amounts of the data were readily
available and were imported directly into the model databases. The
modifications done in the Tel Aviv case were mainly concerned with the
data for the estimation process of the different location choice models and
estimation of land values and improvement values. The land and

improvement values created in the Tel Aviv metropolitan area were initially based on a sample of approximately 1,000 residential, commercial and industrial property transactions collected from the ILA. The data from these transactions were then used for an inverse distance weighted interpolation for the whole of the metropolitan area. When further data became available, we were able to use real rather than interpolated values. The process of location choice model variable estimation for both households and jobs should, preferably, be based on a sample of households and jobs that recently moved. While this form of data is not available for Tel Aviv, we created a Monte Carlo random sample of 5,000 households and jobs (per sector) on which the models were estimated. All statistical discrete choice models were estimated in STATA using either standard or multinomial logit estimation.

3. SIMULATION RESULTS

In this section we report the *UrbanSim* results for two very different scenarios. The first relates to a 'policy scenario' where a major regulatory restriction is imposed on metropolitan development. We simulate the effects of an Urban Growth Boundary (UGB) for metropolitan Tel Aviv. The second case presents results relating to the simulated outcomes of 'event scenarios' in the Tel Aviv metropolitan area. These are micro-level interventions that have a more limited direct land-use impact. However, they can have significant indirect impacts in terms of the socioeconomic composition and land values of the areas in which they are located.

3.1 Macro-level simulation: urban growth boundary imposition

This policy scenario imposes a UGB within the Tel Aviv metropolitan area as outlined in the Israeli National Outline Plan #35 (TAMA35). This UGB is a non-continuous series of boundaries around the main urban clusters within the metropolitan area (Figure 12-4). The testing of this scenario included running a 'business as usual' (BAU) baseline case beginning in the year 1995 (the base year for the Tel Aviv model) and ending in the year 2020. This scenario simulated a non-intervention policy allowing unregulated development of all forms of land use.

A second ('with UGB') case was run stipulating the UGB and prohibiting development beyond its limit. Each of the grid cells within the model was allocated a marker which located them within or outside the UGB. Within the UGB no restrictions were imposed on development. Outside the UGB, the model was directed to prohibit any development from the year 2005 (the year the plan became legally binding) to the target year 2020.

The scenario results are described in two parts. The first part describes the simulated land-use patterns, densities and land values given the UGB scenario. The second part presents the sociodemographic implications of this forecasted land-use change. These data describe the characteristics of the households within the metropolitan region in the year 2020 under the UGB scenario.

3.1.1   Land-use patterns

The results shown in Tables 12-2 and 12-3 clearly show that the UGB has an effect on the development of residential and commercial land use as well as on the land value. The UGB results in a very clear decline in the amount of development in the metropolitan area. Both total and average residential units and commercial floorspace decline in the metropolitan area due to the UGB. However, when we look at the results *within* the UGB area, we observe that average density and the sums of both residential units and commercial floorspace increase. The UGB scenario results in the development of approximately 200,000 less residential units in comparison to the BAU scenario but the number of residential units within the UGB is higher by approximately 13,000 units. When we analyze commercial land use, the results show a similar pattern. The UGB results in an average commercial m^2 per grid cell which is approximately 30% less than the BAU scenario. However, when we look at this average within the confines of the UGB, the results show a higher density of commercial land use compared to the BAU scenario.

These simulations also indicate that imposing a UGB results in a change in both residential and commercial land values in the metropolitan area. The average residential unit value in the Tel Aviv metropolitan area is approximately 30% lower in the UGB scenario than in the BAU scenario. The average improvement value, i.e. the difference in value between built and un-built commercial land use, is approximately 7,000 NIS (New Israeli Shekel i.e. $1,500) per m^2 higher in the BAU scenario than in the UGB scenario. Although the average commercial and residential land value is lower in the metropolitan area overall, opposite results are forecast *within* the UGB area. The value of an average residential unit within the UGB is roughly 55,000 NIS higher and the average improvement value for commercial land use (per m^2) is roughly 1,400 NIS higher.

These initial results suggest that imposing a UGB results in an overall decline in residential and commercial land-use development as well as commercial and residential land value in the metropolitan as a whole.

However, within the confines of the UGB, an opposite effect can be noted. The UGB serves to increase the divide between the areas within the UGB and those outside this boundary.

Table 12-2. BAU and UGB scenario results for the whole metropolitan area: total and mean per grid cell values compared for 2020

Theme	Type	BAU	UGB	Change
Residential units	sum	1,448,567	1,230,299	-218,268
	mean	54	46	-8
Commercial area [m^2]	sum	81,035,036	54,175,886	-26,859,150
	mean	3008	2011	-997
Residential unit value [NIS]	mean	1,046,450	70,933	-975,516
Commercial area value [NIS/m^2]	mean	19,332	12,055	-7,277

Table 12-3. BAU and UGB scenario results within the UGB: total and mean per grid cell values compared for 2020

Theme	Type	BAU	UGB	Change
Residential units	sum	906,207	919,333	13,126
	mean	101	102.5	1.5
Commercial area [m^2]	sum	25,801,111	26,917,741	1,116,630
	mean	2877	3002	125
Residential unit value [NIS]	mean	1,187,432	1,243,371	55,939
Commercial area value [NIS/m^2]	mean	20,893	22,364	1,471

Maps created using the grid cells exported for the year 2020 reiterate the results shown in the tables above. Figure 12-4 shows that imposing the UGB produces a clearly different outcome from permitting uncontrolled development in the BAU scenario. The UGB case results in a controlled form of commercial development with high commercial densities in the UGB and low densities elsewhere. The BAU scenario reflects the possible effects of unregulated development with high density commercial development across most of the metropolitan area. Note that by 2020, the eastern and southern parts of the metropolitan area display a level of commercial density not far from the density levels attained in the core of the metropolitan area.

In terms of land value, Figure 12-5 shows that the average commercial improvement value per grid cell in the metropolitan area overall is lower in the UGB than in the BAU scenario. In the latter scenario, continued development outside the UGB has resulted in higher commercial improvement value. Imposing a UGB does not seem to have driven up land prices within the confines of its boundaries. Rather, releasing development restrictions serves to keep commercial land markets buoyant everywhere. The commercial improvement value in the metropolitan core and around the city centres (which are located inside the UGB) remains high under both scenarios. Finally, the increased commercial development under the BAU scenario results in regional differences within the metropolitan area with a particularly pronounced rise in commercial improvement value in the southern sections.

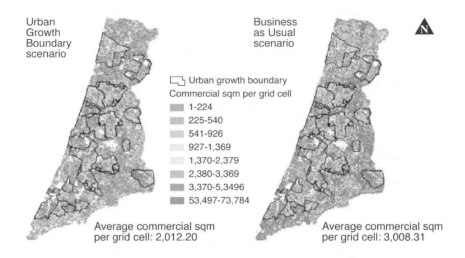

Figure 12-4. Commercial land-use density, 2020, UGB and BAU scenarios compared. (See also Plate 8 in the Colour Plate Section)

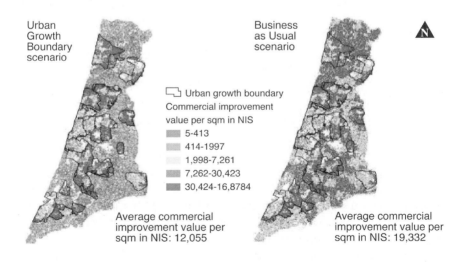

Figure 12-5. Commercial improvement value per grid cell, 2020, UGB and BAU scenarios compared. (See also Plate 9 in the Colour Plate Section)

3.1.2   Household socio-demographics

As each grid cell in the *UrbanSim* model is linked to a string of socio-demographic attributes of the households occupying the cell, simulated land-use change can also be examined in these terms. Tables 12-4 and 12-5 outline the effects of the UGB on the whole metropolitan area and the area within the UGB. In both cases key social, economic and demographic attributes and their differences under the two scenarios are highlighted.

Table 12-4. BAU and UGB socio-economic scenario results for the whole metropolitan area: total and mean per grid cell or household values compared for 2020

Theme	Type	BAU	UGB	Change
Households	Sum	1,397,364	1,230,299	-167,065
	Mean	51.9	45.7	-6.2
Cars	Sum	1,020,053	833,612	-186,441
	Mean	37.9	31.0	-6.9
	per household	0.73	0.68	-0.05
Children	Sum	501,753	448,921	-52,832
	Mean	18.6	16.7	-1.9
	per household	0.36	0.36	0.00
Persons	Sum	4,118,121	3,641,941	-476,180
	Mean	152.9	135.2	-17.7
	per household	2.95	2.96	0.01
Workers	Sum	1,397,220	1,231,713	-165,507
	Mean	51.9	45.7	-6.2
	per household	1.00	1.00	0.00
Income	Mean	10,268	12,147	1,879

Table 12-5. BAU and UGB socio-economic scenario results within the UGB: total and mean per grid cell or household values compared for 2020

Theme	Type	BAU	UGB	Change
Households	Sum	872,075	919,333	47,258
	Mean	97.3	102.5	5.2
Cars	Sum	614,209	594,155	-20,054
	Mean	68.5	66.3	-2.2
	per household	0.70	0.65	-0.05
Children	Sum	305,000	322,392	17,392
	Mean	34.0	36.0	2.0
	per household	0.35	0.35	0.00
Persons	Sum	2,521,689	2,636,592	114,903
	Mean	281.2	294.0	12.8
	per household	2.89	2.87	-0.02
Workers	Sum	872,835	920,595	47,760
	Mean	97.3	102.7	5.4
	per household	1.00	1.00	0.00
Income	Mean	₴ 720	₴ 559	-161

As can be seen, the UGB scenario results in a metropolitan population which is approximately 500,000 persons smaller than the BAU scenario. In terms of household density, as expected, the UGB scenario results in higher density of households within the UGB (74.7%) than the BAU scenario (62.4%). The latter scenario results in particularly high levels of density in the metropolitan core and near the city centres. In terms of household size, the simulated results show that under the BAU scenario the distribution of all household sizes is evenly distributed across the metropolitan region. But in contrast, the UGB scenario results in a concentration of the smaller households inside the UGB, whereas the large households concentrate outside. The UGB seems to force the larger households to seek residential opportunities outside its borders.

The spatial distribution of workers per grid cell under both scenarios seems almost identical (Table 12-4) in the whole of the metropolitan area. The highest concentration of employed persons, in both scenarios, is in the core of the metropolitan area and around the major city centres. However, this obscures some geographic detail that shows that under the UGB scenario 74.7% of the workforce is located within the UGB where under the BAU scenario this figure is only 64.5%.

The simulation results also report similar average household car ownership rates within the overall metropolitan area under both scenarios. However, per household and per geographic unit rates differ considerably with higher rates under the BAU than under the UGB scenarios. This suggests that regulated development, created by a UGB, limits the need to travel as employment, shopping and recreation possibilities are contained within the UGB.

3.2   Micro-level simulation: shopping mall development

The micro event scenario simulates land use, land price and sociodemographic effects associated with the development of shopping malls in different rings (inner, middle and outer) of the metropolitan area. We were particularly interested in observing the differential effects of location in the different rings given that the core area contains a highly developed and competitive retail market while the intermediate and outer rings offer greater opportunity for non-residential land-use development.

We simulate the effects of three similar sized malls of approximately 90,000 gross m^2 each (Figure 12-6). In each case, we simulate a 'with shopping mall' scenario (SM) that is compared with a baseline 'without mall' (BAU) scenario. We observe differences within 1 kilometre and 5 kilometres distances. The inner ring mall (Givatayim) is an existing development that started operation in 2005. The middle ring mall (West

Raanana) is a simulated (fictitious) development on a land parcel zoned for commercial development but with no approved plan. Finally, the outer ring mall (Modiin) is presently under construction and due to open in 2008. The malls were inserted into the *UrbanSim* model using the development events table. The baseline year for the inner ring mall was taken as 2005 and for the middle and outer rings 2008 was the starting year. In all three cases, the model was run to the year 2020.

Figure 12-6. Tel Aviv metropolitan ring structure and shopping mall locations.

The results show that locating the shopping malls in the various rings produces differential impacts. Using spline interpolation, we simulated the effect of mall location on other retail and commercial land uses. In all three cases, the effect of the shopping malls decreases with distance and is negligible at a distance of 5 kilometres (Figure 12-7). This effect remains invariant over time.

However, within the immediate vicinity of the shopping malls, we observe very different effects. Table 12-5 summarizes the simulated results for all three malls within a 1 kilometre radius. Each column in the table shows the 'with' and 'without' mall simulations. The effect of the mall on average residential prices is particularly interesting. In the inner ring scenario, the mall has a negative effect on average house prices but this is overturned in the intermediate and outer rings. This probably suggests that the externalities associated with adding further commercial activity in an already developed area, are in the main negative.

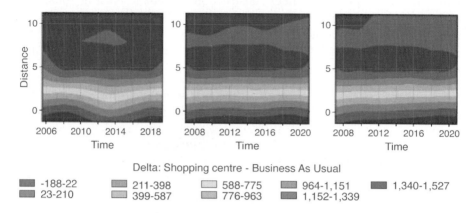

Figure 12-7. Differential impacts of (left) Givatayim Mall, (middle) Raanana Mall and (right) Modiin Mall on commercial land use to 2020. (See also Plate 10 in the Colour Plate Section)

In terms of effect on commercial land use, a new mall in both the inner and middle rings has only negligible effects on stimulating other commercial activity (adding a further approximately 5,000 m^2 within a 1 kilometre radius up till 2020). However, in the outer ring where the commercial land market is less developed, a mall of 90,000 m^2 simulates a further 12,000 m^2 of commercial activity, on top of the 90,000 m^2 by the mall, by the target year. Perhaps not surprisingly this extra supply of commercial floor space serves to pull down the commercial area improvement values. The result is that the 'with mall' land prices in 2020 are forecasted to be lower than those in the 'without mall' situation, but they still remain considerably higher than in the baseline year of 2008.

Table 12-5. Simulated micro impacts within a 1 km radius for three shopping malls (SM) in metropolitan Tel Aviv, 2020: total and mean per gridcell values compared

Theme	Type	Givatayim Mall (inner ring)		Raanana Mall (middle ring)		Modiin Mall (outer ring)	
		SM	BAU	SM	BAU	SM	BAU
Residential units	sum	30,590	30,571	6,338	6,314	2,124	2,101
	mean	369	368	76	76	26	25
Res.unit value [NIS]	mean	974,686	994,557	1,804,680	1,801,181	699,892	696,087
Commercial area [m^2]	sum	375,031	279,307	241,676	146,529	387,122	284,971
	mean	4,518	3,365	2,912	1,765	4,664	3,433
Comm.area improvement value [NIS/m^2]	mean	7,314	6,589	43,231	42,346	4,590	5,243
Residential land use	[%]	94.5	93.4	74.0	74.0	52.4	52.3

4. CONCLUSION

On the empirical level, the results of the simulations highlight two implications for the planning of metropolitan areas. The first relates to the need to disaggregate the metropolitan area into its constituent parts (inner, intermediate and outer rings) for any analysis of deconcentration. As evidenced from the simulations, employment deconcentration in the inner ring elicits very different land-use impacts to those emanating in the outer ring. Very different and distinct processes may be going on in different parts of the metropolitan area. The second issue relates to the policy response to employment deconcentration. If this is perceived as a negative process, the usual policy response (and the one illustrated in this chapter) is to impose regulatory restrictions through growth management (UGBs) or taxation (impact fees). An alternative response exists, however, that relates to redistributing (rather than regulating) the benefits of employment deconcentration. This can be achieved, for example, through encouraging public sector housing in the outer metropolitan area or via reverse commuting. These policy responses can also be accommodated within the simulation capabilities of the *UrbanSim* system and remain a challenge for future work.

Our experience in applying *UrbanSim*, suggests both opportunities and constraints associated with microsimulation as a tool for analyzing metropolitan growth patterns. The simulations presented above show the potential for dynamic analysis at a variety of scales and with agents operating according to different time schedules. We are able to provide answers to a whole score of 'what-if' scenarios in different temporal, spatial and market settings. This, coupled with the assumption of disequilibrium conditions that underpins the *UrbanSim* modelling strategy, means that a constant process of re-adjustment between economic agents takes place based on short-term time schedules. This serves to ground the microsimulation in a much more plausible picture of reality where markets are not perfectly competitive, resources not perfectly mobile and agents do not have full information.

On the constraints side, the data requirements and their limitations cannot be over-stated. Microsimulation requires considerable investment in assembling the initial database. Ideally, this should contain information on the individual agents being modelled: households and their dwellings, workers and their places of employment, developers and the like. In practice however, selection of a finer grid cell level will lead to a simple proportional division of available larger units in order to provide minimum data for the grid cell. Additionally, our experience has been that estimated data sometimes needs to be used in the absence of a comprehensive survey or census source. Similarly, small sample sizes may also require the use of Monte Carlo sampling in order to generate probability distributions of a sufficient size in order to be able to generate decision rules for individual behaviour. A problem therefore arises in that the need to amass data of the right quality and quantity for microsimulation, leads to an ever-increasing 'synthetization' of the data.

Finally, the lure of coupling a microsimulation capacity with a GIS capability means that the analyst is enticed into ever-disaggregated levels of analysis. It takes a veritable leap of faith in order to honestly claim the ability to forecast land-use or land-value changes twenty years into the future at the level of the individual grid cell. In this respect, microsimulation may unwittingly serve as a vehicle for entrapping analysts in their own forecasts.

ACKNOWLEDGEMENTS

This chapter is based on work conducted within the SELMA (*Spatial Deconcentration of Economic Land Use and Quality of Life in European Metropolitan Areas*) research initiative funded by the EU 5th Framework as part of the *City of Tomorrow and Cultural Heritage* programme. Our special thanks go to Chen Greenberg for her assistance with the UGB scenario.

REFERENCES

Bontje, M. (2001) Dealing with deconcentration: population deconcentration and planning responses in polynucleated urban regions in northwest Europe, *Urban Studies*, 38: 769–785.

Ding, C. and Bingham, R.D. (2000) Beyond edge cities: job decentralization and urban sprawl, *Urban Affairs Review*, 35(6): 837–855.

Ewing, R., Pendall, R. and Chen, D. (2002) *Measuring Sprawl and Its Impact*, Smart Growth America, Washington DC, http://www.smartgrowthamerica.org.

Felsenstein, D. (2002) Do high technology agglomerations encourage urban sprawl? *Annals of Regional Science*, 36: 663–682.

Galster, G., Hanson, R., Ratcliffe, M.R., Wolman, H., Coleman, S., and Freihage, J. (2001) Wrestling sprawl to the ground: defining and measuring an elusive concept, *Housing Policy Debate*, 12(4): 681–717.

Hunt, J.D. and Abraham, J.E. (2005) Design and implementation of PECAS: a generalized system for the allocation of economic production, exchange and consumption quantities, in Gosselin, M.L. and Doherty, S.T. (eds) *Integrated Land Use and Transportation Models: Behavioral Foundations*, Elsevier, London, pp. 207–238.

Kratke, S. (2001) Strengthening the polycentric urban system in Europe conclusions for the ESDP, *European Planning Studies*, 9(1): 105–116.

Lang, R.E. (2003) *Edgeless Cities; Exploring the Elusive Metropolis*, Brooking Institution Press, Washington, DC.

Mackett, R. (1990) The systematic application of the LILT model to Dortmund, Leeds and Tokyo, *Transportation Reviews*, 10: 323–338.

McMillen, D.P. and McDonald, J.F. (1998) Suburban subcenters and employment density in metropolitan Chicago, *Journal of Urban Economics*, 43: 157–180.

Persky, J. and Wiewel, W. (2000) *When Corporations Leave Town: The Costs and Benefits of Metropolitan Job Sprawl*, Wayne State University Press, Detroit, MI.

Simmonds, D.C. (2001) The objectives and design of a new land use modelling package: DELTA, in Clarke, D. and Madden, M. (eds) *Regional Science in Business*, Springer, Berlin, pp. 159–188.

Torrens, P.M. and Alberti, M. (2000) Measuring Sprawl, *Working Paper 27*, Centre for Advanced Spatial Analysis, University College, London. http://www.casa.ucl.ac.uk/working_papers.htm.

Urban Audit (2000) The Urban Audit: Towards the Benchmarking of Quality of Life in 58 European Cities, European Union, www.inforegio.cec.int/urban/audit

Waddell, P. (2002) UrbanSim; modeling urban development for land use, transportation and environmental planning, *Journal of the American Planning Association*, 68(3): 297–314.

Waddell, P., Borning, A., Noth, M., Freier, N., Becke, M. and Ulfarsson, G. (2003) Microsimulation of urban development and location choices: design and implementation of UrbanSim, *Networks and Spatial Economics*, 3(1): 43–67.

Wegener, M. (1982) Modeling urban decline: a multilevel economic-demographic model for the Dortmund Region, *International Regional Science Review*, 7(2): 217–241.

Chapter 13

SIMULATION OF POLYCENTRIC URBAN GROWTH DYNAMICS THROUGH AGENTS
Model concept, application, results and validation

W. Loibl, T. Tötzer, M. Köstl and K. Steinnocher
Austrian Research Centres - ARC Systems Research GmbH, Vienna, Austria

Abstract: In contrast to typical urban regions dominated by a core city located in the region's centre and a suburban fringe, the Austrian Rhine Valley, selected as the study area, can be described as a peri-urban region with some medium-sized centres and various rural villages, all developing in different ways. Thus, a simulation approach has to be applied that allows for different settlement growth and densification velocities in order to reach model results which come close to real land-use transitions. The urban growth model presented here is based on a multi-agent system which simulates location decisions of households and company start-ups within a cellular landscape driven by regional and local attractiveness patterns. The model has been applied to generate control runs for the past decade and validated by comparing its results with observed land-use changes at municipality level and at raster cell level. Scenario runs for the future have been conducted assuming suspended zoning regulations leading to a different actors' behaviour.

Key words: Land-use change simulation; urban growth; polycentric development; driving forces; attractiveness criteria; multi-agent systems.

1. INTRODUCTION

Spatial planning requires reliable scenario results for decision making to foresee the effects of spatial planning strategies. A rather new approach to simulate possible future land-use changes is offered by multi-agent simulation (MAS) models that consider the spatial behaviour of individual agents as a basic driver. This approach leads to very detailed, local results and thus provides the information needed by local authorities and inhabitants

E. Koomen et al. (eds.), Modelling Land-Use Change, 219–235.
© 2007 *Springer.*

interested in information about future land-use change consequences for their own neighbourhoods.

Application of MAS models in the spatial sciences started in the mid-1990s (cf. Wegener and Spiekerman, 1997; Portugali, 2000; Torrens, 2001; Batty *et al.*, 2003). MAS models, initiated from sciences like psychology, sociology and biology, did not originally treat space explicitly. The application of MAS in spatial sciences has re-shaped agents' activities from immobile decision making and communication to mobile action within a virtual space, like movement through a landscape or along roads. In land-use change modelling, hybrid models, combining MAS and Cellular Automata (CA) approaches, have often been developed (see Benenson and Torrens 2004). In MAS models, the actors' behaviour is the driving force for land-use change, while CA applies the land-use change based on a cell-by-cell transition function not directly considering actors as drivers. Cells in MAS are applied as entities whose state changes not (only) because of neighbourhood characteristics but also because of agent behaviour changing the state of the underlying cell actively. Both MAS and CA approaches address time through iterative steps, executing state changes of single cells or destination search cycles of single agents, which both perform steady land-use changes inside the study region.

As polycentric regions show differences in settlement development speed, one has to consider agents' destination preferences leading to different choice frequencies where certain municipalities are selected more often as home towns than others. To model these different growth patterns, regional characteristics influencing actors' regional scale choice behaviour have to be investigated in addition to local influences (Loibl and Kramar 2001; Loibl and Tötzer, 2003). The more information regarding actors' behaviour and their spatial consequences is available, the more accurate the simulation results will be.

The MAS approach described in this chapter simulates decisions of several agent classes reacting differently on regional and local characteristics, driving a polycentric settlement development.

2.   MODEL CONCEPT AND REALIZATION

2.1   Overview

MAS-based land-use change models are frequently market- and neighbourhood-oriented (see Portugali, 2000) and focus on the urban local scale. They tend to concentrate on steady growth within a particular city.

The model described here does not consider market aspects explicitly. It is assumed that the real estate market 'works' as an intrinsic general function of regional property development.

The model takes a behavioural approach, where land-use change is the result of many actors' activities in response to general spatial conditions and personal circumstances. The environment is perceived and judged by actors who live in it and who, according to their perceptions, desires and (financial) circumstances, behave differently. Here, emphasis is placed on the simulation of actors' moving and settling decisions to trigger the growth speeds of different settlements. A typology of relevant actor groups is derived and deployed as the model's agent classes, equipped with certain behaviour rules. Six agent classes are defined: four household classes searching for new residences and two entrepreneur classes planning company start-ups. The model simulates only decisions of actors causing the construction of new built-up area or the densification of existing areas. The moving of households or employees in a way of 'exchanging' houses or workplaces without effect on land use is not considered.

To simulate the agents' behaviour properly, it is important to prove and quantify the causal relation of the actors' behaviour with the attractiveness criteria that influence the actors' decisions aiming for built-up area growth or densification (Loibl and Kramar 2001; Loibl and Tötzer, 2003). Therefore, the behaviour rules are verified and thus quantified by relating statistical data describing population and workplace dynamics to spatial attractiveness criteria and to built-up area growth summarized for municipalities.

The model region where the agents act and finally determine land-use change is a cellular landscape realized as a digital land-use raster layer (with 50 metre by 50 metre cells) derived from satellite images and stratified into a set of land-use classes. Additional raster cell layers provide further information like zoning regulations, landscape attractiveness and other characteristics that explicitly relate to local and regional attractiveness. One important layer contains the municipality delineation of the entire model area to enable and control the choice of a certain municipality as part of the destination search process.

2.2 Model concept and structure

The built-up area growth simulation is based on agents' decisions to settle down or to establish a company. These decisions are driven by regional and local attractiveness criteria; regional attractiveness influences agents' municipality choice resulting in different development speeds between settlements; local attractiveness influences the decisions of actors' on the final residential or commercial start-up location within the selected municipality.

During the model initialization, the general increase of household and workplace numbers within the model region is defined, as well as the iteration frequency of the agent classes. The iteration frequency of the four household agents and two entrepreneur agents is firstly calibrated during the model control runs to achieve results, matching the observed distribution and composition of actors. For future scenario runs, the frequency has to be adapted according to the regional development assumptions.

The location of each household and entrepreneur agent is performed by a search routine of two subsequent tasks: (1) *municipality choice*; and (2) *local target area search*. The regional and local attractiveness information for the agent's search has to be provided in advance.

The *municipality choice* refers to the observation that certain municipalities are selected more often as destinations than others, assumed to be driven by certain regional attractiveness criteria. The preference of each agent for a certain destination municipality is controlled by regional choice probability distributions. They are derived in advance applying linear regression models explaining the increase of household and company numbers (using workplaces as proxy for business start-ups) in the municipalities with attractiveness variables. The municipality's absolute growth numbers for households and workplaces, generated by the regression functions, are transformed into relative growth frequencies which are finally applied as municipality choice probability distributions. The municipality selection of the agents is conducted by a random choice of a municipality out of the agent classes' choice probability distribution.

After the selection of a destination municipality, the *local target area search* takes place in the cellular landscape within the municipality considering local attractiveness for housing or for commercial area usage. The search starts in a random cell from where the agent moves to the nearest residential (or commercial) area and, if inside a settlement area, he moves to the cell with the lowest density. Agents accepting higher density settlement areas consider only cells with densities above a certain threshold. The appropriate land-use class (as criterion for a transition into built-up area to be possible) and built-up area zoning (if zoning regulations are considered by the certain scenario) are pre-conditions for further location decisions. To take into account further local attractiveness criteria, additional data layers are integrated. The most attractive cell among the alternatives is selected within the neighbourhood through multi-criteria evaluation, considering the respective cell values from all attractiveness layers. If the selected cell is still open space, a land-use state change to built-up area will take place. If no appropriate cell is found (e.g. because of too high population – or workplace density in the neighbourhood), further search attempts are conducted and if they fail too, a new target municipality is selected.

The additional persons of the moving household or the additional workplaces of the new enterprise are accumulated to the current numbers of the selected cell and additionally to the numbers of the respective municipality. Figure 13-1 gives an overview of the model structure.

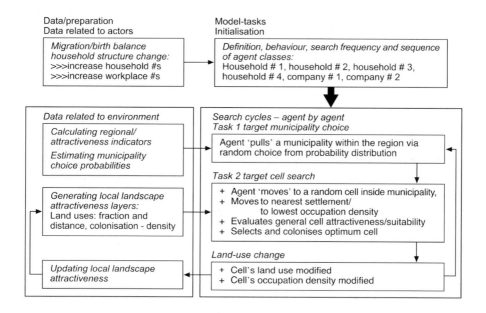

Figure 13-1. Model overview.

The model performs the simulation on an agent by agent basis and thus continuously drives land-use change and change in spatial attractiveness due to growth of building land and increases in cell occupation density. Thus, each agent influences behaviour of subsequent agents. The iterative process introduces temporal aspects and makes the decisions of later settlers dependent on the behaviour of previous ones.

The first MAS model development took place in 2001 using MS Visual Basic as the compiler supports graphical user interface creation and object oriented programming as well as accommodating the authors' FORTRAN programming experience. In 2004, more interactive control has been added to the model relating to, for example, parameter adjustment for 'on-the-fly' model run adaptation and the exchange of input and output data.

3.   MODEL APPLICATION

3.1   Study area and land-use change drivers

The presented model application simulates settlement development within the Austrian Rhine Valley, a peri-urban region with various medium- and small-sized municipalities, south of Lake Constance. Figure 13-2 shows the extent of the Austrian Rhine Valley study area and the observed development of the built-up area between 1990 and 2000.

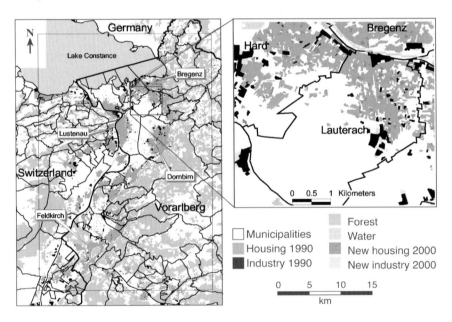

Figure 13-2. Settlement growth in Austria's Rhine Valley, 1990-2000, overview and detail (Satellite images: LANDSAT TM, SPOT; Land-use classification: ARC systems research). (See also Plate 11 in the Colour Plate Section)

The model must be adapted for each region to which it is applied in order to achieve a high 'prediction quality'. This quality can be obtained by calibrating, or tuning, the model with actual observations. Thus an initial model application, or 'control run', applying prior data has to be carried out. As start and end of the control-run period we selected the years 1990 and 2000 for which the necessary data sets are available. These data sets include: a land-use raster layer, population, household and workplace numbers per municipality as well as density raster layers related to building land. The actual simulation results are discussed in Section 4. Here we concentrate on describing how the model is adapted to the study area, which input data are provided and how the model is calibrated to achieve best results for the control run, to provide a reliable model framework for future scenario runs.

The built-up area within the model region (7,330 hectares) was occupied in 1991 by 250,000 inhabitants in 86,000 households and by 97,000 workplaces. The population growth of 15,000 in this period is related to a birth surplus that more than compensates the negative migration balance. The increase of 630 hectares (+10%) of the housing area and 110 hectares (13%) of the commercial area by 2001 is caused by an increase of 16,000 households and 17,000 workplaces. This household number growth is mainly caused by a changing household structure. The share of single and two-person households increased from 22% and 23% in 1991 to 28% and 27% respectively in 2000. The remaining share of larger households decreased. To consider the new household size structure for the destination year, a separation of 13,000 inhabitants from their original households has to be carried out: the inhabitant numbers of the municipalities were reduced by the amount needed to achieve the assumed household size average. This population reduction is distributed to each municipality's housing area cell to reduce the cell's population density and household size: the number of persons per municipality that have left their former household, are divided by the municipality's housing area cells. The person number of each housing area cell is reduced by this share and added to the movers' pool which encompasses 38,000 persons. The 16,000 new households resulting from this movers' pool are divided into the four household agent classes to achieve a certain future household structure.

The impact of an increase in workplaces on settlement growth and densification will be modelled without considering prior workplace losses, taking two entrepreneur agent classes, one that establishes small company start-ups (with few workplaces, accepting densely built up area) and one establishing larger company start-ups with more workplaces demanding new building land along the municipalities settlement borders.

3.2   Attractiveness definition

As described in Section 2.2 the simulation of each agent's action consists of two tasks – municipality choice and local area search. Both tasks are triggered by attractiveness criteria which have to be defined, generated and its effect quantified in advance: regional attractiveness triggers municipality choice, local attractiveness controls the decision to select a final destination cell.

3.2.1 Regional attractiveness criteria

Several regional attractiveness criteria are assumed to influence the destination municipality choice (Loibl and Kramar, 2001). Linear regression analyses were carried out to test the relevant criteria, quantify their contribution to the overall influence on migration or workplace allocation and to examine the statistical performance. The final regression functions for households' and entrepreneurs' destination municipality choice distribution have r^2 values of 0.98 and 0.82 respectively proving that the functions explain 98% and 82% of the total variance. The explanatory variables of the regression functions are:

- numbers of people, households and workplaces in the start year;
- average accessibility (travel time) of district centres and the capital city;
- average share of attractive land-use classes in the vicinity of the municipality settlements (percentage of open space, forest area).

The regression functions allow the calculation of household growth and workplace growth per municipality based on the attractiveness variables. After the transformation of the absolute values into relative search frequencies, they are finally applied as municipality choice probability distributions for the agent classes. Figure 13-3 shows the choice probability distribution for the Rhine Valley municipalities as derived for households. The visual comparison of the municipality choice probability distribution with the migration surplus 1990-2000 proves again that the attractiveness criteria explain the municipality selection quite well. Only the choice frequencies of less attractive municipalities are slightly over-estimated.

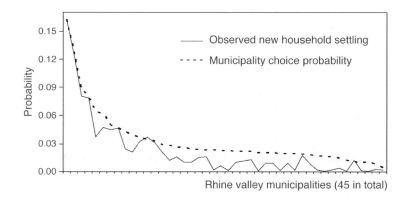

Figure 13-3. Destination municipality choice probability distribution for households.

This approach to model the municipality selection preferences via a set of regional attractiveness criteria allows the incorporation of the impact of large-scale changes in attractiveness; making it possible to simulate, for example, the change in migration and settlement patterns following the improved accessibility due to construction of a new motorway. The effects of different traffic-route alternatives can be tested easily by exchanging the travel-time variable and thus calculating new probabilities. The local search must also be performed as the housing area demand triggered by a new motorway must be confronted with the supply in the respective settlements regarding available new housing area or acceptance of densification.

3.2.2 Local attractiveness criteria

The calculation of local attractiveness is performed in advance on a cell by cell basis, using land-use raster maps and raster GIS functions to estimate the share of more or less attractive land-use classes in each cell's neighbourhood. As part of these calculations, Euclidean distances are estimated to the nearest area of the land-use classes that serve as attractions or obstacles for location choice and thus for land-use change and densification. The cell's population, household and workplace density is calculated through an overlay of census entity polygons on built-up area cells and relating demographic and employment data from official statistics to the underlying built-up area cells to calculate every cell's share on the respective census entity.

The following raster layers with 50 metre cell size have been prepared for the year 1990 to conduct the control run:
- land-use layer with 10 use classes:
  - three urban fabric classes: housing, mixed, industrial/commercial;
  - two semi-urban fabric classes: traffic facilities, urban green;
  - five open space classes: water-bodies, grassland, arable land, forest and natural land cover (wetland, pasture, rock, glacier);
- distance to closest residential area, to closest industrial area;
- distance to closest major road/motorway, to closest motorway exit;
- household density and population density per residential area cell;
- workplace density per settlement area cell;
- zoning regulations: residential-mixed built-up-industrial area; and
- nature conservation: central valley green space protection area.

Some additional layers are prepared for model validation and for future scenario runs: land-use in 2000, density in 2000 and proposed density in 2010. Municipality affiliation, municipality centres and settlement centre layers are also integrated as raster layers for the local search to check whether the agent is still moving inside the selected municipality.

3.3   Agent initialization and parameter definition

3.3.1   Defining the agent composition

To adapt a model run for certain scenario assumptions, the number and composition of the agents' municipality search frequency have to be defined. The agent classes have different behaviour patterns which are deduced from statistical data describing the actors' structure and interactions. Due to limited data availability, agents' stratification is kept simple and additional behaviour assumptions are derived from the literature dealing with migration and settlement motivations with respect to household characteristics (e.g. Bogue, 1969) or about entrepreneurs referring to colonization decisions.

At the start of a model run, the search frequencies of four household agent classes (of 1, 2, 3 and 4 persons) and two entrepreneur agent classes (smaller and larger enterprises) are defined to achieve a certain future composition of household size classes, inhabitant and workplace numbers. The agents' different location preferences for low or higher density settlement areas lead to different search patterns initiating different local developments. The search frequencies of the control run are defined in a way to match the observed population and household distributions and the settlement growth for the destination year. The agents' frequencies can be modified interactively to change the household structure even during the model run, leading to different housing area demand and population density distributions.

3.3.2   Modifying the municipality choice probability distribution

The probability distributions that are based on current observed attractiveness will provide a trend scenario; the control run results for the observation period will match the observed municipality growth pattern. But this probability can also be adjusted individually for each agent class to modify the decision behaviour according to alternative scenario conditions. A weighting and a shifting factor per agent class allow flattening or widening of the shape of the choice probability distribution, leading to different spatial behaviour. Flattening the shape reduces the chances of the attractive municipalities and raises the chances of the non-attractive municipalities, leading to equally distributed settlement growth. Raising the shape leads to further settlement growth concentration. Range thresholds for two regional attractiveness criteria per agent class allow modifying the municipality choice probability distributions by excluding certain municipalities to be selected as a destination.

3.3.3 Modifying the local attractiveness criteria weights

Local cell search considers 10 local attractiveness criteria in a multicriteria analysis as listed in 2.2.3. These attractiveness criteria have to be weighted differently for each agent class to increase or decrease the influence of the particular criteria and thus to modify the behaviour of the respective agent. Default weights are defined as result of the model calibration steps. The weights can be adjusted interactively, even during the simulation runs.

3.4   Simulation cycles

The simulation of settlement growth is conducted through single actions of households and company start-ups of entrepreneurs leading to densification and – in case of open space cells – to land-use change. Change takes place steadily and a destination year is not defined explicitly. It is stated implicitly through certain household, population and workplace growth numbers, expected to be achieved within a certain time range.

3.4.1 Municipality selection

The first step of each search action is to decide on the destination municipality. The regional settling preference of each single agent is conducted by discrete choice: the agent picks by chance a destination municipality from the choice probability distribution. As described in 3.3.2 the probability distribution curvature can be modified to alter the agents' behaviour, reviewing the effects of those changes on settlement growth. Municipalities showing a higher selection probability are more attractive and will be selected more often than others. But a higher probability will only then lead to faster settlement growth in the selected municipality if the demand can be satisfied through appropriate supply, which is available building land or densification acceptance. While regional attractiveness is responsible for the spatial distribution of regional settlement growth, the consequence whether extensive sprawl or moderate growth and some densification will happen is dependent on local attractiveness and on local settlement growth regulations.

3.4.2 Local destination search

The local search takes place in the cellular model landscape. After a target municipality has been selected, the agent starts the final search from a random cell inside the selected municipality. The kind of search depends on the agent's behaviour rules. Agents preferring a low cell density start searching in a random open space cell and move to the nearest settlement. Agents preferring higher density cells start from a settlement centre within the selected municipality and move outwards in random direction seeking for the cell with the lowest density (above a certain density threshold). Single and couple households and small business entrepreneurs show moderate preference for high density cells, while families with children do not except high density cells and show a high preference for low density cells or for vacant building land. The agent class preferences for lower and higher cell occupation densities are predefined through a certain density preference ratio.

Important for the initial local search inside a municipality are population density and distance to housing areas for households, and workplace density and distance to commercial areas for entrepreneurs. These layers can be imagined as elevation surfaces with 'hills' of higher density or greater distance and 'valleys' of low density or small distance, where the agents move along the 'valleys'. The movement of the agents starting in open space is controlled by the distance to the nearest housing or commercial area. The agents follow the decreasing distance values in the cells through the cellular landscape to the outskirts of the next settlement. Inside the settlement the densities of the built-up area cells are examined to perform the further search. After detecting the cell with the lowest density, reaching the 'basin-bottom' inside the 'density valley' system, further alternatives within a neighbourhood are evaluated considering all attractiveness criteria through multi-criteria evaluation (cf. Jankowski, 1995). The weighted linear combination of all criteria allows ranking and selection of the optimal cell; the cell with the highest weighted attractiveness total is finally selected. (cf. Loibl, 2004).

3.4.3 Cell density, land-use and local attractiveness change

The successful agent locates and causes a density change: the household agent increases household density in the respective cell by 1 and the cell's population density by the number of household members; the entrepreneur agent establishes a number of workplaces which increases workplace density in the respective cell. If the cell is unoccupied, the land-use class will be changed to residential, mixed built-up or industrial/commercial land use, depending on the acting agent class, current land-use and zoning regulations. Some local attractiveness criteria (land use, density and built-up area distance) are modified steadily when the cells are occupied. As each agent's action changes the local attractiveness, each agent influences the decision of future moving agents.

If the search fails (e.g. because the densities in the neighbourhood exceed the density acceptance threshold for the particular agent), up to 10 additional search attempts are conducted, starting again from a new random cell inside the selected municipality. If the search remains unsuccessful, a different municipality is picked from the probability distribution and the agent's search starts again.

A 'common search memory' allows simple communication between agents that have already moved and agents searching for a new location within the same municipality: new movers read the entry for their selected municipality on the memory board and search, in the first instance, near cells where the last movers have located successfully. This 'communication' leads to faster search cycles and to a less stochastic and thus less scattered pattern of newly occupied lots. After some 10 successful location actions within one municipality, the memory content for the respective municipality will be erased to avoid too much clustering.

4.   RESULTS AND VALIDATION

To describe the performance of the applied land-use model we carried out a validation on the results relating to municipality choice and local area search. Figure 13-4 allows comparison between the modelled and the observed housing cell growth between 1990 and 2000 for a sample of the Rhine Valley municipalities.

The housing area in each municipality as observed in 1990 and 2000 is shown by hatched bars. The simulated housing area growth in the 1990-2000 period is presented by the black extension of the 1990 bar. The high coincidence of the length of the bars showing the simulated and observed housing area in 2000 indicates that the model is delivering valid results and conducting polycentric development with different settlement growth speeds. It is clear that the good match of simulation and observation owes much to the fact that the calibration and validation of the model are based on the same data and period. In calibrating the model, we used regional macro-scale data to 'tune' the behaviour of the individual agents. The subsequent complex interactions of the individuals, also containing a stochastic element at the local scale are, however, to some extent unpredictable. Capturing this part of the model performance, although not a validation in the strict sense, is considered interesting as it assesses the potential performance of the microsimulation approach. This part of the validation, in our opinion, proves that microsimulation is a suitable technique to model the land-use related effects of complex choice behaviour of a large group of individuals.

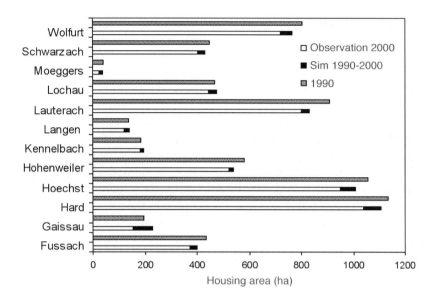

Figure 13-4. Housing-area 1990, observation 2000 and growth simulation 1990-2000 in ha of a sample subset of Rhine Valley municipalities, 1990-2000.

A cell-based validation of land-use change results captures the performance of the local area search and is carried out applying the RIKS Map Comparison Kit using the Kappa statistic (Visser and de Nijs, 2006). The standard Kappa statistic focuses on cell by cell comparison and accepts only total coincidence of location which cannot be expected here, as the microsimulation contains a stochastic element. Therefore, a fuzzy Kappa index is applied which evaluates spatial deviations of land-use change with a fuzzy membership function, considering the gradient from minor to major deviation. (cf. Hagen, 2003).

For this fuzzy Kappa validation, an exponential distance decay function is applied, where large deviations are weighted more strictly and thus judged to have lower coincidence. Using a four-cell neighbourhood extent applied on a 50 metre by 50 metre cell raster, similarity is considered within a distance of 200 metres. A large part of the study area consists of cells that cannot be modified, thus generating a high fuzzy Kappa index that wrongfully indicates a high validity. Therefore, our validation task considers only occupiable building land. After excluding the cells where modification is not allowed (open space, forest areas, water), the fuzzy Kappa index indicated a 84% coincidence in residential area growth and a 97% coincidence of industrial/commercial area growth, showing a good approximation with the observed changes.

The two upper maps in Figure 13-5 show the observed and simulated land use in 2000, the lower left figure shows the fuzzy Kappa map as validation result. The lower right map, by way of example, presents a scenario result, assuming suspended zoning regulations and a response behaviour of agents that forces sprawl.

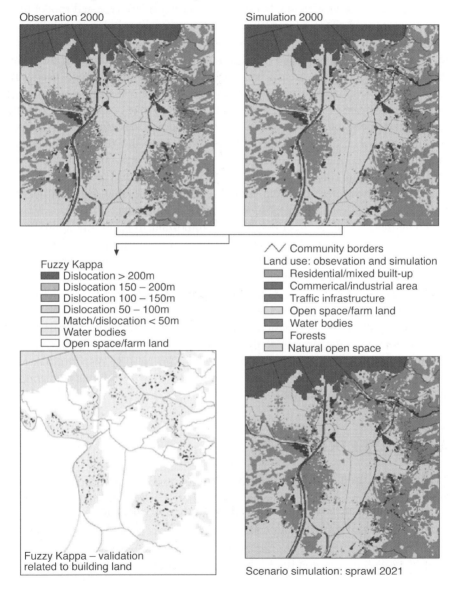

Figure 13-5. Model validation run: observation, simulation runs and validation result. (See also Plate 12 in the Colour Plate Section)

5.   CONCLUSION

The combined multi-agent and microsimulation approach described here shows a high spatial correspondence with the observed changes in the same period, for both municipality choice and local area search. The settlement growth patterns show very stable conditions during many repeated model runs as they refer to the observed migration target municipality choice probabilities.

As the current model calibration forces stable results and a high local accuracy it is not easy to provide scenarios that deviate strongly from the current developments, which is an advantage as non-experienced users are not able to deliver odd results. Sensitivity analysis shows that only a strong modification of the probability distribution shape provokes clear changes in the pattern and speed of urban sprawl. As the model is developed to simulate the effects of population and household growth following current socioeconomic conditions, it has limited ability to incorporate scenarios with radical population changes, e.g. a population decline due to strong ageing. This would, however, call for model modifications that can neither be calibrated nor validated from observed past developments.

To achieve accurate local results, it is important to apply the appropriate attractiveness criteria to quantify the 'true' migration behaviour characteristics at the local scale. Here, further sensitivity analysis is necessary to adapt the weights and parameters in order to achieve proper results. Until now, the goal was to apply variables and adapt the weights in a way to meet the observed development pattern.

Further model development is focussed on the inclusion of interactive control, for example aiming at the modification of regional attractiveness criteria leading to a direct recalculation of the probability distributions and the introduction of specific spatial interventions (e.g. new roads or zoning plans) that affect local attractiveness.

ACKNOWLEDGEMENTS

The model application for the Rhine Valley test case has been conducted within the 'geoland' project, coordinated by Infoterra GmbH, Friedrichshafen, and Medias France, Toulouse, funded by the European Commission in the context of the GMES (Global Monitoring for Environment and Security) initiative of the 6th Framework Programme. Co-funding has been provided by the Austrian Research Centres, Vienna.

REFERENCES

Batty, M., Besussi, E., Chin, N. (2003) Traffic, urban growth and suburban sprawl, *CASA Working Paper Series No.70*, Centre for Advanced Spatial Analysis, University College London. http://www.casa.ucl.ac.uk/working_papers/paper70.pdf.

Benenson, I. and Torrens, P.M. (2004) *Geosimulation: Automata-based Modeling of Urban Phenomena*, Wiley, Chichester.

Bogue, D.J. (1969) *Principles of Demography*, Springer, Berlin.

Hagen, A. (2003) Fuzzy set approach to assessing similarity of categorical maps, *Iinternational Journal of Geographical Information Science*, 17(3): 235–249.

Jankowski, P. (1995) Integrating geographical information systems and multiple criteria decision making methods, *International Journal of Geographical Information Science*, 9(3): 251–273.

Loibl, W. (2004) Simulating suburban migration: moving households, social characteristics and driving forces on migration behaviour, in Feichtinger, G. (eds) *Vienna Year of Population Research 2004*, Austrian Academy of Sciences Press, Wien, pp. 201–223.

Loibl, W. and Kramar, H. (2001) Standortattraktivität und deren Einfluss auf Wanderung und Siedlungsentwicklung, in Strobl, J., Blaschke, T. and Griesebner, G. (eds) *Angewandte Geographische Informationsverarbeitung XIII*, Wichmann Verlag, Heidelberg, pp. 309–315.

Loibl, W. and Tötzer, T. (2003) Modeling growth and densification processes in suburban regions – simulation of landscape transition with spatial agents, *Environmental Modelling and Software*, 18(6): 485–593.

Portugali, J. (2000) *Self Organization and the City*, Springer, Berlin.

Torrens, P.M. (2001) Can Geocomputation Save Urban Simulation? Throw some agents into the mixer, simmer and wait ..., *CASA Working Paper Series No.32*, Centre for Advanced Spatial Analysis, University College London. http://www.casa.ucl.ac.uk/ working_papers/ paper32.pdf.

Visser, H. and de Nijs, T. (2006) The Map Comparison Kit, *Environmental Modelling Software*, 21(3): 346–58.

Wegener, M. and Spiekermann, K. (1997) The potential of micro-simulation for urban modelling, in *Proceedings of the International Workshop on Application of Computers in Urban Planning*, Kobe University, Kobe, Japan, pp. 129–143.

Chapter 14

PUMA: MULTI-AGENT MODELLING OF URBAN SYSTEMS

D. Ettema[1], K. de Jong[1], H. Timmermans[2] and A. Bakema[3]

[1]*Faculty of Geosciences, Utrecht University, The Netherlands;* [2]*Urban Planning Group/EIRASS, Eindhoven University of Technology, The Netherlands;* [3]*Netherlands Environmental Assessment Agency (MNP), Bilthoven, The Netherlands*

Abstract: It is increasingly recognised that land-use change processes are the outcome of decisions made by individual actors, such as land owners, authorities, firms and households. In order to improve the theoretical basis of land-use modelling and to represent land-use changes in a behaviourally more realistic way, we are developing *PUMA* (Predicting Urbanisation with Multi-Agents), a fully fledged multi-agent system of urban processes. *PUMA* will consist of various modules, representing the behaviours of specific actors. The land conversion module describes farmers', authorities', investors' and developers' decisions to sell or buy land and develop it into other uses. The households module describes households' housing and work careers in relation to life cycle events (marriage, child birth, ageing, job change *et cetera*.) and also their daily activity patterns. The firms module includes firms' demography and their related demand for production facilities leading to (re)location processes. The chapter describes the conceptual model, the first phase of operationalisation and initial results.

Key words: Multi-agent models; integrated land-use transportation models.

1.   INTRODUCTION

For more than four decades, social scientists have developed simulation models of land-use conversion processes to assess social and environmental effects of land-use changes as well as the sustainability of land-use policies

E. Koomen et al. (eds.), Modelling Land-Use Change, 237–258.
© 2007 *Springer.*

or to examine the interaction between transportation and land-use development (Lowry, 1963; Wegener *et al.*, 1991; Waddell *et al.*, 2003).

Traditionally, land-use changes and urban development have been modelled using aggregate models, based on zonal information (see Timmermans, 2003 for a review). Notwithstanding their usefulness in many applications, a drawback of these models is their weak theoretical basis. In particular, they describe land-use patterns as the outcome of an allocation process at the zonal level, which is weakly linked to the behaviour of relevant stakeholders. Even if behavioural models, such as discrete choice models, are used, they are commonly applied at an aggregate level. Moreover, these models are focused on the existence of equilibrium, whereas in reality, land use is an ongoing process rather than an end state. This is especially problematic since spatial policies usually have a long temporal stretch, affecting citizens throughout the process (Batty, 2005). Finally, the theoretical basis of these aggregate models becomes increasingly less appropriate due to the dominance of the service sector and information technologies, and the shift from regulatory planning to developmental planning. As a consequence, the models should not treat zones, tracts and grid cells as the decision makers shaping a particular city, but actors such as households, firms, institutions and developers. Also, policy makers have become increasingly interested in such issues as regeneration, segregation, polarisation, economic development and environmental issues (Batty, 2005), which require analyses at the individual level.

This chapter is based on the contention that agent-based models with their focus on individual actors deserve exploration as they potentially do not share the theoretical weaknesses of conventional models and offer considerable flexibility in modelling behavioural processes by applying validated theories and calibrated models. The use of agents offers the opportunity to apply advanced behavioural models to represent agents' behaviour in a more realistic way. In addition, it is possible to model agents as more advanced cognitive units, which are able to display pro-active behaviour, engage in long-term planning and learn about their environment. Another advantage of the use of agents is that interactions and feedback effects at various levels can be modelled (Section 2).

Having noted the potential advantages of agent-based modelling of urban systems, it should be noted that operational agent-based models of larger (metropolitan) urban systems are still scarce, or limited in scale and scope (e.g. Benenson *et al.*, 2002; Mathevet *et al.*, 2003) especially since the increased level of detail required for agent-based modelling implies many new challenges in terms of computational algorithms, data organisation and model architecture. This chapter describes ongoing work on the development of an agent-based model *PUMA (Predicting Urbanisation with Multi*

Agents), aiming at the operationalisation and application of an agent-based model of urban systems at the metropolitan scale (the Northern Dutch Randstad).

This chapter describes the model specification, calibration and first results of the households module which was implemented and tested for the North Wing of the Dutch Randstad, including about 1.5 million households and 1.6 million dwellings.

2. CONCEPTUAL MODEL

2.1 Objectives and scope of the *PUMA* model system

We start from the principle that changes in land use take place in response to individual and/or societal needs, such as the need for housing, commercial buildings, recreational facilities, infrastructure *et cetera*. In turn, these needs arise from activities that individuals, households, firms and institutions want or need to realise, implying that a model of urban development should in some way represent changes in the populations of these agents, changes in their intended activities and changes in their need for physical facilities. The objective in developing *PUMA* is to represent these changes in a theoretically and behaviourally sound way, using state-of-the-art models of individual and institutional choice behaviour.

In particular, the *PUMA* system includes various processes that in one way or another influence the urban system (Miller *et al.*, 2004):

- The evolution of the population through demographic development (birth, death, marriage, divorce *et cetera*), but also through both internal and external migration. Population development is considered a basic driver for land-use development, since it determines the demand for dwellings of various kinds. Also, the spatial distribution of population may determine the location of commercial activities that need proximity to a market or a labour force.
- The evolution of firms (and organisations in general) in terms of their 'birth', location decisions and development of number of jobs. Not only are firms, through their demand of facilities, a driver of land-use change. Through the supply of jobs, they also influence the spatial distribution of population.
- The evolution of the land-use system. That is to say, the conversion of land (farm land, nature areas) into other uses (residential, commercial), which is the outcome of decisions made by owners (farmers, real estate owners, authorities) and buyers (developers, investors, authorities). We

hypothesize that the decisions of these actors are at least to some extent based on the demand for dwellings and commercial buildings, stemming from the first two processes identified above.

- Daily activity and travel patterns of individuals and (workers in) firms and institutions. These activity patterns are important for various reasons. First, they have to be carried out within the current spatial system and, in that sense, generate demand for facilities (stores, work places, recreation, schools) and transportation infrastructure. If one of these is insufficient (or if the demand of an individual/household/firm changes), this may lead to adaptations such as relocation or suppression of activities. In the latter case, a demand exists for facilities or infrastructure, which may trigger changes in the physical spatial system (e.g. additional development of residential area). Second, it is through the generation and execution of daily activity patterns that mismatches between demand and supply become evident, changing the perceived quality of the urban system and possibly leading to adaptations. For instance, if demand for road space is too high, congestion will occur, leading to deterioration of accessibility and possibly causing households or firms to relocate.

A more detailed analysis of these fundamental processes, described in Section 2.6, illustrates that the mutual interaction between these processes occurs through interactions between individual agents or between agents and higher-level components of the system. We will, however, first introduce the main model components and discuss the concepts for modelling the behaviour of the agents.

2.2   Components

The *PUMA* system consists of various components: a grid-based land-use system with spatially fixed sites and facilities, a transportation system and spatially non-fixed agents deciding about the development and use of the urban system.

The spatial land-use system is represented as a grid-based system, which serves as the spatial reference point of agents. Each grid cell is defined by the coordinates of its centre point. In addition, the grid cell contains spatial characteristics, such as the number of inhabitants, dwellings, firms, jobs *et cetera* and the accessibility level of the cell to jobs or population. The grid cell may also inherit properties of larger spatial units (such as Traffic Analysis Zones (TAZ) in which the grid cell falls. In *PUMA,* the grid cells serve as containers of smaller, spatially fixed, units, such as dwellings, commercial buildings, social and recreational facilities and jobs. Each dwelling, commercial building and job is defined as an individual agent, which carries attributes such as dwelling type, market value, size (in case of dwellings), market value and functionality (in case of commercial buildings) and sector, required education level and salary (in case of jobs).

To simulate daily activity and travel patterns, a transportation system is required that connects the potential activity locations, and defines the travel times between locations, accounting for the effect of traffic intensity on travel speed. For this purpose, transportation networks for car traffic, public transport and slow modes are used in a very similar fashion as in numerous transportation studies. The connection between the spatial system and the road network takes place on the grid cell level.

The drivers of the urban system are active, spatially non-fixed agents such as individuals (organised in households), firms/institutions, land owners and developers, who either use the urban system through their activity patterns and (re)location behaviour or directly change the spatial or functional characteristics of it (see Sections 2.3–2.5). The agents are connected to a number of base locations, such as dwellings or work places.

2.3 Households/individuals' behaviour

As noted before, households are one of the main drivers of urban development. Through their emergence and evolution they create a demand for dwellings and facilities like schools, jobs, stores, recreational facilities *et cetera* which triggers the development of such facilities in reality, either through developing undeveloped areas or through redevelopment of existing urbanised areas. Also, by residing or working in certain areas, the households determine the characteristics of neighbourhoods and the attractiveness of such areas for other households. As a consequence, the relevant behaviours of households include demographic events, residential choice and work location choice. In addition, households' daily activity patterns are of importance, as these determine, on an aggregate level, patronage levels of infrastructure and facilities, leading to negative externalities such as congestion and pollution. Also, these externalities are experienced in the daily activity pattern, for instance in the form of congestion.

2.3.1 (Re)location decisions

Demographic events like getting married or cohabiting, having children and leaving the parental home will be determined by factors such as age, gender, education level, work status and cohort (e.g. attitudes towards marriage, career and parenthood may change considerably within a generation). These demographic processes are modelled using empirical data of life trajectories.

Regarding the (re)location behaviour of households, an elaborate discussion of the approach is given in Devisch *et al.* (2005). In this chapter, we will only discuss some key aspects. We assume that agents will try to optimise their lifetime utility. For instance, if we term the utility experienced in dwelling *d* in year *y* as U_{dy}, households will try to maximise:

$$U_d = \sum_{y=1..n} U_{dy} \rho_y \tag{1}$$

where *n* is the length of a households planning horizon and ρ_y is a discount factor to represent that short-term utility may be more important than longer-term utility. The utility given household and dwelling characteristics, U_{dy} can be defined as:

$$U_{dy} = U_{by} + U_{ly} \tag{2}$$

Where U_{by} is the utility derived from the remaining monetary budget after housing expenditures. This reflects that households make tradeoffs between the budget to spend on housing and the budget to spend on consumption of goods. The utility U_{ly} reflects the direct utility of living in a particular dwelling, depending on characteristics of the dwelling and the surroundings.

Obviously, when a household first chooses a residential location (dwelling), it will choose the dwelling *d* yielding the highest utility U_d. However, if a household already lives in a dwelling, it may occur that an alternative dwelling *e* is available, with a higher utility U_e. The decision to move will then depend on the gain in utility traded off against the transaction costs of relocation.

As noted by various authors, housing relocation is a process consisting of various stages. The first stage, called awakening, implies that a household becomes aware of the fact that it can improve its utility by moving to another dwelling. This awareness can be caused by various factors. An important classification of triggers is into push and pull factors. The push factors are related to changes in the household or in the living conditions, such as a change in household composition or finding a new job elsewhere. Pull factors are related to the opportunity to find a better dwelling elsewhere. Note that the attractiveness of alternative dwellings may increase over time, for instance as a result of growing income or changed household circumstances.

The process of relocating to another dwelling is now conceptualised as follows. First, as defined earlier, a household derives a lifetime utility U_d from the current dwelling. In addition, a household will have a perception of the housing market. In particular, it will have some idea of the utility to be derived from the most attractive dwelling available in the market. If we term this utility 'abstract utility' (denoted U_a), a household will decide to start searching for another dwelling if $U_a > U_d + \tau$, where τ represents transaction costs. An important implication is that the decision to start searching for another dwelling may be due to a decrease in the current utility U_d, but also to an increase in the perceived abstract utility U_a. It is noted that the availability of an alternative dwelling and its utility depend on the household's perception of the market. Building up this perception may take place in a gradual indirect way, such as by coming across advertisements and newspaper articles, but also in a direct way by receiving a direct offer (e.g. a family member or friend selling his/her house). Once a household perceives that the utility of alternative dwellings is higher than the current utility, it will actively explore its options and possibly move to another dwelling.

2.3.2 Work participation and location choice

An agent's work situation determines his income and through that where he can afford to live. In addition, work location choice is relevant since the spatial distribution of jobs affects the spatial distribution of residents. In our modelling approach, we have treated work status by developing a model that describes the probability that an agent works as a function of gender, education level, age and life cycle.

Work location choice resembles residential choice in that workers will choose one out of a set of available jobs, based on considerations such as salary, job type, distance to the dwelling, and personal preferences regarding type of organisation etc. In the current era where dual income families are the norm, distance between the current dwelling and the job, possibly implying the need to relocate the household, is increasingly important. However, multiple candidates usually apply for one job. This implies that the labour market can be depicted as a real market, with a demand for workers by firms and institutions, and a supply of labour by individual workers. With respect to changing jobs, we hypothesize that workers will trade-off the utility from their current job against the potential utility (in terms of salary or other factors) of another job. Again, this perception is based on job advertisements, job changes by friends and relatives *et cetera*.

2.3.3   Daily activity patterns

Daily activity patterns are of importance, as they constitute the confrontation between the physical urban system and individuals' behaviour. On the one hand, the aggregated individual behaviours lead to system user levels and externalities such as congestion and pollution. On the other hand, it is through their daily trips and activities that individuals/households experience the externalities and respond to them, for instance by relocating their residence or job. Since the objective of the *PUMA* system is to represent changes in the urban system as the outcome of agents' behaviours and to represent these behaviours in a realistic way, we argue that the daily activity and travel patterns should be modelled using activity-based models.

2.4   Firms/institutions

Firms and institutions (non-commercial employers) are of importance as they influence the urban system by locating in a particular place, affecting the spatial distribution of jobs and the use of the transportation system. The main challenge in agent-based modelling is to represent firms' behaviour in a behaviourally sound way. That is to say, firms are represented as individual agents that can be started, develop (in terms of number of employees), search appropriate locations in various stages of development and hire employees. Thus, firms are related to locationally-fixed agents such as business estates and spatially non-fixed agents such as individuals who work for them. De Bok and Sanders (2005) give an example of modelling firms' location behaviour, which depends on accessibility levels, distance to the old location, agglomeration considerations and land uses in the surrounding areas. An example of modelling the demography of firms can be found in Van Wissen (2000). Inclusion of firms' spatial behaviour in *PUMA* is planned in future and will not be discussed in further detail in this chapter.

2.5   Land owners and developers

Changes in the spatial distribution of population and economy can also occur through the conversion of land to other uses. For instance, changes in agriculture may lead to transitions of agricultural land into residential, commercial or recreational use. With respect to the conceptualisation of the land use conversion processes, we posit that the owner of the land has the strongest influence on what will happen to the land and in fact takes the decision about the land use. We assume that associated to each grid cell is a landowner, who is characterized by attributes such as:

- type of owner (farmer, developer, private person, authority *et cetera*);
- spending power (which investments can be made); and

- technological and managerial knowledge.

At each point in time, a landowner can decide to (Hunt *et al.*, 2004): leave his land as it is; develop his land by changing the land use and exploit it; develop his land by changing the land use and sell it; and sell his land to another owner.

However, for some owners the options are more limited. For instance, a farmer cannot develop his property into residential area, as he lacks the necessary investment power and skills. Also, not all actions may be allowed given planning regulations. To start, we distinguish between three types of owners with specific options: farmers (options: exploit, sell or buy), authority (options: keep, sell to farmer, sell to developer or develop and keep), and developer (options: develop and sell, (re)develop and exploit, sell).

Ultimately, the decision about which action to take depends on the expected utility of each alternative to the owner. In the case of commercial owners, utility will coincide with profitability; the action that delivers the highest profit will be taken. In the case of authorities, social benefits will also play an important role, whereas for farmers personal and emotional reasons may affect their decision.

An important factor in deciding whether or not to sell the land (with or without developing it) is the market price. According to the agent-based perspective chosen in this study, the market price depends on the willingness-to-pay (WTP) of other agents. Also in this case, the profit that can be made with the land will be an important factor for the WTP. However, also non-commercial values, such as the natural quality of land, can play an important role. The factors that we assume to influence owners decisions to sell, exploit or develop their WTP are the profitability of land in its current and new use, conversion costs of the land, the land price, the demand for dwellings and real estate, the price level of dwellings and real estate, land uses in environment, environmental and liveability concerns and characteristics of the firm or organisation.

The fact that the owners' decision to take particular actions also depends on other agents' WTP, suggests that in fact a market for land transactions exists, where sellers (e.g. a farmer) and buyers (e.g. a developer) negotiate about the price of a transaction. In this respect, both parties will base their negotiations on their perception of other transactions in the market, and buyers will also base their decision on their perception of the WTP of potential buyers after development.

2.6   Interactions and timescales

PUMA will encompass behaviours by various agents on varying time scales. Most of the behaviours (residential and job location choices by households, location decisions and growth processes of firms and land use development decisions) take place on a longer-term time scale and can be updated annually. Obviously, daily activity and travel patterns take place (and should be simulated) on a daily basis. When simulating urban systems over various decades, the simulation of daily activity and travel patterns will take place only for a subset of all days, but with sufficient frequency to represent non-daily activities such as social visits.

Interactions can, in *PUMA*, occur between system components on various scales (Figure 14-1). We distinguish between (1) individual agents, and (2) the aggregate urban system containing every micro-level agent or component and markets, on which supply and demand of land, buildings and facilities takes place.

Various interactions between system levels are hypothesized for households and individuals. Through residential and job location choice, households and individuals influence the physical spatial system in terms of the composition of neighbourhoods and concentration of workers. These can be considered aggregate system characteristics, which may in turn influence the behaviours of other households and firms. However, households display also short-term (or daily) behaviour in the form of daily activity and travel patterns. These may affect the physical system in a dynamic way, leading to temporary concentrations of congestion and pollution. Although temporary, these effects may affect other households' (re)location decisions. Another important aspect is that the experience of daily activity patterns may affect households' demands. For instance, the experience of limited options for recreation may lead to a relocation of a household and a demand for a dwelling in a particular area (or with particular characteristics). In a similar vain, demand for jobs may arise through daily experience. Finally, it is noted that also demographic events affect demands. For instance, the (anticipated) birth of a child may lead to a demand for a different dwelling, as may the desire from someone to leave the parental house.

Firms (or institutions) affect the physical system by their location choice. Through that, they affect the land use, the spatial distribution of jobs, but possibly also the spatial distribution of pollution and noise. Also, the flows of persons and goods to and from the firms/institutions result in the use of the spatial system with associated aspects such as congestion and pollution. Like households, if the utility of activity and travel patterns is insufficient (e.g. too much congestion or too few customers), the firm may decide to relocate to a more advantageous location, leading to a demand for commercial buildings and possibly an actual relocation.

To conclude, we note that an important form of interaction takes place on markets, where buyers and sellers (of land or buildings) or employers and employees negotiate the price and eventually try to make a deal. Modelling this process adequately is one of the main challenges in urban systems modelling. Important in this process are professional intermediates (brokers, recruitment agencies, real estate brokers *et cetera*), which have access to databases of potential buyers and sellers and of transactions, and play a role in connecting buyers and sellers and setting the price. Therefore, we argue that higher-level agents such as brokers should be included in the model system.

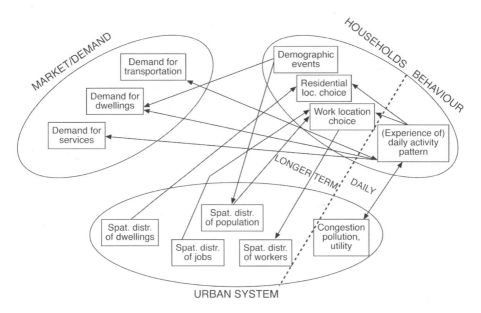

Figure 14-1. Interactions between system components for the households agents class.

3. OPERATIONAL MODEL

3.1 Scope of the operational model

This chapter presents work in progress. To operationalize the *PUMA* system, we have decided first to develop a simple system, allowing for the possibility to exchange simple agents and agent behaviour with more advanced agents over time. In particular, the operational model focuses on households and individuals, as being the drivers of spatial development. Land-use changes, brought about by developers, authorities and firms/institutions (and changes in the number of dwellings and jobs) are still exogenous to the model. Their daily activity patterns are also not simulated yet. Plans are to incorporate an updated version of Aurora (Joh *et al.*, 2003; Arentze and Timmermans, 2005).

3.2   Study area and input data

The study area consists of the northern part of the Dutch Randstad, containing the major cities of Amsterdam and Utrecht as well as Schiphol airport. The area contains about 1.5 million households and 3.16 million inhabitants and is densely populated (950 inhabitants per km^2). In the next decades, some 100,000 dwellings need to be constructed in the Randstad, a considerable part of which will be located in the study area. Due to the concentration of population and economic activity the area potentially suffers from negative feedback effects caused by congestion, noise and pollution.

Given the scale of the area and the potential effects, the area is an interesting test bed for the application of multi-agent models on metropolitan scale. The necessary data to run the model includes:

- a grid-based spatial system, containing necessary spatial information: land uses/densities, facilities, accessibility levels, job availability;
- a transportation network, connecting zones/cells, and providing a travel time matrix;
- the specification of individual dwellings: dwelling type, price and location (grid cell); and
- a synthetic population of individuals organised in households, living in dwellings.

With respect to the spatial representation, the study area is divided into 500 by 500 metre grid cells. For each cell, data is available with respect to the available services, the percentage of public space and non-built up area, the distance to arterials and highways, the number of jobs in various sectors and the accessibility to jobs and population. In addition, an origin-destination matrix with travel times between each pair of cells was available.

For the base year 2000, a synthetic population was generated for the study area using Monte Carlo simulation as described in Veldhuisen *et al.* (2005). The synthetic dataset specifies households by the number of adults and children and the age of the household head. Using distributions of age differences between spouses and between mothers and their children taken from CBS statistics, an estimate was made of the ages of each household member. Based on the same statistics, the education level was drawn randomly.

To initialise households' residential location and individuals' work location discrete choice and regression models were used that were calibrated on the Dutch residential preferences survey (WBO) (Table 14-1). First, using a logistic regression model, the working status of each adult individual was determined. The probability of working is defined as:

$$P_{work} = \frac{1}{1 + \exp(-\sum_j \beta_j X_j)} \tag{3}$$

Where: X_j are factors influencing the probability of working, and β_j are parameters (specified in column two of Table 14-1) indicating the impact of each factor. Next, the income was determined using a regression model. Once these items are specified, the residential choice model (Section 3.4) that is applied in subsequent stages is applied here to assign each household to a particular dwelling. Each dwelling is defined as an individual agent, with characteristics such as dwelling type, price and neighbourhood characteristics. The dwellings are taken from the Address Coordinates Netherlands (ACN) database, specifying each Dutch dwelling by type and exact coordinates. Therefore, each dwelling can be assigned to a particular grid cell and inherits the properties of the grid cell.

Table 14-1. Work status (1= working) and work location choice models

Explanatory variable	Logistic regression model[1]: work status	Explanatory variables	Multinomial logit model[1]: work location choice
Constant	0.867	Commute distance	-0.092
Age < 25	0.462	HighEduc*commute dist.	0.055
Age 40-54	-0.336	Kids*commute dist.	-0.029
Age 55-65	-1.980	Female*commute dist.	-0.050
Mother	-0.401	# jobs within 30 min. by car	0.425*10-5
Male	0.845	# inhabitants within 30 min. by car	-0.156*10-5
Male*Age < 25	-0.785	HighEduc*% office jobs	1.539
HighEduc*Age < 25	0.443	HighEduc*# of jobs	0.938*10-6
HighEduc*Age 25-39	1.128		
HighEduc*Age 40-54	1.081		
Nagelkerke ρ^2	0.272	adjusted ρ^2	0.327

[1] all parameters are significant at $\alpha=0.05$

Once each individual is assigned to a dwelling, the working individuals are assigned to a working location using a discrete choice model. For each worker, a work location was selected by randomly drawing from a limited set of alternative work locations, weighted by the probability of each location. Following the multinomial logit model, this probability is defined as:

$$P_i = \exp(V_i) / \sum_j \exp(V_j), \text{ with } V_j = \sum_k \beta_k X_{jk} \tag{4}$$

Where: X_{jk} is the kth attribute of work location j, and β_k is a parameter (in column 4 of Table 14-1) describing the relative effect of the kth attribute. In this stage of development, we do not work with a one to one relationship between jobs and workers, but simply describe the choice of workplace as the choice of a particular work zone.

3.3   Simulating events

Given the behaviours included in the current version of the model, all households and individuals are updated in time steps of one year. In each period, demographic events, residential relocations and job changes can occur. In *PUMA*, all events are simulated for each agent/household subsequently, in order to ensure consistency within households and individuals. First, the demographic events are simulated, followed by residential location choice and job changes. However, within the category of demographic events, the sequence of events is randomised. The demographic events are ageing, giving birth, leaving the parental home, getting married (used as a unifying term encompassing also a couple that decides to start living together) and divorce (also referring to split-up of a couple that lived together). All events are regarded as binary probabilities. The probability of any event to happen is defined in probability tables as a function of age, gender and (in the case of giving birth) marital status. It is important to note that some events are conditional on household or personal characteristics, such as age, gender, marital status and position in the household (Table 14-2). These conditions imply that earlier events may rule out potential later events. The table also summarises which events are ruled out by events that take place earlier in the year.

We argue that each event is equally likely to take place on any day of the year, implying that each sequence of events is equally likely to happen. This leads to the following approach: (1) determine the relevant events based on state S_t; (2) randomly determine the sequence in which to simulate the events; (3) simulate emergence of events (yes/no), taking into account the fact that earlier events may rule out later events; and (4) define the state of the next year S_{t+1}.

Some events may produce new households. For instance, someone leaving the parental home to marry or live alone creates a new household. Those leaving the home to marry but also singles deciding to get married are collected in two pools of male and female marriage partners. From these pools new couples are created, based on a distribution of age differences between spouses, based on national statistics. All new households (both couples and singles) select a dwelling from the list of vacant dwellings, according to the residential choice models described below.

Table 14-2. Conditions for events and imposed consistency between them

Event	Necessary condition of S_t	Rules out (for the same year)
Giving birth	Female, age 18+	Giving birth, leaving home, marriage, divorce, moving
Dying	No requirements	Marriage
Leaving parental home to live single	Living with parents, age 18+	Divorce
Leaving parental home to marry	Living with parents, age 18+	Divorce
Marriage	Single or living with parents, age 18+	Divorce
Divorce	Being married	Marriage
Moving	No requirements	No conditions

3.4　　Residential choice model

Households' (re)location behaviour consists of two phases. First, a household builds up a perception of available dwellings on the housing market and its opportunity to improve lifetime utility by moving to another dwelling. Next, a decision is made whether or not to start searching for another dwelling and eventually to move. Ideally, the representation of this process would require the modelling of individuals' learning about the housing market through their perception of transactions as well as their negotiation process to acquire a new dwelling. However, in lieu of such advanced models, we decided to model this process as a sequence of three consecutive choices, using discrete choice models that were estimated using the Dutch WBO data set. The model estimates are provided in Table 14-3. A description of the three consecutive choice-models follows below.

The first relevant choice is whether or not to search for another dwelling. This decision will depend on the characteristics of the current dwelling, but also on the expected utility of moving to another dwelling. The choice is modelled using a binary logit model of the form:

$$P_{search} = \frac{V_{search}}{V_{search} + V_{non-search}}, \text{ with: } \begin{array}{l} V_{search} = \sum_j \beta_j X_j \\ V_{non-search} = 0 \end{array} \tag{5}$$

where X_j are characteristics of the household or the current dwelling. The parameters in Table 14-3 suggest that younger people are more likely to search for another dwelling than older people. In addition, people with a low income, with children younger than 12, those living in an apartment with children and people of which the dwelling has a high value are more likely to search for another dwelling. Single people, people living in an apartment and couples without children are less likely to search for another dwelling.

Table 14-3. Residential (re)location choice models

Explanatory variable	Binary logit model: search for new dwelling	Binary logit model: move to new dwelling	Multinomial logit model: choice of dwelling
Constant	-1.090		
Age < 35	0.242	0.262	
Age > 55	-0.446	-0.506	
Double income, no kids	-1.588	-2.736	
Price/Income	0.507	0.503	-0.00156
Income < 1000€/month	0.914	1.782	
Child < 12 years	0.941	0.929	
Lives in apartment	-0.601	-3.460	
Lives in app. With child	1.255		
Single	-0.683	-0.721	
Commute distance male			-0.117
Commute distance female			-0.109
Age > 55 * apartment			-1.195
Age > 55 * row			-1.144
Child * apartment			1.414
Child * semi detached			1.538
Child * detached			1.157
HighInc * row			1.470
HighInc * semi detached			1.288
HighInc * detached			2.357
3+HH * row			1.059
3+HH * detached			1.543
Adjusted ρ^2	0.33	0.32	0.34

The second decision is whether or not to move, depending on characteristics of the current dwelling and on the expected utility of moving to another dwelling. This decision is modelled using a binary logit model of the form:

$$P_{move} = \frac{V_{move}}{V_{move} + V_{non-move}}, \text{ with: } \begin{array}{l} V_{move} = \sum_j \beta_j X_j \\ V_{non-move} = 0 \end{array} \tag{6}$$

where X_j are characteristics of the household or the current dwelling. The parameters in Table 14-3 suggest that the factors that affect the decision to search or not also affect the decision to move or not. The only difference is living in an apartment with a child, which does not affect the decision to move.

The last residential choice is the selection of one out of a set of available dwellings. Following the multinomial logit model, the probability of choosing dwelling i is defined as:

$$P_i = \exp(V_i) / \sum_j \exp(V_j), \text{ with: } V_j = \sum_k \beta_k X_{jk} \tag{7}$$

where X_{jk} is the kth attribute of dwelling j, and β_k is a parameter (in column 4 of Table 14-3) describing the relative effect of the kth attribute. The price parameter clearly indicates that the probability of choosing a dwelling decreases with increasing price (relative to income). The importance of the work location to residential location choice is obvious. Limiting the commute distance of male and female partners apparently is an important factor in residential choice behaviour. Also age, having a child, earning a high income and household size affect the probability that a certain dwelling type is chosen.

New households, which emerge from demographic events such as home leaving and divorce, choose a dwelling based on the residential choice model in Table 14-3. For each existing household, we simulate subsequently whether the household searches for a new dwelling and, (if so) whether the household will move and (if so) what new dwelling is selected. Households are simulated subsequently, each time followed by an update of the available housing stock.

3.5 Work location choice

Work status of individuals is also updated on an annual basis. In particular, based on the Dutch WBO, a logistic regression model was estimated, describing the probability of a person taking part in the work process (Table 14-1). According to the model, workforce participation is a function of age, gender, having children and education level. Since the current model lacks historical dependence, it is likely to overestimate the percentage of transitions, although the aggregate shares of workers and non-workers should be realistic.

If an individual switches from non-working to working, he/she will choose a working location. Also, individuals not changing work status reconsider a change of work location each five years. In both cases, a work location is selected using a destination choice model that was calibrated on the Dutch WBO (Table 14-1). Important to note is that distance from the dwelling is an important factor in explaining work location. This combined with the importance of commute distance in residential location choice suggests that changes in the spatial distribution of jobs and dwellings may both lead to changes in the spatial distribution of population and changes in commute patterns.

4.   APPLICATION

4.1   Technical issues

The operational model is implemented in C^{++}, applying object oriented programming. Since for each simulated year, each household and each individual is updated, the simulation is quite time consuming. In particular, a total model run comprising of 30 years takes about 12 hours on a Pentium 4 PC. For each simulated year, data is stored on the individual level, allowing for the calculation of numerous statistics on various aggregation levels in post-processing procedures. An important tool in this respect is the visualisation of model outcomes per year in maps, based on grid statistics. These maps yield valuable insight into the spatial distribution of population segments in various policies.

4.2   Results and future applications

Figure 14-2 displays some demographic processes simulated by *PUMA*. The figure suggests that the total population remains rather stable for 40 years and then starts decreasing, especially due to aging of the population, leading to lower birth rates and higher mortality rates. The birth rate is rather high in the first 20 simulated years, due to the high proportion of population in the fertile cohorts (20-35 year). Given the low proportion of population in the cohort 0-20 year at the start of the simulation, birth rates drop gradually, with a temporary increase around 2035, when the first baby boom reaches the fertile age. Figure 14-2c also suggests that demographic events can be affected by urban system characteristics. Until 2030, the number of households that is created each year increases, especially, due to the number of people leaving the parental home. After 2030, the available housing stock can no longer accommodate the number of new households, leading to a drop in the number of newly created households. This is reflected in the drop in all demographic events leading to new households, such as leaving home, marriage and divorce. Overall, the demographic events appear to be logical given the input data, although we did not yet have the opportunity to check input data and outcomes against external data. Also, the results indicate the importance of interaction between demographic processes and spatial developments, such as regional housing stock development.

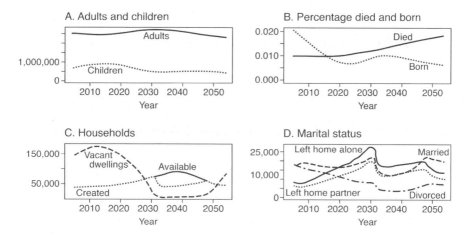

Figure 14-2. Simulation of demographic developments.

Figure 14-3 illustrates the development of the spatial distribution of the population in relation to the urban system. On the left, the figure shows the number of vacant dwellings per grid cell, with high concentrations in the cities of Amsterdam, Utrecht and some smaller cities. The right-hand side of the figure shows that the concentrations of vacant dwellings have almost disappeared, due to the increase in the number of households that was also clear from Figure 14-2. In a similar vein, analyses can be made of the spatial distribution of for instance individuals in particular age cohorts, the proportion of working population or other socio-demographic segments.

The main conclusion from these first results, however, is that they support the feasibility of modelling urban processes on the ultimate disaggregate level of households/individuals at the metropolitan level.

Although the system in its current form still lacks crucial processes such as firms'/institutions' behaviour, these processes will be computationally and memory wise far less burdensome, as the number of involved agents is much smaller. The feasibility of the approach in terms of computation time and data handling implies that a potentially very detailed and behaviourally sound model of urban dynamics can be developed, allowing for an the analysis of a variety of spatial effects at a much more detailed level than is possible in current aggregate models.

In the coming months the model will be tested more extensively. In particular, various scenarios, in which changes are made to the spatial distribution of housing stock or the distribution of jobs, will be evaluated to test whether *PUMA* produces plausible results.

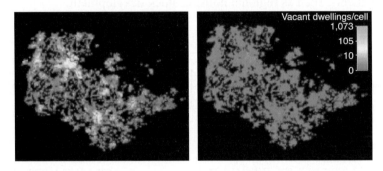

Figure 14-3. Spatial distribution of vacant dwellings in the base year 2004 (left) and according to the PUMA predictions in the year 2030 (right). (See also Plate 13 in the Colour Plate Section)

5.   CONCLUSION

In this chapter, we have reported some results of the current implementation of an agent-based system of land-use dynamics, entitled *PUMA*. Due to space limitations, we could only very briefly summarize the motivation underlying the system and the kind of models that are used. It is important to note, however, that the process of developing the system is to specify the scope and architecture, exploring the system first on the basis of easy to implement, well-known models and gradually replacing these with richer, new, behavioural models. Agents to be included in the future are, for example, cognitive agents capable of activity-scheduling and rescheduling behaviour, learning the environment and capable of adjusting their behaviour, agents for simulating housing search and choice, incorporating negotiation between developers and potential buyers in a dynamic context, and agents simulating life trajectories and their impact on transport decisions. These models have already been conceptualised and their implementation is now being tested.

Other agents, especially for firm demography and land-use change need much more thinking through and development. This task is more difficult as firms and organisations are quite different and distinct. Moreover, data collection may be more difficult.

Despite the early stage of development, the simulation results reported in this paper, demonstrate the potential of an agent-based approach in terms of computation time and data handling. This implies that a potentially very detailed and behaviourally sound model of urban dynamics can be developed, allowing for the analysis of a variety of spatial effects at a much more detailed level than is possible in current aggregate models.

REFERENCES

Arentze, T.A. and Timmermans, H.J.P. (2005) A cognitive agent-based micro-simulation framework for dynamic activity-travel scheduling decisions, Paper presented at the *Knowledge, Planning and Spatial Analysis* Conference, Pisa, June.

Batty, M. (2005) Agents, cells and cities: new representational models for simulating multi-scale urban dynamics, *Environment and Planning A* 37, 8: 1373–1394.

Benenson, I., Omer, I. and Hatna, E. (2002) Entity-based modeling of urban residential dynamics: the case of Yaffo, Tel Aviv, *Environment and Planning B*, 29: 491–512.

De Bok, M. and Sanders, F. (2005) Modeling firms' location behaviour, Paper presented at the 2005 TRB Annual Meeting, Washington D.C.

Devisch, O., Arentze, T.A. Borgers, A.W.J. and Timmermans, H.J.P. (2005) An agent-based model of residential choice dynamics in non-stationary housing markets, Paper CUPUM Conference, London.

Hunt, J.D., Abraham, J.E. and Weidner, T. (2004) The land development module of the Oregon2 modeling framework, Paper presented at the 2004 TRB Annual Meeting, Washington D.C.

Joh, C.-H., Arentze, T.A. and Timmermans, H.J.P. (2003) Understanding activity scheduling and rescheduling behaviour: theory and numerical simulation, in Boots B., et al. (eds) *Modelling Geographical Systems*, Kluwer Academic Publishers, Dordrecht, pp. 73–95.

Lowry, I.S. (1963) Location parameters in the Pittsburgh model, *Papers and Proceedings of the Regional Science Association*, 11: 145–165.

Mathevet, R., Bousquet, F., Le Page, C. and Antona, M. (2003) Agent-based simulations of interactions between duck population, farming decisions and leasing of hunting rights in the Camargue (Southern France), *Ecological Modelling*, 165: 107–126.

Miller, E.J., Hunt, J.D., Abraham J.E. and Salvini P.A. (2004) Microsimulating urban systems, *Computers, Environment and Urban Systems*, 28: 9–44.

Timmermans, H.J.P. (2003) *The saga of integrated land use-transport modeling: how many more dreams before we wake up?* Paper presented at the 10th International Conference on Travel Behaviour Research, Lucerne, 10–15 August .

Van Wissen, L. (2000) A micro-simulation model of firms: applications of concepts of the demography of the firm, *Papers in Regional Science*, 79: 111–134.

Veldhuisen, K.J., Timmermans, H.J.P. and Kapoen, L.L. (2005) Simulating the effects of urban development on activity-travel patterns: an application of Ramblas to the Randstad North Wing, *Environment and Planning B*, 32(4): 567–580.

Waddell, P., Borning, A., Noth, M. Freier, N., Becke M. and Ulfarsson, G. (2003) Microsimulation of urban development and location choices: design and implementation of UrbanSim, *Networks and Spatial Economics*, 3: 43–67.

Wegener, M., Mackett, R.L. and Simmonds, D.C. (1991) One city, three models: comparison of land-use/transport policy simulation models for Dortmund, *Transportation Reviews*, 11: 107–129.

Chapter 15

INTEGRATING CELLULAR AUTOMATA AND REGIONAL DYNAMICS USING GIS
The Dynamic Settlement Simulation Model (DSSM)

K. Piyathamrongchai and M. Batty
Centre for Advanced Spatial Analysis (CASA), University College London, UK

Abstract: The evolution of cities at an urban-regional scale reflects complex relationships between ways in which urban structure develops in response to local decisions involving land development which is set within the more aggregate pattern of urban and regional structure. There is a mutual interaction between physical development and the urban hierarchy which is not often accounted for in the new wave of cellular models that have appeared in the last ten years. This chapter describes an implementation of a simulation model that is based on integrating these local and regional dynamics. We call it the *Dynamic Settlement Simulation Model (DSSM)* and we develop the integration using two different cell-based modelling techniques: cellular automata (CA) and raster GIS. The model is implemented using an object-oriented programming approach, and after we describe its rudiments, albeit briefly, we show its application to real data from Chiang Mai, a major city in Thailand. Finally, this chapter indicates how the model can be used as a part of a spatial decision support system (SDSS) generating predictive outcomes that represent possibilities for implementing predictive and scenario-based applications in urban and regional planning and related fields.

Key words: Cellular automata; dynamic simulation model; GIS; nodal regions; spatial interaction and diffusion; spatial decision support systems.

1.   INTRODUCTION

In recent years, there has been an increasing number of applications of cellular automata (CA) techniques applied to simulating the growth and form of human settlements. Many researchers have applied CA models to both

E. Koomen et al. (eds.), Modelling Land-Use Change, 259–277.
© 2007 *Springer*.

developed and developing countries' situations where they have attempted to integrate such models with other techniques such as GIS. Such models have followed the tradition of aggregate spatial modelling as well as incorporating more disaggregate agent-based modelling structures that seek to simulate development at the level of the individual and agency. The *Simland* model due to Wu (1998), for example, integrates CA with multi-criteria evaluation (MCE) techniques in order to derive behaviour-oriented rules for the transition from one state of development to another, in each cell. Li and Yeh (2000) have taken this approach further by integrating CA models with the spatial analysis of sustainable urban development as developed using constrained transition rules. The *SprawlSim* model by Torrens (2002) has been developed using traditional modelling techniques such as spatial interaction, discrete choice, econometric and input-output models. In fact, most researchers in this field have attempted to take advantage of several complementary theories to create such models. Moreover, these models have attempted to combine the dynamic and self-organizing behaviour which CA models generate with a variety of spatial functions so that such models might yield more relevant and realistic results.

There are three aspects which the research reported in this chapter will focus on. First, in the case of urban and regional development, the model developed here is based on integrating cellular representations of land development with regional spatial structure through the measurement of interactions between cities, and how these interactions affect urban and regional expansion. The focus of the model is thus to simulate 'interaction' which we consider essential in mirroring key differences between the cities to be measured and this leads quite naturally to methods for estimating interaction between them.

To achieve this first goal, two other related issues emerge which we treat as our second objective. This involves the way in which several static and dynamic spatial variables reflecting factors which clearly affect a regional space can be efficiently organized and analysed. To visualize the process and to monitor changes in these key variables, spatial simulation based on a dynamic simulation modelling (DSM) structure was developed. Two technical approaches were integrated in this model construction, namely the CA model structure representing the local level and the DSM reflecting more spatially aggregate regional dynamics. Thus, the second main purpose of the study was to implement a hybrid model which we call the *Dynamic Settlement Simulation Model (DSSM)* which is based on a fusion of CA and DSM. This hybrid model depicts the complexity of urban and regional development, building on well established urban and regional theory which couples local with more global spatial structure and processes.

The third and last aim we set out to accomplish is to ensure that this hybrid model is consistently applied, and thus rooted in a real world situation. Actual datasets obtained from real urban situations in Thailand have been collected and as well as the various theoretical experiments that we designed for testing the model prior to its application, such data were used for validating the fit of the model to real urban growth patterns. In this chapter, we develop the model for the city of Chiang Mai but we have also developed applications for the city of Phitsanulok (Piyathamrongchai, 2006) and as the program for the model is quite generic, it is being adapted to other cities at the present time.

This chapter briefly reports the entire modelling process from a sketch of the model structure through to its full implementation and use in generating urban scenarios for Chiang Mai. The model implementation is essentially based on coupling urban development represented as changes of state in urban land use where land parcels and districts are presented as cells and the more aggregate regional structures that imply various patterns of spatial dependence, in turn reflecting diffusion and spatial interaction. The cellular model reflects the development of local factors in local neighbourhoods influencing state change in the spirit of work by Wolfram (1984; 2002) and the regional analysis is in the tradition of work on the spatial analysis of nodal regions by Haggett (1965; 1975) and Haggett *et al.* (1977). The integration is effected by representing the local level by cells and the regional level factors by points and areas which, in turn, are translated and linked together through the concept of raster GIS. In one sense, the model builds its own GIS to accomplish this coupling and as such it is a contribution to the wider context of developing models within GIS. These techniques have been consistently developed into a suite of computer programs which are supported and elaborated both by arbitrary spatial data sets and by real world spatial data which pertains to urban growth in small and medium sized cities in developing countries. This chapter finally reports how the model can be applied to predict urbanisation of Chiang Mai City in the next twenty years.

2. CONCEPTUAL MODEL STRUCTURE

In integrating the CA model structure with the dynamic simulation modelling (DSM) structure, various concepts of spatial organization (Haggett, 1965; 1975; Haggett *et al.*, 1977) such as *interaction, node, hierarchy* and *diffusion* form the essence of the model structure that we are designing. In essence, development is simulated at the local level according to various factors that pertain to the land development process. At each time step of development, the aggregate pattern of development is then used to structure the regional pattern from which the nodal spatial structure is extracted. At this regional level, nodes or poles of development are reinforced or weakened depending upon the local development pattern but these structures are also used to influence local development at the cellular level. The way aggregate patterns are used in the model is by converting nodal and spatial structure at the more aggregate levels of cells into a raster GIS, thus cementing a firm link from local to global. This is more or less a process of positive feedback between the scales but in the presence of various constraints at both levels and in the presence of various triggers and thresholds that are needed to structure the appropriate polarity and spread of the regional system. In short, if development occurs in one area of the city and its hinterland, then new nodal points might be established at the regional level but this will depend on various thresholds and the whole range of variables that affect accessibility and population density across the region. A simplified structure for the model system is shown in Figure 15-1.

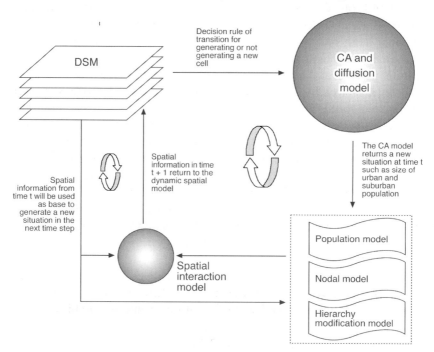

Figure 15-1. Conceptual framework of the *DSSM.*

Figure 15-1 illustrates various flows of activity within the model where it is clear that the structure is built around the DSM with the CA framework being used to control diffusion which is a central part of the way urban and regional growth and development is handled. The DSM is the main module which creates regional development which we treat and represent as active thematic map layers. Traditional GIS normally has little ability to process spatial data in a dynamic sense, and thus the DSM contains the various operators for manipulating diverse GIS functions to handle the analysing and synthesising of the dynamic spatial data which drives the spatio-temporal structure of the model (Yuan, 1995; Egenhofer and Golledge, 1998; Langran and Chrisman, 1988; Couclelis, 1999; Peuquet and Duan, 1995). Spatial variables which measure urban and regional growth are thus defined as the key data sources for the DSM. All spatial and non-spatial data are classified and analysed by several GIS functions at each time slice and in every time period. Cartographic modelling is used to synthesize all the various map layers which are modified at each time instant or over each time period (Tomlin, 1990; DeMers, 2005). Outputs from the DSM also represent important local constraints on the operation of the CA and diffusion models in successive stages. Outcomes from this process are also represented as map layers as well as statistical information which are essential for visualisation and quantitative measurement of the performance of the model as well as its predictions. The way these functions are put together is illustrated briefly in Figure 15-1.

On the other hand, the CA model is based on a more conventional structure in which cells are developed according to what happens in their neighbourhoods. The model thus enables 'cells' to be generated from transition rules that use the values or attributes of neighbouring cells. Cells are also associated with 'births', 'survivorship', 'decline' and 'death' during the simulation period although the controls on the amount of development are determined by constraints and simulations at the DSM level. A diffusion model is applied at the cellular level and this enables random seeds to be generated within the cellular space letting the processes of transition occur with dynamic probability. This reflects a modest degree of randomness in spatial decision making, and such randomness, as in all such models, is constrained within strict limits (Batty, 2005).

The processes in the model focus on spatial organization based on the concepts of node, hierarchy, and spatial interaction. First we need to be clear about the measurement and definition of nodes and hierarchies. Here a node is assumed to provide the 'centrality' of an urban area, and nodes are generated using an algorithm which searches for centres within any given search radius. Furthermore, an existing node can be replaced by a new location that is more 'central' than an old one. To define the order or hierarchy of nodes, one or more variables that are able to separate nodes from each other need to be considered. As population is a major driving force of the urban growth process, we define population data as the main variable in evaluating and ranking the hierarchy. Various techniques can be used at this point to project the population growth in each area as well as for constructing a map that represents the population density with respect to land use. The hierarchy of a node can then be ranked using population density over the land-use classification which we discuss in more detail elsewhere (Piyathamrongchai, 2006). The hierarchy of a node is not only useful for visualisation of the evolving importance of centres within the emergent urban structure, but this is also essential for quantitative measurement of the degree of spatial interaction in terms of accessibility during the next stage of the model simulation. In essence, the nodal structure of the system represents the structure of the region within which cellular development takes place at the lower level but with both levels in mutual interaction with one another through positive feedback.

Spatial interaction is the most important module in representing and visualizing urban-regional relationships. Users or decision makers need to measure spatial interaction for any kind of planning which involves urban growth, for transportation is key to the way urban growth diffuses. As mentioned earlier, population density is another key variable in this study which is set up as 'mass' in a series of gravity type model functions which formalise the way interaction is computed (Haynes and Fotheringham, 1984). Two different variables are output from the spatial interaction models which involve the key centrifugal and centripetal forces which drive the regional dynamics of the system. Centrifugal forces originate from the push factors that flow from the centre to hinterland while centripetal forces are influenced by pull factors that agglomerate economic activities into a centre. The models used here are modified spatial interaction models which generate these two forces whose outputs from this module are then conceived of as layers within the DSM.

All these processes are implemented using relatively simple concepts and techniques. However, they all generate a degree of unpredictability in their feedbacks in each time period. We thus are unable to predict in advance the complexity inherent in their mutual and simultaneous operation within the model structure. The model is therefore based on intricate and subtle processes with inputs and outputs where all these systems consistently work together, and this enables the model framework to replicate the kind of complexity that is observed in real urban and regional growth processes.

3.   MODEL DEVELOPMENT

Building a model using cellular automata enables us to take advantage of the fact that CA models are well established in their operation. Thus various operations such as neighbourhood effects, state changes, spatial rules and so on accord to the standard processes embodied in CA. These represent the dimensions of spatial complexity that we consider important and relevant to the growth of modern cities in most parts of the world (Xie, 1994; Batty and Xie, 1997). However, spatial and temporal complexity in standard CA, as noted earlier, is based on predefined states and transition rules in which implementations of standard CA applications have neglected other crucial static and dynamic spatial information. It is this information that the theories alluded to in the section above relate to.

Probabilistic and constrained CA applications, on the other hand, have been developed to fill this gap. Many CA researchers developing urban applications since the 1990s have increasingly introduced more and more factors which are both spatial and non-spatial, incorporating effects from local to global levels (local cells to regional and thence to the whole space) (Batty *et al.*, 1997; Batty, 2005). These factors cover many aspects such as accessibility, topographic features, and definable policies, e.g. sustainable development, land suitability, and protected or restricted areas (Li and Yeh 2000; 2001; Yeh and Li, 2001; White and Engelen, 2000; Wu and Martin, 2002; Clarke *et al.*, 1997; Ward *et al.*, 2000; Loibl and Toetzer, 2003). To integrate these factors into a CA model, it is necessary to utilise one or more functions of GIS, at least, in collecting or preparing spatial data using GIS, since the functions in GIS are able to support such integration easily and efficiently. Moreover much data now comes in a form that requires a GIS to unlock it.

Even though developing stochastic formulations and adding constraints using relevant GIS operations within models are now quite well developed procedures, it is still important to measure how close (or far) a CA simulation is from complex real world phenomena. In short, CA models need to be calibrated to get some sense of how close they are to reality. Some essential questions spring from this, namely:

- how has recent work developed the capability and functionality of GIS operators for manipulating spatial and non-spatial data in creating new and more useful information within the processes of CA modelling?
- are there changeable constraints within recent CA applications as reflected in the dynamic aspects of CA?
- have any model components or modules been developed based on dynamic spatial analysis which have then been used to manage static and dynamic spatial information within such models?

Much research in urban CA simulation which has focused on urbanisation in real cities or major urbanised areas or on development in fictional cities, has used a wide range of land-use classifications from binary (urban or non-urban) to multilevel developments which contain a complex array of urban land uses (Xie, 1994; Batty and Xie, 1994; Batty, 2005). Some work has included a regional dimension, developing simulations in larger areas within higher resolution spaces, thus representing regional growth simultaneously along with local growth simulated at the cell level (Li and Yeh, 2000; White and Engelen, 2000). However, many of these studies have overlooked the interactions between these different levels and thus the importance of how central a town is compared to the others in the hierarchy has been neglected. This is where the notion of *nodes and hierarchies* comes into its own; how interrelated such towns are within the same regional space is a function of *spatial interaction*, and how these interactions evolve dynamically impacting on the growth of cities and towns through *diffusion* (Hägerstrand, 1953). As we have noted, this is developed within *DSSM* using various developments of the nodal region concept, which concentrate on urban and regional interaction as mentioned in the previous section. The *DSSM* has been developed to respond to the questions noted above and thus focuses on integrating the techniques of CA and GIS with the key elements in the nodal regional system.

It would take too long to describe all the details of the way the model is developed technically through its equation structures but readers are referred to other publications where the model is spelt out in greater detail (Piyathamrongchai, 2006; Piyathamrongchai and Batty, 2007). We do however need to give the reader some sense of how the model is programmed and an illustration of its working is available at http://www.casa.ucl.ac.uk/kampanart/home.htm. The *DSSM* has been fully developed using the JAVA programming language that is built using the object-oriented programming (OOP) approach (Wood, 2002; Horstmann and Cornell, 2005a; 2005b; Wiener and Pinson, 2000). The model structure was defined as 21 main classes in JAVA programming syntax so that the various concepts introduced above could be incorporated in a consistent and modular fashion, making it relatively easy to extend and adapt the model. These classes have their own functions which perform various operations in the model as in Figure 15-1. The key classes are summarized in Table 15-1 below.

Most classes in the *DSSM* consist of many model parameters which are crucial in driving the simulation. The *DSSM* interface allows users who want to set up applications to flexibly adjust all the parameters which then generate results from different scenarios that in turn can be easily and efficiently visualised. These parameters can be classified into three groups:

1) parameters for the CA simulation such as thresholds for CA transition rules and diffusion; 2) parameters for the constrained map dynamics which affect the construction of the dynamic maps at each time state or period; and 3) parameters for the hierarchical node and potential surface map generation such as distance measures and functions which are needed to create nodes and fashion the measurement of potential. The way these are used in the calibration process is described in more detail elsewhere (Piyathamrongchai, 2006; Piyathamrongchai and Batty, 2007).

Table 15-1. Description of some key classes in the *DSSM*

Class	Description
CAmodel	The main *DSSM* program.
CaSim	CaSim is created in order to perform cellular automata.
Dasy	Dasy is the dasymetric application in this model. This class contains some methods that are used to calculate the dasymetric population density
DasyPanel	An inherited panel from JPanel (JAVA swing) used in order to display graduated colour maps: population density map and potential surface map, for example
Diffuse	A Hägerstand diffusion model is applied in this class
GraphPanel	A real time graph panel is built from numbers of urban cells in the urban simulation space and time state
Node	The node class represents all modules used for generating nodes, computing node sizes, locating nodes on the space and generating the centrifugal and centripetal potential maps

4. APPLICATIONS TO THE CITY OF CHIANG MAI

Due to limitations posed by the model computation and the visualisation of its inputs and outputs, it is not possible to picture the entire area of the Chiang Mai province within the model interface. This is due to the fact that there are limits on the size of problem that can be effectively handled due to the number of individual cells required in processing and of course, limits on the size of map that can be easily visualised. The model is designed to study the growth of urbanising areas through their interactions in their surrounding hinterlands but for this pilot demonstration, the Chiang Mai case study was restricted to a fixed distance from the existing city centre. For purposes of model visualisation, a trial 12-kilometre radius was fixed from the centre and all the administrative territories (sub-districts) that fell within the main city and this hinterland were taken as the area for simulation as a cellular space. Figure 15-2 depicts the region selected within 12 kilometres from the city centre. The Chiang Mai case study covers approximately 864 square kilometres with 58 sub-districts covering the key districts of Muang Chiang

Mai, Hang Dong, Mae Rim, San Sai, Doi Saket, San Kamphaeng and
Saraphi.

Figure 15-2. The Chiang Mai simulation area.

In preparing the necessary datasets, thematic maps from the GIS database
in GIS vector format were produced for administrative areas/boundaries, the
transportation routes, and the water bodies and streams. These were
classified and subsequently converted to the GIS raster format. The vector
and raster image maps were prepared using the same coordinate system, and
this ensured that all maps could be properly superimposed. At this stage, all
the raster maps were again corrected and finally resized to a suitable
dimension for use in the simulation model. Note that the limits set in the
computer program were that any application should not be greater than
122,500 pixels or 350x350 pixels. All Chiang Mai data were thus resized to
340x322 pixels (about 120 metres: four times less than the resolution of the
original data). The slope layer was constructed from a raster DEM (digital

elevation model) which in turn was created from the contour data of 1:50,000 maps. Based on 3D GIS functions, the slope layer was directly created, and it was masked and resized to the same scale as the converted GIS raster maps. The last two data layers required are the urban seeds used to drive the simulation and the land-use layer that shows the areas and uses of development. The urban seed layer was constructed by extracting only the urban cells from the classified 1989 image while most of the features in the land-use layer were also obtained from this classified image. Two classes – transportation and the rivers – were added to the land-use data from the GIS database. Finally all the data were converted to the ASCII image format which is the format used in *DSSM.*

4.1   Simulating Chiang Mai 1989-2003

Prior to starting the simulations and engaging in analysis that will lead us to an evaluation of the model, the simulation parameters and the time variables will be reconsidered for the purpose of defining the scope and range of the sets of model parameters by their values. The simulation iterations will also be matched up to the real time structure in the data. After we have defined these parameters and time iterations, we are in a position to run the model to produce outcomes. In case of model parameters, four scenarios were defined with different sets of parameters. Results from these scenarios are visualized and compared in an evaluative way.

Once we assign the parameters, one other general problem of all spatio-temporal models such as *DSSM* involving iteration in the allocation of activities to space is to measure or to match the simulation time with the real time. In particular, the time scale of the simulation is set by the 'iterations' which are associated with the set of program algorithms that will be executed in one cycle, with all the changes sequentially appearing after the cycle has been completed. The key question involves how many days or months or years one iteration represents, and this of course will define a cycle. To resolve such definitional issues of temporality, it is necessary to find out what the relationships are between these two temporal scales. To do so, this study first quantified the urban development between 1980 and 2003 from the two available satellite images for the city of Chiang Mai. The changes in the urban area evaluated from these images are then used to set a parameter threshold which measures each cycle in the simulation. We ran the simulation ten times and gathered the numbers of iteration when the simulation reached the numbers of urbanized cells as measured from the multi-date images. These were used for computing the average number as 137 iterations over 14 years (1989–2003); this is approximately 10 iterations per year and we used these estimates in simulating each of our scenarios.

Figure 15-3 shows the results visually at each five year simulation for each scenario in terms of the urbanised areas. It clearly shows that the urbanised cells in the first scenario are distributed more across the space, whereas in the others, they are more concentrated and close to the city centre. Only in the last scenario is there a slight trend towards their distribution further out from the existing city. Compared to the pattern of urbanised cells (red cells) in the 2003 classified image in Figure 15-4A, it appears that the most similar is our baseline, benchmark or archetypal scenario, the first one.

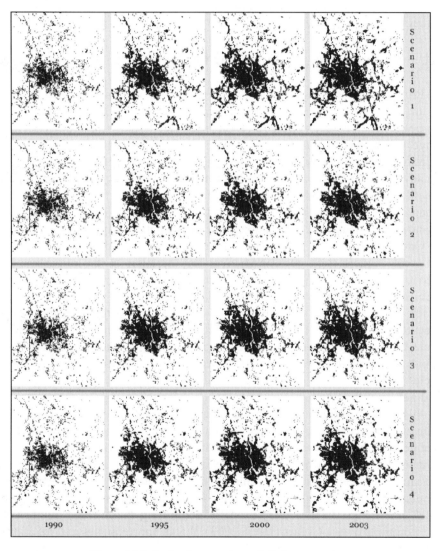

Figure 15-3. A visual comparison of all four scenarios.

To evaluate the model, it is important to compare these simulation results to the real world data in validating and calibrating the model. The land use maps were classified into five types based on urbanised areas (red), agriculture or vacant land (yellow), forest area (green), road cells (black), and water bodies (blue). After 140 iterations of the simulation, two large forest areas remained in the north and the west of the city, which are close to the forest features classified from the satellite image in 2003 (see the green areas in Figure 15-4A). The urbanised areas, on the other hand, were distributed around the city centre and developed along the main transportation routes in all directions especially through to the south and the east of the city. From the actual data, the image clearly shows that the urbanised area has developed to the south along the Chiang Mai-Hang Dong road and this is consistent with the model. Moreover, the land-use maps, generated from simulation, were compared to the actual data using a technique called 'confusion matrix', brought from digital image processing, to measure the accuracy indexes of simulated land-use map. Furthermore, the *DSSM* model also provides other comparable spatial information: the nodal distribution maps as shown in Figure 15-4B and surface maps for the centripetal forces in terms of potential surface-like representations.

Table 15-2 shows that all the results are acceptable in terms of the confusion indexes. The overall accuracy and the Kappa coefficient average 82 percent and 0.67 respectively. This gives an acceptable performance in terms of the overall simulation generated by the model which we judge to be satisfactory. Nevertheless, when considering more detail in each class, we find some differences between the scenarios. In case of agriculture or the bare land and the forest classes, all the scenarios show a very high percent accuracy; approximately 80% for agriculture and 90% for forest class (see the producer accuracy in Table 15-2 below).

Table 15-2. Summary of the accuracy measures, in %, and Kappa statistics for all scenarios for the Chiang Mai area

Scenar.	Overall accuracy	Kappa	Producer accuracy			User accuracy		
			Urban	Agricult.	Forest	Urban	Agricult.	Forest
1	81.4	0.66	50.4	80.9	90.8	39.1	87.9	84.6
2	82.8	0.67	42.2	84.2	91.3	44.8	87.8	84.2
3	82.7	0.68	44.8	83.8	91.2	44.6	87.6	84.3
4	82.0	0.67	47.6	82.4	90.6	41.5	87.7	84.3

In summary, this section has demonstrated various experiments with the model which enable us to get some sense of how well the model performs in simulating urban growth in the Chiang Mai case study. It is clear that the model is able to generate somewhat different results according to the definable parameters. Evaluation of these results using the confusion matrix shows that the best results with respect to the scenarios involves the first scenario which has an overall 81% accuracy, a Kappa coefficient of 0.66 and about 51% producer accuracy in the urban class.

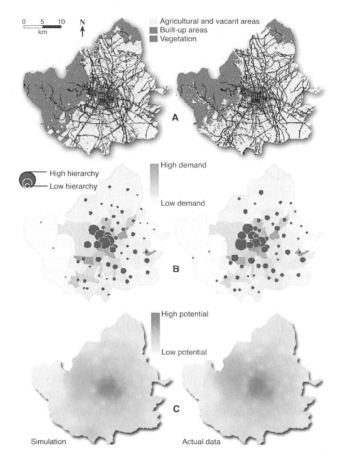

Figure 15-4. Simulated versus real data: (A) land-use map, (B) nodal distribution and land demand and (C) potential surface maps for the centripetal force surface, all for data and model predictions in 2003. (See also Plate 15 in the Colour Plate Section)

4.2 Forecasting Chiang Mai 2003-2023

This last substantive section will demonstrate a simple application to the city of Chiang Mai City by predicting 20 years of urban change beginning at the year 2003. This will result in various visual images of how Chiang Mai will look in the near future from which we can make a rapid assessment of the scale of urbanisation and the rate of growth. In order to perform the simulation, initial data sets have to be prepared, and the urban seed layer and land use were thus based on the urbanised areas classified from the satellite image in 2003. Note that we did not take the outcome of the 1989 to 2003 simulation as the starting point for we are conscious that all that this would reveal would be a compounding of errors over almost 40 years. We consider that with the development of this model at this point, it is better to simply take the best model parameters from the calibration and apply these to the actual 2003 data, not the simulated results. Population growth for each future year has to be computed and we base this on a linear regression which we in fact used to fill in the missing data from 2002 and 2003. This is much preferable to taking random growth rates defined for each time state. Transportation data had to be assumed with no changes in any of the road features from the year 2003, notwithstanding that there are bound to be changes in the road infrastructure to 2023.

Figure 15-5 represents a prediction of the urbanised areas simulated from the model over this 20 year period. Considering the urbanised areas around the city centre in the year 2003, the urbanisation is predominantly along the superhighway and inner ring road. As the simulation proceeds, there is urban expansion along the routes to the east and south of the city. After the year 2012, some nodal points where the radial roads intersect the second ring appear as very clear urban clusters on every side of the city. A little later in the simulation, urban areas at the edges of the city grow up and are finally connected to urban clusters at these nodal points. Some development links the city to nearby small towns, especially in the east and south of the city as shown in Figure 15-5. The eastern outskirts of the city and a small town in the east previously separated from the city, are obviously connected with urbanised areas along the road by 2010. This also happens in other areas to the far south side of the city.

During the 20 year forecasting period, the results show visually that many urban cells are generated around junctions of main roads with the middle and outer ring roads radiating from the city centre. This has occurred, for example, around the second ring road during the years 2003 to 2015 shown in Figure 15-5. These results indicate the possibility that these areas will become urban in the future.

Although the model generates urbanised cells in the space based on probabilities calculated from the dynamic constraints which adapt themselves during the simulation process, running these simulations with the same parameter sets several times generates outcomes that imply a fairly similar pattern of urbanisation and a standard well-structured urbanisation process. This application however does demonstrate another way in which the model can perform in predicting possible spatial scenarios. Although we are sure that the model is not generating accurate futures, we do consider it is able to generate rather well informed spatial trends in population distribution which in terms of the urbanisation in Chiang Mai, would appear to be quite plausible. This means that we consider the model to be a good reflection of the existing processes of land-use change in this region and these forecasts give us more confidence in the model's predictive ability.

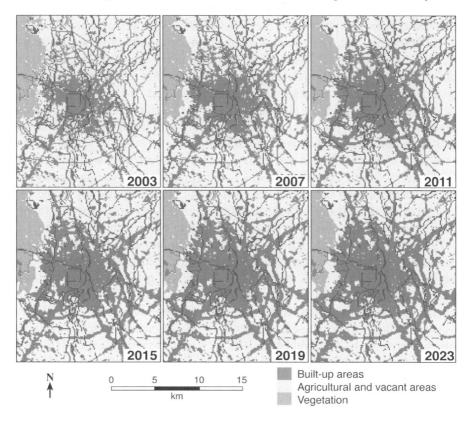

N
↑

0   5   10   15
km

Built-up areas
Agricultural and vacant areas
Vegetation

Figure 15-5. Chiang Mai land use 2003–2023. (See also Plate 14 in the Colour Plate Section)

5.   CONCLUSION

This chapter has presented a sketch of the conceptual development and the implementation of the hybrid *DSSM* which is based on integrating two developed traditions techniques: CA and DSM which reflect the local and more aggregate methods of simulating growth processes in the urban-regional system. The model was developed using the object-oriented programming language, JAVA and it can be adapted to a variety of circumstances as it is in modular form. Its contains its own GIS functions which enable the user to go smoothly from local cellular representation to surface representation and back again thus showing how cities develop at two different scales – at the scale of the land parcel and plot and at the scale of its regional organisation. As observed here in this chapter, the *DSSM* generates results which represent the complex behaviours of an urban-regional system and it yields useful spatio-temporal information, such as urbanized areas, land use, centrifugal and centripetal surfaces, and nodal distribution maps. These kinds of maps clearly support information for urban and regional planning and development.

We only sketched the experiments we have undertaken with this model, but these were essential in bringing the conceptual simulation model to reality. The case study area, Chiang Mai, was explored in terms of both calibrating and validating the results from the model. In order to use the model for real applications, a major consideration which is somewhat new and specific to this model is how to synchronise a 'year' in the real world with 'iterations' in the simulation. Once time becomes synchronous, the second issue is to choose the initial parameters for running the simulation, since there are various ways to mix parameters as mentioned previously.

The model worked reasonably well and we have been successful in getting it to return visual outcomes as sequential sets of map layers. These, we feel, provide the model builder and user with a much richer set of visualizations than any that we have seen hitherto. The confusion matrices provide good results as they indicate that generally the model is able to yield fairly accurate results across all three land-use classes: urban, agriculture and forest. This chapter has also sketched the use of the model in forecasting applied to Chiang Mai. The results clearly indicate that urbanised areas tend to expand to the east and the south of the city, also representing the increasing importance of all three ring roads in the future. These experiments have ensured that it is possible to use the model in 'what-if' scenarios and to adapt it to observing and learning about the city. From its use in defining different scenarios, it is possible to integrate the model as part of a spatial decision support or planning support system which embraces both trend projections, 'what-if' scenarios, learning about the city, and using the model to visually communicate ideas about its present and future.

REFERENCES

Batty, M. (2005) *Cities and Complexity: Understanding Cities through Cellular Automata, Agent-based Models, and Fractals*, The MIT Press, Cambridge, MA.

Batty, M., Couclelis, H. and Eichen, M. (1997) Urban systems as cellular automata, *Environment and Planning B: Planning and Design*, 24(2): 159–164.

Batty, M. and Xie, Y. (1997) Possible urban automata, *Environment and Planning B: Planning and Design*, 24: 175–192.

Clarke, K.C., Hoppen, S. and Gaydos, L. (1997) A self-modifying cellular automaton model of historical urbanisation in the San Francisco Bay area. *Environment and Planning B: Planning and Design*, 24: 247–261.

Couclelis, H. (1999) Space, time, geography, in Longley, P.A., Goodchild, M.F., Maguire, D.J. and Rhind, D.W. (eds) *Geographical Information Systems: Volume 1; Principles and Technical Issues*, Wiley, New York, pp. 29–38.

DeMers, M.N. (2005) *Fundamentals of Geographic Information Systems*, 3rd ed., Wiley, New York.

Egenhofer, M. and Golledge, R. (eds) (1998) *Spatial and Temporal Reasoning in Geographic Information Systems*, Oxford University Press, New York.

Hägerstrand, T. (1953) *Innovation Diffusion as a Spatial Process*. postscript and translation by Allan Pred (1967), University of Chicago Press, Chicago.

Haggett, P. (1965) *Locational Analysis in Human Geography*, Edward Arnold, London.

Haggett, P. (1975) *Geography: A Modern Synthesis*, 2nd ed., Harper & Row, New York.

Haggett, P., Cliff, A.D. and Frey, A. (1977) *Locational Models*, Arrowsmith Ltd, Bristol.

Haynes, K. E. and Fotheringham, A.S. (1984) *Gravity and Spatial Interaction Models*, Sage Publications, Beverly Hills.

Horstmann, C.S. and Cornell, G. (2005a) *Core Java 2, Volume I: Fundamentals*, 7th ed., Prentice-Hall, Englewood Cliffs.

Horstmann, C.S. and Cornell, G. (2005b) *Core Java 2, Volume II: Advanced Features*, 7th ed., Prentice-Hall, Englewood Cliffs.

Langran, G. and Chrisman, N.R. (1988) A framework for temporal geographic information, *Cartographica*, 25(3): 1–14.

Li, X. and Yeh, A.G. (2000) Modelling sustainable urban development by integration of constrained cellular automata and GIS, *International Journal of Geographical Information Science*, 14: 131–152.

Loibl, W. and Toetzer, T. (2003) Modelling growth and densification processes in suburban regions—simulation of landscape transition with spatial agents, *Environment Modelling & Software*, 18: 553–563.

Piyathamrongchai, K. (2006) A dynamic settlement simulation model: Applications to urban growth in Thailand, PhD Thesis, University of London, Senate House, London.

Piyathamrongchai, K. and Batty M. (2007) Coupling cellular automata models to urban spatial structure through GIS, *CASA Working Paper 120*, Centre for Advanced Spatial Analysis, University College London, London.

Peuquet, D. J. and Duan, N. (1995) An event-based spatiotemporal data model (ESTDM) for temporal analysis of geographical data, *International Journal of Geographical Information Systems*, 9(1): 7–24.

Tomlin, C.D. (1990) *Geographic Information Systems and Cartographic Modelling*, Prentice Hall, New Jersey.

Torrens, P.M. (2002) SprawlSim: modelling sprawling urban growth using automata-based models, in Parker, D., Berger, T. and Manson, S. (eds) *Agent-based Models of Land-Use and Land-cover Change*, LUCC International Project Office, Belgium, pp. 72–79.

Ward, D.P., Murray, A.T. and Phin, S.R. (2000) A stochastically constrained cellular model of urban growth, *Computers, Environment and Urban Systems*, 24: 539–558.

White, R. and Engelen, G. (2000) High-resolution integrated modelling of the spatial dynamics of urban and regional systems, *Computers, Environment and Urban Systems*, 24: 383–400.

Wiener, R. and Pinson, L.J. (2000) *Fundamentals of OOP and Data Structures in Java*, Cambridge University Press, New York.

Wolfram, S. (1984) Cellular automata as models of complexity, *Nature*, 311: 419–424.

Wolfram, S. (2002) *A New Kind of Science*, Wolfram Media, Champaign, IL.

Wood, J. (2002) *Java Programming for Spatial Science*, Taylor & Francis, London.

Wu, F. (1998) Simland: a prototype to simulate land conversion though the integrated GIS and CA with AHP-derived transition rules, *International Journal of Geographical Information Science*, 12: 63–82.

Wu, F. and Martin, D. (2002) Urban expansion simulation of Southeast England using population surface modelling and cellular automata, *Environment and Planning A*, 34: 1855–1876.

Xie, Y. (1994) Analytical models and algorithms for cellular urban dynamics, PhD Thesis, University of New York at Buffalo, The Graduate School, Buffalo, NY.

Yeh, A.G. and Li, X. (2001) A constrained CA model for the simulation and planning of sustainable urban forms by using GIS, *Environment and Planning B: Planning and Design*, 28: 733–753.

Yuan, M. (1995) Temporal GIS and spatio-temporal modelling. Available online at: http://ncgia.ucsb.edu/conf/SANTA_FE_CDROM/sf_papers/yuan_may/may.html.

PART V: OPERATIONAL LAND-USE SIMULATION MODELS

Chapter 16

A LAND-USE MODELLING SYSTEM FOR ENVIRONMENTAL IMPACT ASSESSMENT
Recent applications of the LUMOS toolbox

J. Borsboom-van Beurden, A. Bakema and H. Tijbosch
Netherlands Environmental Assessment Agency (MNP), Bilthoven, The Netherlands

Abstract: Changes in land use may have important implications for nature, environment and water management. Thus, insights into future land-use patterns are needed for explorations of the future state of the environment and subsequent policy making. The Netherlands Environmental Assessment Agency, in co-operation with other partners, funded a consortium with the aim to invest in further development of the *Land Use MOdelling System* (*LUMOS*). This system contains a cellular automata-oriented model and an economics-based model. Both models are discussed here as well as a number of their applications. This chapter ends with an evaluation of these applications and the model performance to date.

Key words: Policy-making; land-use models; environment; nature.

1.   INTRODUCTION

The Netherlands Environmental Assessment Agency (MNP), chair and founder of the *LUMOS* consortium, is one the four independent assessment agencies in the Netherlands. These agencies all have a role to play in operationalising the World Bank's 'People-Planet-Profit' concept, with the Social and Cultural Planning Office of the Netherlands (SCP) dealing with 'People', the Netherlands Bureau for Economic Policy Analysis (CPB) handling 'Profit', and the Netherlands Environmental Assessment Agency (MNP), along with the Netherlands Institute for Spatial Research (RPB), being responsible for the 'Planet'.

E. Koomen et al. (eds.), Modelling Land-Use Change, 281–296.
© 2007 *Springer.*

The primary task of the MNP is to advise the Dutch Government on a wide variety of environmental issues from a scientific base built on knowledge and expertise. Policy makers use MNP research findings to develop, implement and enforce environmental and ecological policy. MNP underpins policy through its monitoring, modelling and assessment of possible risks and impacts. Operating within the bounds of the Environmental Protection Act and the Nature Conservation Act, the MNP has assumed the role of charting the current status of the environment and nature in collaboration with a range of scientific institutes and other national assessment agencies to support a broad, but ecologically-based political and social discussion.

Various current policy issues in the Netherlands are strongly related to changes in the way our physical environment is used and will be used in future. Urbanisation pressure is high in the central part of the Netherlands and is not expected to drop in the coming decades. Apart from the persistent quantitative housing shortage that can be ascribed to population growth and household fission, there is a qualitative shortage as well, since the demand for low density residences in rural or semi-rural environments is not satisfied. New building locations have to be found but may conflict with water management or environmental and ecological policy goals. Besides, several traffic problems in this congested part of the Netherlands have to be addressed through the construction of new infrastructure or the enlargement of the existing networks. Whilst this may affect habitats (fragmentation) and valuable landscapes, it is doubtful whether it will adequately solve the congestion problems in the long term. The major problem of exceeding noise and particulate matter thresholds in residential areas appears to worsen if mobility continues to rise as it has done in recent decades. At the same time, in the countryside, further scale enlargement in Dutch agriculture is commonplace because revenues are increasingly under pressure from further liberalisation of world trade and the reform of the EU Common Agricultural Policy (CAP). In traditionally strong sectors such as greenhouse horticulture, cattle breeding and dairy farming, this scale enlargement is seen as a precondition for firm continuation in the long term, which requires further rationalisation of agricultural practices. This rationalisation may, however, lead to a loss of historic patterns and cultural features in the landscape. Besides, biodiversity problems resulting from the transport of nutrients by air and water from agricultural activities to nearby nature areas have not yet been solved.

Apart from these 'autonomous' developments, several policies are having a profound impact on land use currently and will continue to do so in the near future. The realisation of a National Ecological Network and the establishment of robust corridors linking nature conservation areas together, involves a considerable area of land. The introduction of explicit spatial measures in water management poses severe restrictions to further urbanisation and influences the perspectives for agricultural sectors as greenhouse horticulture. The designated national landscapes in the National Spatial Strategy (VROM *et al.*, 2004) add further restrictions to spatial developments.

It is clear that the developments sketched above all have consequences for the physical environment. For this reason, insights into the spatial effects of long-term demographic, societal, technological and economic developments are needed to formulate robust policies anticipating potential problems in the future. Maps of future land use are one of the means used by MNP to support environmental and ecological policy making. These maps are usually based on the geographical output of land-use simulations.

This chapter focuses on the way land-use models are applied for exploring the environmental and ecological effects of possible future developments in the Netherlands. It shows that information provision on current and future land use is an indispensable intermediate step in integrated assessment studies.

2.   THE *LUMOS* LAND-USE MODELLING SYSTEM

LUMOS, the *Land Use MOdelling System* (www.lumos.info) is a toolbox for land-use modelling. The *LUMOS* consortium is a national partnership of Governmental institutes, universities and commercial R&D parties that conduct research on land-use change and related spatial modelling. These partners have a long history of cooperation that has now been formalized in the consortium, which is chaired by MNP. The participants exchange scientific knowledge about the factors determining land-use changes, and cooperate in research in this field. The emphasis of the research lies on spatial modelling of these factors, the relationships with thematic models and the evaluation and visualization of model results. The *LUMOS* toolbox is a product of the *LUMOS* consortium that consists of two land-use models, the *Land Use Scanner* and the *Environment Explorer*, and various tools for data analysis such as the *Map Comparison Kit*.

The *Environment Explorer* is a dynamic spatial model, in which land use and the effects on social, economic and ecological indicators are modelled in an integrated way. It is similar to the *MOLAND* model described in Chapter 17 of this book. The model contains indicators concerning economic activity, employment, social well-being, transportation and accessibility, and the natural environment (de Nijs *et al.*, 2001). The *Environment Explorer* is characterized by three scale levels. At the national level, the model combines national economic demographic and environmental growth scenarios and passes them onto the regional level. At the regional level, the *Environment Explorer* models the location of new activities as a result of national and regional growth and determines the relocation of activities and people within the 40 administrative (COROP) regions of the country. The resulting regional developments can be limited to a minimum and maximum level per region or set to a fixed value. At the local level, the Netherlands is represented on a grid composed of cells, in which each cell represents land use for a parcel of land of 500x500 metres. A constrained cellular automata (CA) model carries out the dynamic allocation of the regional claims for space. At the local level, global information about the availability and quality of land for further expansion is returned to the COROP and national levels, thus influencing the dynamics between these levels. An example of a recent study is de Nijs *et al.* (2005).

The *Land Use Scanner* (Figure 16-1) takes the demand for land by the various land-use functions as input and distributes that based on the suitability of each location for each function. This suitability is determined by the present land use at the location and its surroundings and by Government stimuli or restrictions for a specific land-use function at that location. For instance, the proximity of housing locations may increase the suitability of a location for building new housing, but restrictions due to nature conservation policies may decrease the suitability at the same location. The weight attached to these different factors may also vary for each scenario, depending on the world view that underlies the scenario, as will be described shortly. Based on the demands for land and the suitability of each location for each land-use function, the *Land Use Scanner* distributes the land-use functions in such a way that each function is located optimally at its most suitable location, thus deriving future land-use maps for each scenario. Chapter 20 of this book contains a more detailed description of the model.

In addition, the *LUMOS* toolbox contains a number of tools that help in analysing the results of both models, of which the *Map Comparison Kit* is the most important. Comparison of maps is useful for identification of the spatial differences between socioeconomic scenarios, detection of temporal changes, calibration/validation of land-use models, hot-spot detection, and uncertainty analysis. The *Map Comparison Kit* enables map comparison and quantification of similarities and dissimilarities, especially for 'categorical' or 'nominal' maps, either in a cell by cell comparison or in a 'fuzzy' way, where the exact location of a specific feature is considered of less importance. The software, commissioned by MNP, has been designed by the Research Institute for Knowledge Systems (Hagen, 2002; Visser, 2004). Other tools available in the *LUMOS*-toolbox are the Site Seeker (de Nijs and

Kuiper, 2006), a tool for automatic calibration and validation and a tool for spatial optimisation. The latter tool uses heuristic methods such as genetic algorithms to optimise land use from spatial policy perspectives and is further discussed in Chapter 9 of this book and Loonen *et al.* (2006).

The next sections describe three recent applications of the *Land Use Scanner* relating to the construction of a number of national future scenarios, a regional simulation of future land use for the province of Zuid-Holland and the possible impact of land use on nature and the environment.

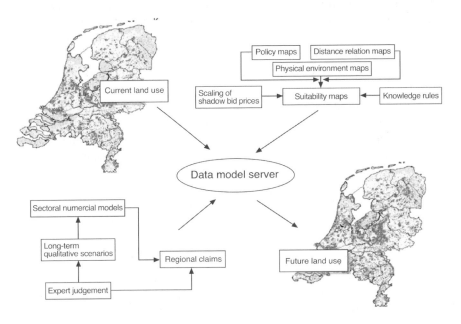

Figure 16-1. Components of the *Land Use Scanner* Model.

3. SPATIAL IMPRESSIONS OF THE FUTURE

The aim of the Spatial Impressions study was to create a consistent set of spatially elaborated scenarios for *ex-ante* evaluations of the environmental and ecological effects of land-use changes. A detailed description of the study can be found in Borsboom-van Beurden *et al.* (2005). The *Land Use Scanner* played a central role here. Before the model could be applied, however, the spatial processes anticipated had to be determined for each scenario on the basis of the qualitative storylines developed for the

'Sustainability Outlook' and 'Environmental Outlook'. In the Sustainability Outlook (MNP, 2004a), two axes are considered as key uncertainties: 'efficiency' versus 'solidarity', and 'globalisation' versus 'regionalisation'. The trend towards more 'efficiency' means that decision making is increasingly based on economic rationality and market forces. The limited Government intervention is mainly directed to facilitate market processes. The trend towards more 'solidarity' involves decision making which is determined by values on social equity and solidarity, cultural identity and sustainability. Government coordination is important here and not restrained. 'Globalisation' on the vertical axis means intensification of the physical, economic and sociocultural exchange between societies, resulting in an increased uniformity. 'Regionalisation', at the other end of the axis, implies that globalisation slows down or stops. The resulting framework is shown in Figure 16-2.

For the Sustainability Outlook, these scenarios have been enriched by adding values, goals (preferences as well as concerns), means to achieve these goals and different Government philosophies. This results in four world views addressing the quality of life and the way this quality should be realised from a specific perspective of sustainability.

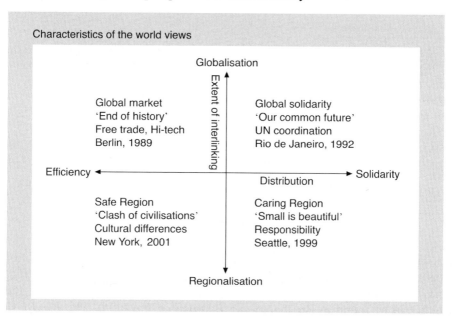

Figure 16-2. Scenario framework for the Sustainability Outlook with selected keywords.

Apart from the description of future trends in housing, employment, traffic and transport, agriculture, energy provision, and nature, the world views contain assumptions on population number, population composition and economic growth. Population growth diverges from a decrease of 1 million to an increase of 4 million people; economic growth ranges from very low to high.

Thematic experts were questioned about possible changes in driving forces, their resulting implications for land use, their physical appearance and their relationship to changes in other land-use types. This information was translated into a base map containing 28 land-use types and a set of allocation rules and suitability maps. Meanwhile, information on the demand for land in the future was collected by gathering land-use claims from specialised models as *PRIMOS* for demography and housing (de Bok *et al.*, 2004) and *BLM* for employment (CPB, 1999). Sometimes the demand for land was estimated by specialists on the basis of policy documents on ecological development and water management. After pre-processing, the data were used as input in the *Land Use Scanner* model. The various claims were allocated with the model and the results were checked and approved by project teams and thematic experts. Apart from the visual interpretation of the maps, the *Map Comparison Kit* was used to explore the differences between base year and projection, as well as those between different scenarios.

The resulting land-use maps (Figure 16-3) show noticeable differences in urbanisation patterns, not only as a result of variation in population increase, household fission and formation, but also due to the varying degree of urban containment in spatial planning. Apart from considerable further urbanisation in all scenarios, rural areas also showed significant changes. Arable farming disappears to a large extent in some regions because it is superseded by dairy farming. In some scenarios, greenhouse horticulture is booming, whereas in other scenarios it diminishes. In Global Solidarity, explicit spatial measures are taken to counter water problems. Examples are the creation of new surface water to retain excess water for dry periods, groundwater protection by prohibiting new construction and intensive cattle breeding and far-reaching restrictions to further extension of built-up areas, greenhouse horticulture and intensive cattle breeding in the winter beds of large rivers such as the Meuse, Waal and Rhine.

It appeared that not only the resulting maps of future land use were of interest for various policy fields, but surprisingly also 'by-products'. These included suitability maps depicting the most attractive locations, suitability maps containing all policy restrictions for particular land-use types, and GIS analyses such as overlays with a number of policy contours. As mentioned earlier, these maps were predominantly created to serve as an intermediate step in model chains.

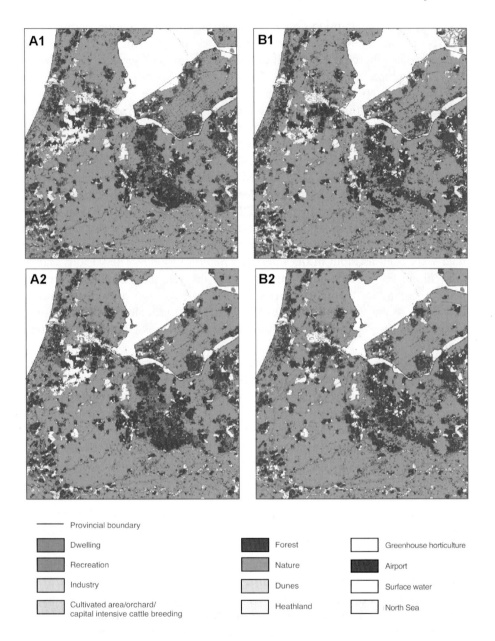

Figure 16-3. Land-use patterns in 2030 in the Randstad area for the Global Market (A1), Safe Region (A2), Global Solidarity (B1) and Caring Region (B2) scenarios. (See also Plate 16 in the Colour Plate Section)

4. APPLICATION ON A REGIONAL LEVEL

By request of the province of Zuid-Holland, MNP developed a more detailed regional level elaboration of the Spatial Impressions of the Future study (Bouwman *et al.*, 2006). The largest cities of the province (The Hague and Rotterdam) are part of the urbanised Randstad area, whereas a considerable part of the countryside is located in the unique cultural landscape of the Green Heart. Urbanisation pressures and congestion affect environment, nature and water quality. The spatial impressions served to gain insights into the most important long-term dilemmas in the spatial organisation for Zuid-Holland and supported in this way the provincial policy plan 'Green Space, Water and the Environment'.

This study elaborated further on the Spatial Impressions study developed for the Sustainability Outlook by using provincial data and applying guiding principles that describe specific aspects of a desired spatial development that were formulated by the staff of the Department for Green space, Water and the Environment. Examples of these principles are:

- prohibition of new construction in the National Ecological Network;
- preferably no residential, industrial, commercial and office development in peaty areas;
- urban expansion situated near high quality public transport junctions; and
- concentration of new greenhouse horticulture in locations designated by Government.

In fact these guiding principles can be compared to the suitability maps in the *Land Use Scanner*. In this particular application these were largely based on datasets provided by the province. A number of datasets are actually refined versions of data already included in the Spatial Impressions configuration, such as the provincial version of the National Ecological Network. Most datasets, however, added new information representing the location factors that play an important role in regional decision making (Table 16-1). Unfortunately this material is not available on a national scale.

This study also used more recent datasets on the demand for space, mainly coming from the joint study of long-term economic, demographic and residential developments by the different assessment agencies. The additional demand for residential land was decreased following the most recent adjusted population projections (de Jong and Hilderink, 2004; Hilderink *et al.*, 2005). Besides, demand for agricultural land was lower because of the reform of CAP. Another major difference with the Spatial Impressions study is that an adjusted allocation algorithm was used in combination with more detailed grid cells of 100 by 100 metres containing only one land-use type per cell.

Table 16-1. Specific regional level datasets and related policy goals in the Zuid-Holland study

Dataset	Policy goal
Robust ecological connections between existing and planned nature areas	Counter defragmentation of nature areas and provide large habitats
Non-polluted seepage water	Possibilities to develop new wetlands
Underground infrastructure	No development permitted for safety
Sand extraction	No building, economically important
Geological relicts	Conservation
Historical landscapes	Conservation
Historical elongated villages	Conservation
Proposed water retention areas	Counter flooding and dehydration
Industrial noise zones	No new houses, densification of employment
Supporting power of the soil	Prevent subsidence of buildings
Possibilities for subterranean construction	Save space
Possibilities for heat conservation in soil	Sustainable use of energy
Low polders	No construction to prevent water inconvenience

Figure 16-4 shows the results of the Zuid-Holland study for two scenarios. In the Global Market scenario, high economic growth and a substantial population and household growth result in a strong increase in the built-up area at the expense of agriculture. Further urbanisation takes place near the large cities in valuable landscapes and within the National Ecological Network. Rotterdam Airport and the acreage of greenhouse horticulture also expand substantially. The amount of land used by industrial and commercial sectors and offices increases as well. The port of Rotterdam also needs extra space in this globalizing scenario calling for additional extensions. New surface water is created to meet the demands of outdoor leisure.

The Global Solidarity scenario shows a very different picture. Economic and population growth are moderate. Societal values with respect to sustainable development make the Government pose many restrictions from nature conservation and water management to new urban development. Urbanisation shifts to the current flower bulb-growing areas and an island in the south. The use of space of greenhouse horticulture decreases to the benefit of residential development. Large new surface water areas are created, mainly to serve water retention and incidentally leisure purposes. In addition, this water can prevent the growing together of urban areas, articulating them in this way. The Safe Region and Caring Region scenarios are more similar to current land use. For this reason they will not be discussed further here.

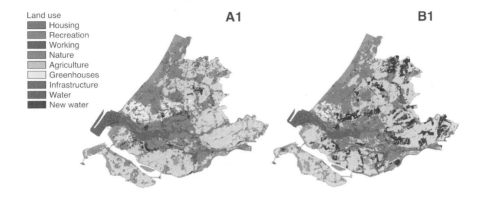

Figure 16-4. Land use in Zuid-Holland in the Global Market (A1) and Global Solidarity (B1) scenarios. (See also Plate 17 in the Colour Plate Section)

As in the Spatial Impressions study, the results show how diverging spatial developments may be in the long term, depending upon the degree of urbanisation pressure and spatial restrictions through Government intervention. In the Global Solidarity and Caring Region scenarios, the province is too small to meet all demands for space, resulting in an overspill to nearby provinces. In all scenarios, urbanisation continues, albeit in a different form: concentrated or deconcentrated, near the large cities or at a greater distance. Open spaces disappear in the market-oriented scenarios. Current policy plans for green recreation near the large cities will not materialise.

The Dutch province of Zuid-Holland is not large enough to provide the space needed for housing and employment, and at the same time take care of optimal nature and landscape conservation and water protection. It can be concluded that further urbanisation requires clear choices with respect to:

- concentrated or deconcentrated urbanisation?
- location of port-related activities near Rotterdam or in other provinces?
- housing and leisure or air transport and employment near Rotterdam Airport?
- extension of greenhouse horticulture or flower bulb-growing industry or relocation to other provinces?
- reserving space for nature areas and water management or other functions?

5.   IMPACT OF LAND USE ON NATURE AND ENVIRONMENT

In integrated assessment studies, the effects of current and proposed policies are analysed (see, for example, MNP, 2004b). Maps of future land use are often only an intermediate step in a chain of models to determine ecological and environmental effects. RIVM and Stichting DLO (2001), for example, analysed the effects of existing and proposed policies of land-use planning on the preservation of valuable landscapes, quality of nature areas and the protection of specific groundwater infiltration areas. Previous studies omitted the possible influence of surface water and groundwater on the quality of surrounding nature areas. Therefore, the aim of this case study was to analyse the potential for improving water quality in the EU Bird and Habitat Directive (BHD) areas by optimising the spatial allocation of intensive land use functions from the perspective of hydrological impact.

In the Netherlands, 141 BHD areas have been designated as part of the European network of protected nature areas, Natura 2000. Within these BHD areas, it is obligatory to take measures to maintain protected habitats and/or species and to make arrangements to prevent disturbance and loss of ecological quality (MNP, 2004c). Apart from protecting BHD areas themselves, it is also important to reckon with surrounding land use functions that might influence them via hydrological processes. By taking this impact into account in the process of spatial planning, water quality in BHD areas may be substantially improved. Several hydrological processes affect BHD areas but, in this case study, focus was on seepage and changes in groundwater and surface water flows in relation to other types of land use in the vicinity. The land-use optimisation aimed at excluding the adverse land-use types from the locations that possibly influence the BHD areas by restricting their allocation. In doing so, alternative future land-use configurations can be simulated that result in improved water quality in BHD areas.

Land-use types that have an adverse effect on surface water and groundwater quality (e.g. through nutrient emissions) and groundwater level (e.g. through artificial lowering) include urban or semi-urban land-use types such as residential areas, holiday centres, business estates and infrastructure and certain types of agriculture such as arable farming, livestock and dairy farming, greenhouse horticulture, capital-intensive cattle breeding and orchards. The adverse effects of parks and recreational areas, nature areas, and extensive forms of livestock farming (grassland) are assumed to be much smaller. The actual impact of the land-use functions on the nearby BHD areas is determined by the land-use intensity and the hydrological system. The latter especially influences the distance at which impacts are to be expected. These relationships are as yet not exactly known for every BHD area. We, therefore, defined the BHD-influencing areas as the total area that may provide a hydrological impact. Large water resources such as rivers, the central lake (IJsselmeer) and the North Sea, have been left out of consideration because their quality is determined predominantly by foreign countries.

The *Land Use Scanner* configuration for the Safe Region scenario from the Spatial Impressions study served as a starting point for this analysis. In an additional land-use simulation we strived to exclude the adverse land-use types from BHD-influencing areas by assigning low or even negative suitability values to these areas. Figure 16-5 shows desirable and undesirable land-use types within the areas that potentially have a hydrological influence on the BHD areas in the central part of the country. The figure shows the original land-use map from the Spatial Impressions study (left) and the optimised patterns resulting from the BHD analysis (right).

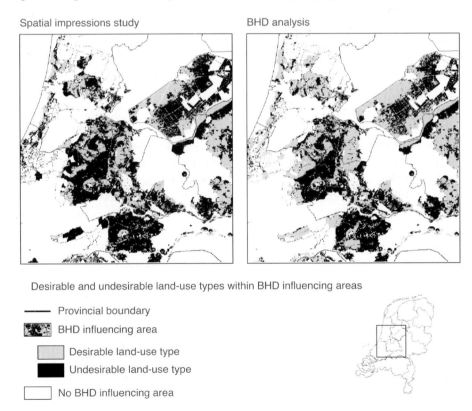

Spatial impressions study   BHD analysis

Desirable and undesirable land-use types within BHD influencing areas

— Provincial boundary

BHD influencing area

Desirable land-use type

Undesirable land-use type

No BHD influencing area

Figure 16-5. Distribution of desirable and undesirable land-use types within BHD-influencing areas for the western part of the Netherlands.

The abundance of undesirable (black) land-use types in BHD-influencing areas has decreased by about 13% when compared to the original Safe Region land-use map. These land-use types have now been allocated outside the influential zone. Considering the current size of the restricted BHD influencing areas, it is not possible to exclude all adverse land-use types completely from these areas because of the large demands for land and the limited availability of suitable alternatives.

More accurate results may be obtained when the hydrological influence of BHD areas is specified per land-use type. In this case, the extent of the impact areas will most probably decrease, resulting in a lower amount of land that needs to be allocated and therefore better possibilities to exclude adverse land-use types from BHD-influencing areas. Following this analysis, it can be concluded that it is possible to simulate more optimal future land-use configurations from the perspective of nature conservation by taking into account possible adverse effects of neighbouring land-use types. The method shows promise in pinpointing the desired locations of these land-use types. This would require a further refinement of the exact influence of each land-use type.

6.   CONCLUSION

What can be concluded on the usability of the land-use models in the *LUMOS* toolbox for support in policy making? Considerable progress has been made in the last years at several points. Not unimportant, the underlying database containing suitability maps has become very extensive. Data collection and updating have become more or less a routine task, and recent versions and spatially related policy changes are quickly incorporated into the models. Information on the demand for land is coming from recent studies, for example the joint study by the national assessment agencies mentioned earlier. Thematic experts have become heavily involved in the elaboration of spatial processes and provided ample information on allocation rules. The *LUMOS* models are operational and have been applied in various projects: apart from the applications discussed in this chapter, see also de Nijs *et al.* (2002), Groen *et al.* (2004) and MNP (2004b; 2006). In most cases, the results have been used to prepare policy making.

In addition, the models have been adapted and improved. The *Environment Explorer* has been calibrated and validated and extended with a traffic module (see Engelen *et al.*, 2005, for detailed description). Difficulties in interpretation of land prices per cell and 3D visualisation (Borsboom-van Beurden *et al.*, 2006) when using mixed land use in 500 by 500 metres necessitated the extension of the allocation mechanism of the

Land Use Scanner with discrete allocation, that uses only one land use per cell of 100 by 100 metres. Another extension has been the ability to sketch on top of maps, for example representing new residential areas, while at the same time the demand for land is corrected with this area.

In spite of these improvements, the *LUMOS* models are not yet finished. The mutual relation between land use and infrastructure is not yet worked out at a sufficiently detailed level. Although the *Environment Explorer* contains such a link with the traffic module, this link is still too coarse to be suitable for the assessment of a number of policy options in traffic and transport, such as road pricing. In the *Land Use Scanner,* this link is hitherto absent, although at the moment the possibilities for a connection with the TIGRIS XL traffic model are being investigated by the Rand Europe company, who developed the latter model for the Ministry of Transport and Water Management. This will require making the static model dynamic, a significant change in the allocation mechanism. Another discussion is whether the concept of transformation should not be placed centrally instead of as the estimation of a probability or optimisation by linear programming.

Land prices used in the *Land Use Scanner* should be improved too by incorporating regional variations and assumptions about their development in future. Further, both the *Environment Explorer* and *Land Use Scanner* ignore the international context of the Netherlands. At least a zone of neighbouring countries should be included because many spatial processes extend beyond the border. The link with the specialised models on housing, employment and agriculture providing the demand for space for these functions can be ameliorated. An example is the relationship between a lack of space to accommodate housing and the substitution of residential preferences in the housing market. The same goes for the link between land-use models and the models that evaluate environmental and ecological effects. Optimisation from the perspective of nature preservation seems promising, but an integrated assessment requires a more systematic overview of possible environmental and ecological effects of all land-use types.

REFERENCES

Borsboom-van Beurden, J.A.M., Boersma, W.T., Bouwman, A.A., Crommentuijn, L.E.M., Dekkers, J.E.C. and Koomen, E. (2005) Ruimtelijke Beelden. Visualisatie van een veranderd Nederland in 2030, *RIVM rapport 550016003*, RIVM, Bilthoven.
Borsboom-van Beurden, J.A.M., van Lammeren, R.J.A., Hoogerwerf, T. and Bouwman, A.A. (2006) Linking land use modelling and 3D visualisation. A mission impossible? In van Leeuwen, J. and Timmermans, H. (eds) *Innovations in Design and Decision Support Systems in Architecture and Urban Planning*, pp. 85–102.

Bouwman, A.A., Kuiper, R. and Tijbosch, H.W. (2006) Ruimtelijke beelden voor Zuid-Holland, *Rapportnummer 500074002.2006*, Milieu- en Natuurplanbureau, Bitlhoven.

CPB (1999) BLM-Regionale verkenningen 2010–2020: in gesprek met de regio's. *Werkdocument no.112*, 's-Gravenhage.

de Bok, M., Heida, H. and Brouwer, J. (2004) *Vier Lange Termijn-scenario's Wonen voor het Milieu- en Natuurplanbureau*, ABF Research, Delft.

de Jong, A. and Hilderink, H. (2004) Lange-termijn bevolkingsscenario's voor Nederland, CBS/MNP/CPB/ RPB/SCP/NiDi, Voorburg/Bilthoven/Den Haag.

de Nijs, T., Engelen, G., White, R., van Delden, H. and Uljee, I. (2001) De LeefOmgevingsVerkenner. Technische documentatie, *RIVM rapport 408505007/2001*.

de Nijs, A.C.M., Crommentuijn, L.E.M., Farjon, H., Leneman, H., Ligtvoet, W., Niet, R. de and Schotten, K. (2002) Vier Scenario's van het Landgebruik in 2030, Achtergrondrapport bij de Nationale Natuurverkenning 2, *RIVM rapport 408764 003*, RIVM-MNP, Bilthoven.

de Nijs, A.C.M., Kuiper, R. and Crommentuijn, L.E.M. (2005) Het Landgebruik in 2030. Een project van de Nota Ruimte, *Rapport 711931010/2005*, Milieu- en Natuurplanbureau, Bilthoven.

de Nijs, A.C.M. and Kuiper, R. (2006) De Locatiezoeker. Lagenbenadering voor verstedelijking, *Rapport 500074001/2006*, Milieu- en Natuurplanbureau, Bilthoven.

Engelen, G., Hagen-Zanker, A., de Nijs, A.C.M., Maas, A., van Loon, J., Straatman, B., White, R., Uljee, I., van der Meulen, M. and Hurkens, J. (2005) *Kalibratie en Validatie van de LeefOmgevingsVerkenner*, Milieu- en Natuurplanbureau, Bilthoven.

Groen, J., Koomen, E., Piek, M., Ritsema van Eck, J. and Tisma, A. (2004) Scenario's in kaart, Model- en ontwerpbenaderingen voor toekomstig ruimtegebruik, Ruimtelijk Planbureau, NAi Uitgevers, Rotterdam.

Hagen, A. (2002) *Comparison of Maps containing Nominal Data*, Report commissioned by National Institute for Public Health and the Enironment (RIVM), Research Institute for Knowledge Systems, Maastricht.

Hilderink H., den Otter, H. and Jong, A. de (2005) Scenario's voor huishoudensontwikkelingen in Nederland, *Rapportnr. 550012005*, MNP/ABF Research/CBS/CPB/RPB/SCP, Bilthoven/Den Haag.

Loonen, W., Heuberger, P.S.C., Bakema, A.H. and Schot, P.P. (2006) Application of a genetic algorithm to minimize agricultural nitrogen deposition in nature areas, *Agricultural System*, 88(3): 360–375.

MNP (2004a) *Quality and the Future. Sustainability Outlook*, Netherlands Environmental Assessment Agency, RIVM, Bilthoven.

MNP (2004b) Milieu- en Natuureffecten Nota Ruimte, *RIVM-rapport 711931009*, RIVM, Bilthoven.

MNP (2004c) *Natuur Balans 2004. Milieu- en Natuurplanbureau*, RIVM, Bilthoven.

MNP (2006) Nationale Milieuverkenning 6 2006–2040, *Rapport 500085001*, Milieu- en Natuurplanbureau, Bilthoven.

RIVM and Stichting DLO (2001) Who is afraid of red, green and blue, Toets van de Vijfde Nota Ruimtelijke Ordening op ecologische effecten, *RIVM-rapportnr. 711931005*, Wilco BV, Amersfoort.

Visser, H. (ed.) (2004) The Map Comparison Kit. Methods, Software and Applications, *RIVM report 550002005/2004*, Rijksinstituut voor Volksgezondheid en Milieu, Bilthoven.

VROM, LNV, V&W and EZ (2004) Nota Ruimte, Ministerie Volkshuisvesting, Ruimtelijke Ordening en Milieubeheer, Den Haag.

Chapter 17

THE *MOLAND* MODELLING FRAMEWORK FOR URBAN AND REGIONAL LAND-USE DYNAMICS

G. Engelen[1], C. Lavalle[2], J.I. Barredo[2], M. van der Meulen[3] and R. White[4]

[1]Flemish Institute for Technological Research (VITO), Mol, Belgium; [2]Joint Research Centre of the European Commission (JRC), Ispra, Italy; [3]Research Institute for Knowledge Systems (RIKS), Maastricht, The Netherlands; [4]Department of Geography, Memorial University of Newfoundland, St. John's, Canada

Abstract: The Joint Research Centre (JRC) has implemented the following elements for a growing and representative sample of cities and urbanised regions: (1) high resolution spatial databases of land-use change over approximately 40 years, (2) urban sustainability indicators, and (3) a modelling framework enabling the implementation of hybrid land-use models representing urban regions ranging from the single, average-sized city to complex hierarchies of cities. The core of the latter framework enables the configuration of dynamic spatial models operating at a macro- and a micro-geographical scale. Both levels are intimately linked and permit an integral and detailed representation of the evolving spatial system. The models are complemented with: a series of spatial indicators, tools for interactive design, analysis and evaluation of policy measures or scenarios and a set of routines enabling the semi-automatic calibration and validation of the applications developed.

Key words: Land use; cellular automata; spatial dynamics; modelling framework; spatial indicators; calibration.

1.   AIMS OF MOLAND

Cities are growing worldwide at a pace faster than ever. Currently, one in two human beings lives in a city, and it is expected that most future population growth will occur in cities. Growth has been and will be associated with an increased complexity in the institutional, functional and

E. Koomen et al. (eds.), Modelling Land-Use Change, 297–319.
© 2007 *Springer.*

technical organization of cities. These changes should be managed to preserve or enhance the quality of life for billions of people, and to ensure economic and environmental sustainability. Effective planning and management increasingly requires data on past and current conditions, an accurate understanding of the cause and effect relations responsible for urban and regional change in time and space, and an ability to foresee the likely consequences of proposed projects and policies.

The DG Joint Research Centre of the European Commission, Institute for Environment and Sustainability, Land Management and Natural Hazards Unit in Ispra, Italy, initiated in 1998 the MURBANDY (Monitoring Urban Dynamics) project, later followed-up by the MOLAND (Monitoring Land Use Changes) project, with the objective to monitor the developments of urban areas and identify trends at the European scale (Lavalle *et al.*, 2002). The aim also was to provide a spatial planning tool for monitoring, modelling and assessing the development of urban and regional systems. The work includes the creation of land-use databases for various cities and the definition of indicators enabling the assessment of the impacts of anthropogenic stress factors (with a focus on expanding settlements, transport and tourism) in and around urban areas, and along development corridors.

To date, the MOLAND methodology has been applied to an extensive network of cities and regions (Figure 17-1; http://moland.jrc.it/) for an approximate total coverage in Europe of 70,000 km^2. Originally, a representative sample of major cities was tackled. However, with the advent of EU enlargement, the work focused on bordering areas anticipating potential effects of the development of local towns and economic corridors. A new activity, initiated in 2004, involves spatial planning and mitigation of hazards (floods, droughts, land slides, forest fires) in urban areas in response to the dramatic effects of climate and human induced catastrophes that Europe has witnessed recently. MOLAND thus contributes to the assessment of the impacts of extreme weather events, in the frame of research on adaptation strategies to cope with climate change. In addition, applications outside Europe have been targeted at mega cities including Lagos and Delhi (Barredo and Demicheli, 2003; Barredo *et al.*, 2004).

MOLAND defined, applied and validated a methodology in support of sectoral policies with territorial and environmental impacts. The reference framework is provided by, among others, the Thematic Strategy on the Urban Environment, the European Spatial Development Perspective (ESDP), the strategic environmental assessment of the Trans-European Network for transport (TEN), and the environmental evaluation of EU Structural and Cohesion Fund programmes. The following three interrelated component projects are implemented: *'Change'*, *'Understand'* and *'Forecast'*.

Figure 17-1. European urban areas investigated by MURBANDY/MOLAND.
Source: *EU-JRC-IES*

'Change' measures changes in the urban spatial extent and structure for a representative sample of cities and urbanised regions in Europe. A database of land-use change over approximately the last 40-50 years is generated, involving the creation of detailed geographical information system (GIS) datasets of land-use types and transport networks at a mapping scale of 1:25,000 and a mapping unit of 1 hectare. For major cities, datasets are gathered for typically four dates (early-1950s, late-1960s, 1980s, 2000s). For larger urbanised regions, this is the case for two dates (mid-1980s, 2000s). The land-use classification used is an extension of the CORINE land-use/land-cover database which significantly extends the number of the artificial surfaces classes and thus enables a more detailed study of typical urban land uses and their dynamics. In addition to the land-use and transportation information, ancillary data are collected in either a GIS or

other format related to demography (e.g. population size, age cohorts, educational level), economy (e.g. production per sector, employment, income), master and zoning plans (e.g. protected natural areas), transportation, energy consumption, environment (e.g. air quality, noise levels), geology and topography.

'Understand' identifies, develops and computes territorial and urban environmental indicators aimed at measuring the level of sustainability of urban and peri-urban areas. These provide a synthetic assessment of urban and regional development comparable across Europe in terms of social, economic, environmental and landscape characteristics. The methodology emphasizes the use of a large variety of indicators: high resolution and spatially-explicit versus non-spatial, dynamic versus static, descriptive of a momentary state versus aimed at change detection.

'Forecast' develops a dynamic, high-resolution modelling tool suitable for the exploration and forecasting of urban futures as the consequence of the inherent autonomous dynamics of the urban system, scenarios, trends and intended or unintended policies. This tool is applied to a wide set of cities and urbanised regions and is available for the realistic exploration of urban futures under a variety of planning and policy scenarios, involving: current and planned extensions of transportation networks, trends in population growth, immigration trends, economic trends, zoning measures and legislation, energy use, emissions and waste management.

In the remainder of this chapter, we focus our attention on *'Forecast'*, the third component of the MOLAND methodology.

2.   MOLAND: A MODEL OR A MODELLING FRAMEWORK?

MOLAND aims at providing a methodology applicable to any urbanised area in Europe and the world beyond. It defined for its *'Forecast'* component, analytical and functional requirements which require more than simply 'a model'. It should enable the development of a wide range of applications, all in a similar urban analytical context, yet each specifically targeted and differing in a number of essential characteristics. These are requirements pointing at the need for a flexible and generic modelling framework, rather than a model as such. This argument is based on the following considerations:

- Applications vary considerably in *geographical extent* and *size*. Cities such as Dublin are modelled alongside with Leinster Province and Northern Ireland covering close to one third of Ireland. Beyond this, future work in MOLAND will certainly address urbanisation problems of growing complexity set in ever larger areas, such as traffic congestion or emission control in urban corridors like those in the pentagon shaped by London, Paris, Munich, Milan and Hamburg.

- The *spatial resolution* and *detail* at which problems need to be addressed changes with the problem at study. Typically the *'Change'* component provides for time series of land-use data at the 1:25,000 scale and 1 hectare resolution purposely generated. Thus, a unique basis is laid for developing and calibrating powerful, high-resolution land-use models. However, this same quality is not necessarily available in the ancillary data collected from local authorities. Models should ideally have a spatial resolution and level of detail adapted to the problem and the available data.

- The land-use classes mapped in *'Change'* are standardised throughout the MOLAND project and results are comparable over time and among applications. However, the drivers of the land-use changes are essentially demographic, social, economic and environmental in nature. The MOLAND models (Section 3), exploiting the interlinked processes at the micro- and macro-level, require a *crisp mapping* between economic land-use classes and economic sectors as well as residential land-use classes and population categories. This is far from straightforward and is highly dependent on the classification schemes adhered to by the national and regional authorities providing the ancillary data for the application. Flexible mapping between land uses (at the micro-level) and sectoral information (at the macro-level) needs to be supported.

- Urban *policy* and *planning problems* vary considerably and so should the MOLAND applications developed to confront them. Examples in this case include urban sprawl, its effects on land abandonment and forest fire incidence in the Algarve region in Portugal, flooding problems associated with new residential developments in flood-prone areas exacerbated by climate change in Pordenone province in Italy (Barredo *et al.*, 2007), and economic development in the context of the EU enlargement in the Dresden-Prague corridor. This requires a high level of flexibility in providing more or less detail in representing economic sectors, population categories and environmental drivers.

- Extending the previous point, picking up the relevant signals associated with growth and land-use change and detecting the main desired and adverse consequences of planning and policy actions requires targeted *indicators* in the social, economic, environmental or landscape domains.

- With respect to policy and planning exercises, a MOLAND modelling tool should support the rapid and interactive design, implementation and extensive analysis of very diverse sets of *scenarios* and *measures* as means to intervene in the autonomous dynamics of the system studied.

- Monitoring and analysis of urban change and the associated spatial planning problems is set in a dynamic context in the sense that the models need *frequent adaptations* as new knowledge and data become available and the urgency of problems and themes requiring attention shifts. Again, this asks for a high level of adaptability at the level of MOLAND.

- Finally, almost any change in a model or its databases necessitates at the least a partial *recalibration* and *revalidation*. Calibration and validation are prerequisites for a model to be reliably used in practical planning and policy-making. However, these are time consuming tasks because of the complex nature of the dynamic models used. Ideally, a MOLAND modelling tool therefore should be equipped with facilities supporting calibration and validation.

These are requirements favouring the availability of a flexible *modelling framework*. The prime aim of such a framework is to build and run a class of applications defined primarily by the problem and the available data to address it rather than by the particulars of an existing model, most often a research product. The latter seems an obvious conclusion. In reality, however, the opposite seems to be the rule and models are often used to address problems for which they are only partly suited. As a result, overly sophisticated models are applied because they seem to hold the implicit advantage of being widely applicable. Apart from delivering the wrong type of output in terms of the embodied variables, the temporal and spatial resolution represented, and the levels of detail covered, this approach is inefficient in the amount of resources required to build, feed, calibrate and apply the model. Provided a sufficiently powerful modelling framework is available, it is much more efficient to create or adapt a model attuned to the precise problem within the scientific boundaries set by the approach.

This aim has been largely achieved with the venue of MOLAND Version 3 (Engelen *et al.*, 2004), which is set up as a framework enabling the configuration of an application in a variety of ways. In the next section, its theoretical, functional and technical characteristics will be briefly described. Next, the levels of freedom allowed in configuring problem specific applications will be discussed.

3.   THE MOLAND MODEL: THEORETICAL BASIS

The goal of developing a generic modelling framework – one that will be applicable to essentially any European urban area – implies the rather strong assumption that, at some level, cities are fundamentally similar, evolving by the same processes. Especially in the European context, where cities grow out of, and express, a wide variety of cultural, economic and historical contexts, this seems a bold hypothesis. However, the fractal analysis of land-use patterns provides evidence in its favour (Frankhauser, 1994; White and Engelen 1993; 1997), and suggests that the substantial differences among cities are largely due to city-specific boundary conditions. In particular, we hypothesise that it is such city-specific factors as location and relative attractiveness in a regional system of cities, the geometry of the transport network, local topography and local planning regulations (which themselves represent a particular local intervention in what might otherwise be a generic process of urban development) that determine the observed differences among cities. The MOLAND project in effect constitutes a test of this hypothesis.

3.1 Coupled models at three geographical levels

Inherent in the MOLAND approach is the systems view that urban areas evolve as the result of endogenous processes combined with exogenous events including policy-induced changes. Therefore, it is essential to incorporate in the model a sufficient description of the processes making and changing the urban system and to represent policy and other constraints as elements interacting with these, in order to form integral pictures of possible futures of the modelled urban area.

The core component of the MOLAND modelling framework is essentially a dynamic land-use-transportation model. The model represents processes operating at three geographical levels (Figure 17-2): the global (one spatial entity, typically representing a contiguous group of countries, a country, or an administrative or physical entity), the regional (n administrative entities, typically representing NUTS 2, NUTS 3 or NUTS 4 regions) and the local (N cellular units). At each level, the user can decide on a representation adapted to the needs of the application in terms of the degree of sophistication and spatial resolution of the models used.

3.2 Global level

At the Global level, the scale of the entire modelled area, the model integrates data taken from economic, demographic and environmental growth scenarios. Hence, at the global level, the model essentially runs on exogenously provided time series. Sophistication in producing these remains external to the MOLAND modelling framework proper.

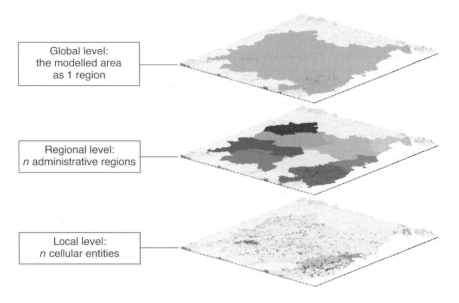

Figure 17-2. MOLAND representation of processes at global, regional and local levels using the application of the Dublin Metropolitan region. (See also Plate 18 in the Colour Plate Section)

The main economic activities are typically condensed into three to 10 aggregated sectors, including: farming, mining, industrial, commercial, services, socio-cultural, and recreational activities. The population is typically represented in two to four residential categories, for example: high- and low-density residential. While economic and population figures are posed in terms of numbers of people, jobs, or economic output, the natural land-use categories are expressed in terms of area occupied. They include, typically, wetlands, forests, shrub vegetation and extensive grasslands. The choice of the categories is based on the difference in spatial behaviour, the availability of data at the three geographical levels of the model, and the end-use requirements. Generally speaking, the best model is one incorporating sectors with distinctly different spatially-dynamic behaviour and not one with the highest number of sectors.

3.3   Regional level

At the regional level, consisting of administrative regions (typically a couple of dozen), the national growth figures are a constraint for models catering for the fact that the national growth will not evenly spread over the modelled area, rather that regional inequalities will influence the location and relocation of new residents and new economic activity and thus drive regional development. Models of different levels of sophistication may be applied.

In the simplest of cases, there is no regional level, or, better stated, there is only one region incorporating the entire modelled area. A most sophisticated model at the regional level is a dynamic spatial interaction based model or a dynamic gravity based model (see for example: Wilson, 1974; White, 1977; 1978; Allen and Sanglier, 1979a; 1979b). Such a model arranges the allocation of national growth as well as the inter-regional migration of activities and residents based on the relative attractiveness of the regions. One of two principles is applied:

For the allocation and relocation of the *population* and *economic* activities, a standard potential based model is applied: each region competes with all the other regions for new residents and new activities in each economic (and population) sector on the basis of its geographical position relative to the other regions, its employment level, the size of its population, the type and quantity of activity already present, and its location relative to the public and the private transportation systems. In addition to these, and novel in the context of interaction based models, summarized cellular measures, obtained from the model at the local level, are characterising the space within the regions and are factors determining the relative regional attractiveness. Three local level measures are the abundance of good quality land, the zoning status of that land and its accessibility relative to transportation or other infrastructure.

Four sub-models can be distinguished:

- A *regional economic sub-model*, applied per economic sector, calculates the amount of production and employment in the sector, its spatial allocation and re-allocation among the regions.
- A *regional demographic sub-model*, applied per population category, calculates the growth of the regional population, its allocation and re-allocation among the regions and the demand for housing.
- A *transportation sub-model*, possibly a complete four-stage transportation model, linked dynamically in the modelling framework, calculates the changes in the characteristics of the transportation infrastructure, the flows of people and goods travelling over it, and their impact on interregional distances and accessibility.
- A *land-claim sub-model* translates the regional growth numbers into spatial claims. The latter are passed on to the model at the local (cellular) level for a detailed allocation. Two principles may be applied: (1) a claim for land is fixed and passed on as a hard constraint. This principle reflects the fact that for particular activities policies determine the amount and location of land that is to be created or to be preserved in a region. It applies mostly for the natural land categories and recreational land. Alternatively, (2) the principle of supply and demand is applied to

regulate the densification of land use as well as its spatial allocation. This principle applies in particular to housing and most of the economic activities.

Simpler models can be implemented too at the regional level. Examples are the constant share model, in which the proportional shares of the state variables among the regions is preserved over time, and models based on the shift and share principle, in which additional assumptions are made relative to the regional specialisation in one or the other economic activity, population or natural category.

3.4   Local level

At the local level, the detailed allocation of economic activities and people is modelled by means of a cellular automata based land-use model (Couclelis, 1985; White and Engelen, 1993; 1997; Batty and Xie, 1994; Engelen *et al.*, 1995). The modelled area is represented as a mosaic of grid cells typically representing parcels of land covering anything from 0.25 hectare to 4 km^2. Each cell is modelled dynamically and together the cells constitute the changing land-use pattern of the area. Land use is classified into a (software-technical) maximum of 32 categories, subdivided in 'feature states' (fixed land uses that do not change dynamically), 'function states' (that change dynamically as the result of the local and the regional dynamics) and 'vacant states' (that change dynamically due to the local dynamics only). The function states are by all means the most important land uses in the model and are chosen with a view to guarantee as much as possible a one-to-one relation with the economic and residential categories at the regional level. In principle, it is the relative attractiveness of a cell as viewed by a particular spatial agent, as well as the local constraints and opportunities, that cause cells to change from one type of land use to another. This model is driven by the demands for land per region generated at the regional level. Four elements determine whether a cell is taken in by a particular land use function (Figure 17-3):

- Physical *suitability* is represented in the model by one map per land-use function modelled. The term suitability is used here to describe the degree to which a cell is fit to support a particular land-use function and the associated economic or residential activity. It is a composite measure, prepared in a GIS, on the basis of physical, ecological and environmental maps. It includes such factors as: elevation, slope, soil quality and stability, agricultural capacity, air quality, and noise pollution.
- *Zoning* or institutional suitability is also characterized by one map per land-use function. It is another composite measure based

on master plans and planning documents available from national or regional planning authorities including: ecologically valuable and protected areas, protected culturescapes and buffer zones. For maximally three consecutive planning periods, determined by the user (example: 2005-09, 2010-14 and 2015-30), each zoning map specifies which cells can and cannot be taken in by the particular land use. For the analysis of policy and planning alternatives, it is of paramount importance that suitability and zoning can be handled separately.

- *Accessibility* for each land-use function is calculated relative to the transportation infrastructure. It is an expression of the ease with which an activity can fulfil its needs for transportation and mobility in a particular cell. It accounts for the distance of the cell to the nearest link or node on each of the networks, the importance and quality of that link or node, and the particular needs for transportation of the activity or land-use function.

- *Dynamics at the local level* are the most important factor representing the impact of land uses in the area immediately surrounding a location. This is no longer the domain of abstract planning, rather that of the reality on the ground, representing the fact that the presence of complementary or competing activities and desirable or repellent land uses is of great significance for the quality of a location and thus for its appeal to particular activities. For each cell, the model assesses the quality of its neighbourhood: a circular area with a radius of eight cells. For each land-use function, a set of rules determines the degree to which it is attracted to, or repelled by, each of the other functions present in the neighbourhood. They articulate the strength of the interactions as a function of the distance separating the different functions within the neighbourhood. If the attractiveness is high enough, the function will try to occupy the location; if not, it will look for more attractive places. New activities and land uses invading a neighbourhood over time will thus change its attractiveness for activities already present and others searching for space. The spatial interaction rules determining the interplay between the different functions – the inertia, the action at a distance, the push and pull forces, and economies of scale - are defined as part of the calibration of this cellular automata model.

On the basis of these four elements, the model calculates for every simulation step the *transition potential* for each cell and each function. Cells will change to the land-use function for which they have the highest transition potential, until regional demands are satisfied. Thus, the transition potentials are a proxy for the land rent reflecting the pressures exerted on the land. As such they constitute important information for those responsible for the design of sound spatial planning policies.

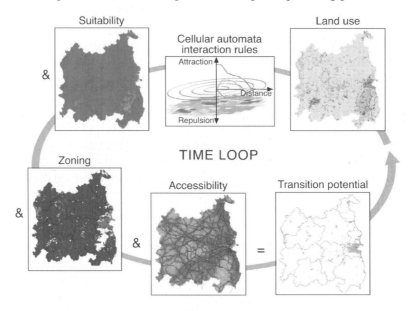

Figure 17-3. Different elements determine the spatial dynamics of the urban system. (See also Plate 19 in the Colour Plate Section)

The linkage between the models at the global, regional and local levels is bi-directional and very intense: the global growth figures are imposed as constraints on the regional models, the regional models distribute and allocate the global growth to the regions and impose the regional growth numbers on the cellular models. Finally, the cellular model determines growth at the highest level of detail. Then the cellular model sends to the regional model information on the quality and availability of space for each type of economic or residential activity. It is an input into the spatial interaction calculations at the regional level and influences the relative attractiveness of the individual regions. As regions gradually run out of space for an activity, they lose part of their competitive edge and exert less attraction. Growth is consequently diverted to other, more attractive regions. This framework constitutes a powerful instrument for representing non-linear spatial dynamics operating across a range of scales.

4.   SETTING-UP AND CALIBRATING NEW MOLAND APPLICATIONS

4.1   Configuring a new MOLAND case

In a number of past and ongoing projects, MOLAND has proven to be a generic and very flexible modelling framework applicable to cases varying greatly in complexity, detail, spatial and temporal resolution (e.g. Van den Steen, 2005; Van Delden *et al.*, 2005; Barredo *et al.*, 2007) and for areas equipped with the appropriate set of data. Configuring a new case remains a complex operation due to the large amount of information required and stored in a number of files and directories. MOLAND is not a GIS, hence manipulation and preparation of the base data has to be finished beforehand. No programming is required to set up new applications, rather, the configuration is supported by a wizard and requires the successful completion of four consecutive steps. The flexibility of MOLAND resides in the Steps 2 and 3 pertaining to the local and the regional level respectively:

- Step 1: *defining the working directories* informs the model about the directory structure used to store the files constituting the application;
- Step 2: *defining or changing the local-level characteristics* involves entering information in the *Simulation settings* window relative to the modelling area, including: cell size, the precise number of rows and columns of the cellular representation, and the number of regions modelled. Further, the characteristics of the cellular automata model applied at the local level are defined including in particular the respective numbers of vacant, function and feature states. Also the specification of the simulation interval is provided by entering the initial year of the simulation (and the date for which the base data apply) as well as the simulation interval. Essential input involves also the specification of the number of regions modelled (at the regional level) as well as the model applied to that effect. The latter is specified by providing the name of the particular DLL executing the model. In the example given in Figure 17-4, the *MolandModel.DLL* is specified, which implements the model described in Section 3.3.

  As explained in the text, other models can be used at the regional level: it suffices to specify the name of the DLL, software coded according to a template fitting MOLAND's technical interfaces, implementing it. For example, the simplest case described in the text, the one with one region only at the regional level, is invoked by specifying the *MurbandyModel.DLL*. Finally, reference is entered relative to the files containing essential information pertaining to the land use (map and legend), the transportation system (map, legend and number of networks modelled) and the regions (map representation and names in the legend file) modelled.

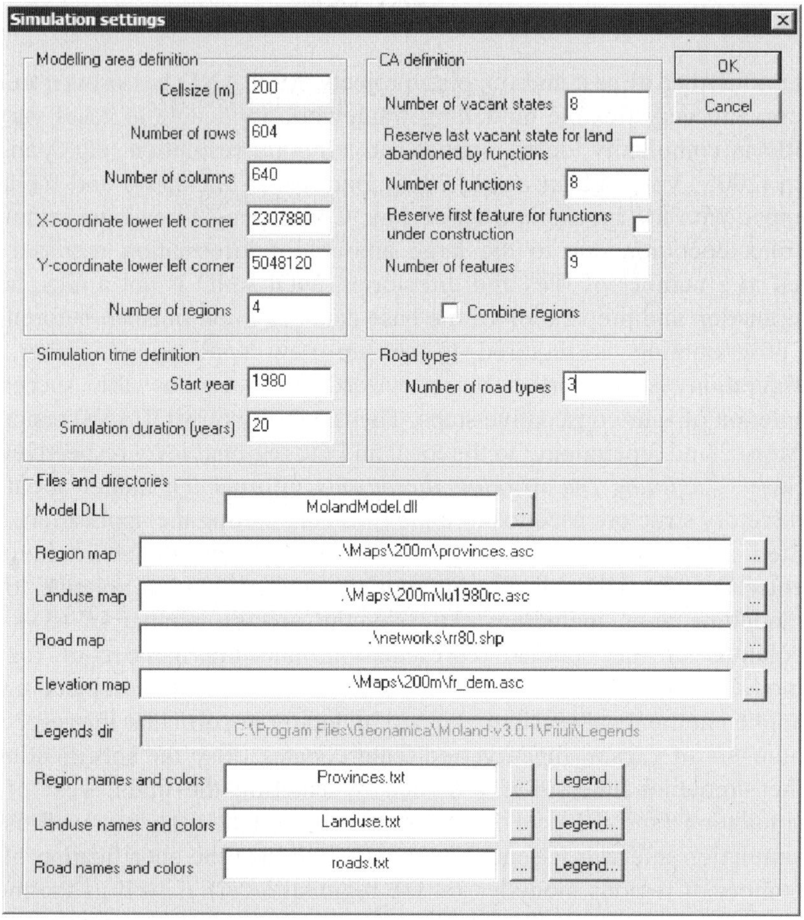

Figure 17-4. Dialogue window to be completed in Step 2 while setting up a new application with the MOLAND modelling framework.

• Step 3: *defining or changing the regional-level characteristics* specifies the model at the regional level. In the case where no regional level is applied, this step is skipped in the set-up procedure. If however, the user has decided to configure an application with a full regional level model, hence has not checked the *Combine regions* check box, has specified a *Number of Regions* larger than one, and has referred to the appropriate .DLL in Step 2, then the *Model configuration* dialogue window (Figure

17-5(1)) will enable him to provide the two essential items required for specifying a regional model.

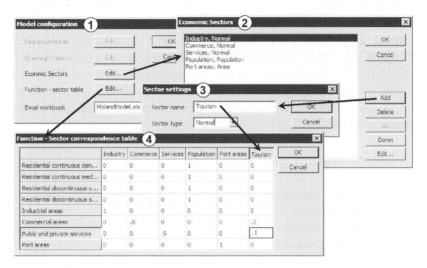

Figure 17-5. Cascade of dialogue windows to be completed in Step 3.

First of all, the sectors need to be specified. By default, the model assumes four sectors: Industry, Commerce, Services, and Population (Figure 17-5(2)). By means of delete and add operators, this list can be reduced to one sector, or expanded beyond four (Figure 17-5(3)). In the model described in Section 3.3, sectors can be of three different types: *Normal*, *Population* or *Area*. Normal and Population refer to sectors that are part of the economic sub-model and the demographic sub-model respectively and are modelled accordingly, while Area refers to a sector which is modelled in a simplified manner on the basis of its spatial extent only. Secondly, the *Function – sector table* needs to be filled-out. By entering proportions in the range 0 to 1 in the cells of the table, the user creates the desired mapping between the sectors of the regional level – in the columns – and the land uses at the local level – in the rows – (Figure 17-5(4)). Row totals should add up to one exactly. At the global level, a new application is automatically primed with a no-growth scenario produced on the basis of cell counts performed on the land-use map for an application without regional model and on the basis of values per sector totalled over the regions for a model with a regional model. These values can be replaced consecutively by actual scenario values obtained from forecasts, stakeholder analysis, *et cetera*.

- Step 4: *finalising the application* involves saving the case and providing the information that can be entered via the user-interface of the application. At set-up, a default cellular automata algorithm is selected replaceable by another one selected from the library, if not by a user-defined one specified by means of the in-built script language. Similarly, at set-up, the state variables and parameters are primed with default values having a neutral effect on the behaviour of the models. These values are replaced interactively by the actual population, production and employment figures, technical parameter values, and cellular automata interaction rules.

4.2   Calibration and validation

Calibration and validation of spatially-dynamic models is not a trivial problem. It is so because in principle every region or cell modelled represents at the least one state variable in the model. Thus, the models consist of tens of thousands, if not millions of coupled dynamic equations resulting in a potentially extremely rich behaviour. However, in the calibration process, the task is precisely to ensure that the model behaves in a realistic manner and is able of generating realistic, observed and/or existing spatial patterns. Typically a historic calibration is carried out in which the model is iteratively tested and tuned to generate a spatial behaviour observed in a past period documented in available maps, time series and other data (Engelen *et al.,* 2005). A historic calibration will require a sufficiently long calibration period, typically some 10 years, so that the underlying processes in the system have time to manifest themselves in a representative manner. Three types of measures are applied. First, and at the highest level of abstraction, the spatial patterns generated are measured and compared with those observed in the actual map. *Fractal* measures (White and Engelen, 1993; 1997) or *landscape structural analysis* measures (McGarical *et al.*, 2002; Barredo *et al.*, 2007) are used. At the next level of abstraction map comparison methods are applied to detect *qualitative* similarities between maps. Such measures allow for small spatial or categorical deviations at the cellular level. Methods utilized are the *fuzzy inference system* developed by Power *et al.* (2001) and the *fuzzy map comparison* method developed by Hagen (2003). Finally, the most strict comparison involves a *cell-by-cell comparison* method and the associated *Kappa statistic* (see, for example, Hagen, 2003; Monserud and Leemans, 1992). The three types of techniques are typically used in a chronological order so that with the advances made in fine-tuning the model, the measures become more strict as well.

The MOLAND modelling framework is equipped with automated procedures for calibrating the regional model, the local model and the linked global-regional-local model. The algorithms use state of the art hybrid optimization methods based on principles of simulated annealing, genetic algorithms, golden section search, and random search methods (Van Loon, 2004). In particular for the cellular automata model, the calibration routine produces the set of interaction rules capable of generating the spatial patterns observed.

Currently, the calibration procedures are available in a prototype version and none could be called 'fully automated' since they still require a substantial user involvement. However, when used in semi-automatic mode, the methods speed up the calibration of both the regional and the local model considerably: a good parameter set is found in a matter of hours, typically one or two days, rather than one or two person-months of hard and tedious work by a specialist. Moreover, when started from an initial parameter set delivered by a specialist or taken from another application, the methods substantially outperform the specialist in the quality of the parameter and rule sets generated (Engelen *et al.*, 2005).

5. STATE VARIABLES AND INDICATORS

From the model description, it will be clear that a MOLAND application is driven by scenarios and other inputs at the global, regional and/or local levels and that it generates output at the regional and the local levels. Typically the model is run for a 30-year period into the future, but shorter or longer time intervals are possible too. Results are calculated and visualized on a yearly basis. At the regional level, the population, as well as the employment and production figures in each economic sector are calculated. At the local level the resulting new land-use map is generated and presented for every simulated year (Figure 17-6).

In addition to these and conforming to the aims of the *Understand* component of MOLAND (Section 1; Lavalle *et al.*, 2002), the model calculates a number of spatial indicators expressing changes in the economic, social or environmental status of the area modelled, its regional and cellular entities. Indicators are interactively and custom designed by the user on the basis of five built-in generic algorithms performing operations on the local output of the model. In fact, an indicator consists of a *generic algorithm*, and its *parameter set*. The latter are specified by means of a dedicated dialogue window and determine the land uses, ancillary maps, threshold values, search radii, weights, *et cetera*, on which the spatial algorithm is to perform its tasks. Within a model and a run, a single algorithm can be used in a different configuration as often as required. For example, the indicators 'urban values at stake in flood prone areas' and 'adverse agricultural activity in landslide prone areas' are calculated by means of the same *disturbance algorithm*.

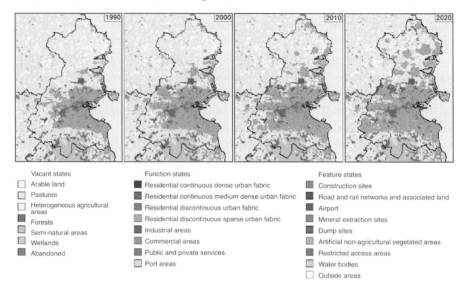

Vacant states
☐ Arable land
☐ Pastures
☐ Heterogeneous agricultural areas
▪ Forests
▪ Semi-natural areas
▪ Wetlands
▪ Abandoned

Function states
■ Residential continuous dense urban fabric
▪ Residential continuous medium dense urban fabric
▪ Residential discontinuous urban fabric
▪ Residential discontinuous sparse urban fabric
▪ Industrial areas
☐ Commercial areas
▪ Public and private services
☐ Port areas

Feature states
▪ Construction sites
▪ Road and rail networks and associated land
▪ Airport
▪ Mineral extraction sites
▪ Dump sites
▪ Artificial non-agricultural vegetated areas
▪ Restricted access areas
☐ Water bodies
☐ Outside areas

Figure 17-6. Snapshots taken from a high growth scenario of the MOLAND application to the Greater Dublin region. (See also Plate 20 in the Colour Plate Section)

All indicators are calculated synchronously with the model and are thus updated yearly. They are presented on a (dynamic) map. Like any other map they can be exported in ArcInfo GRID or IDRISI format. In addition, for every indicator a synthetic value – depending on the algorithm: a sum, weighted sum or average of all cells – is calculated per region modelled and for the whole modelling area. This value is presented as an index value in a time chart. The list of available algorithms consists currently of the following five:

- the *count algorithm* for defining indicators consisting of the ratio of sums of weighted land uses in the vicinity of each cell. This type of indicator can present an image of the supply of, or demand for, certain land uses by other land uses within a specified radius;
- the *cluster algorithm* used to define indicators pinpointing clusters consisting of specific land uses and combinations thereof;
- the *distance algorithm* for defining indicators approximating the smallest distance from a source cell to a cluster of target cells, for all source cells. Each land use can be specified to be either a source, a target, or none.

The minimum size of a target cluster can be specified, as can the measure of distance (Figure 17-7);

- the *distance to map algorithm* for defining indicators approximating the distance from a cell with a certain land use to the nearest target cell, specified on an ancillary binary map. As in the distance algorithm, one can specify the measure of distance. However, the distance is measured to the closest target cell, not the closest target cluster; and

- the *disturbance algorithm* for defining indicators pinpointing (potentially disturbing) land uses and associated activities coinciding with areas of particular interest located on an ancillary map.

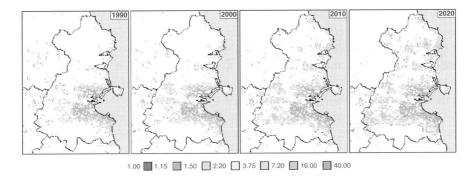

Figure 17-7. The indicator 'access to open space for residential locations' associated with the states shown in Figure 17-6 and based on the 'distance algorithm'. (See also Plate 21 in the Colour Plate Section)

Apart from the above, indicators can be calculated and manipulated as part of a post-processing task, among others, on the basis of spatial output written to file during a simulation. The *Map Comparison Kit* (Visser, 2004) and the *Overlay-Tool* (Engelen *et al.,* 2003), both intimately linked with the MOLAND modelling framework, are dedicated instruments for producing indicators based on crisp or fuzzy map comparison methods, respectively, composite indicators based on the combination and weighing of land use and ancillary maps. Indicator maps, like other output maps, can be animated or stored, opened, viewed and compared by means of the *Map Comparison Kit* enabling the analysis of changes within a single run or in different runs of the model. Animations are available after the simulation as animated GIF files and can be viewed with commercial software packages including: the Windows Picture and Fax Viewer, MS PowerPoint, Internet browsers, or graphical packages like Paint Shop Pro.

6.   DISCUSSION AND CONCLUSION

MOLAND, both the project and the modelling framework with the same name developed in its *Forecast* component, have come a long way in the past seven or eight years. The MOLAND methodology has taken a definite shape and has been applied to a large number of cases of varying complexity by a growing number of users in the EU but also in the mega cities of developing countries. A lot of theoretical work on data acquisition, change detection and indicator development, modelling and calibration has propelled its application by its intended end-users in planning and policy–making institutions. In the course of time, its emphasis has shifted from the fairly theoretical applications on urban areas proper to the regional context and the practical policy problems associated with urban expansion and sprawl, transportation infrastructure and accessibility, the EU enlargement, and most recently, the adverse effects of global change on urbanisation trends and urban life style (Barredo *et al*. 2007). Alongside this, the functionality requirements of MOLAND have gradually shifted from the 'closed and confined model' to the 'flexible and generic modelling framework' enabling the development of models dedicated to the precise case in a scientifically correct and robust manner more rapidly and reliably. The MOLAND modelling framework currently available is equipped with a functionality and flexibility enabling its application with a fraction of the resources required originally. Its user-friendliness facilitates its use in demanding exercises such as interactive and participatory sessions aimed at quantifying, visualising and analysing storylines and scenarios developed by stakeholders. The example in case is the recent PRELUDE project of the European Environment Agency (van Delden *et al.*, 2005; http://scenarios. ewindows.eu.org/reports/fol077184), which is based on an architecture similar to MOLAND.

The scientific achievements of the project have been reported in a growing number of publications. Relative to the use and usability of the MOLAND modelling framework for practical planning and policy making, it is fair to come to the preliminary conclusion that it provides insight in the interconnected nature of different functions, processes and cause effect relations underlying the spatial configuration of urbanised areas. It makes the effects of policy interventions explicit in the specific domain of the user but also in that of its colleagues and counterparts. It enables quick calculation of, and consequently more, alternatives than would otherwise be considered or possible and it enables an objective evaluation of the results generated. However, the model represents a complex reality and the reality proves that it is difficult to keep the system itself from being complicated. Despite the fact that major effort has gone into rendering it as transparent and user-friendly as possible, it will currently only meet the expectations of part of its intended end-users on this subject. Mostly technicians, more familiar with GIS and modelling and having worked their way through the extensive documentation, will be happy with the system as it is. For occasional and non-technical users this is much less the case. That is why training and working sessions involving groups of end-users working on practical problems, facilitated by specialists, are required. Material for the kind of workshops has been compiled and has been used with success among others in Ireland, the Netherlands and Spain. However, it goes without saying that increased user-friendliness of any modelling framework should come with due care and caution for modelling as a scientific discipline.

Many modelling problems remain unsolved in MOLAND. Instruments of the kind depend largely on good quality data. In particular high-resolution current and past land-use maps are required to determine trends, calibrate and validate the model. This is a problem because good land-use maps are still rare. Most often their detail, their number and the choice of the land-use categories are insufficient to be useful. Moreover, definitions and categories change over time and making the land-use maps consistent between years becomes a laborious prerequisite. The MOLAND project is unique in this respect as it compiles the databases required. Moreover, and for all practical purposes, the MOLAND modelling framework should be equipped with a good (meta-) documentation procedure to keep track of the input data used and those generated in the exercises carried out. The modelling framework facilitates handling data files with great ease, yet keeping track of the precise sets used and generated while defining and running scenarios, strategies and alternatives is far from evident.

Also, the match between the land-use categories of the land-use map (at the local level) on the one hand and the activity classes (at the regional and global levels) represented in economic and census tables on the other hand, is not necessarily one to one. In practice, it will often be necessary to go back to the most fine-grained representation of the census data in order to establish a workable match. *Vice versa*, land-use categories are seldom pure and devoted to a single use. For example, a farming cell refers to an economic activity, but it is most often also the residence of farmers and their families. Elements like these require due care when establishing the match between the local and regional level of the MOLAND models. More user support and consistency checking at this level are required.

The calibration and validation of the type of model is far from easy or fast. This is partly due to the limited availability of good quality data, but also to the integrated nature of the model: all linked processes need to be calibrated in isolation and in combination in order to generate reliable results. It has been the experience with similar models (such as Environment Explorer, De Nijs *et al.*, 2001) that they require re-calibration regularly when new or better data become available. Automated or semi-automated calibration procedures are much needed and much wanted to take care of this problem. The MOLAND modelling framework is one of the few equipped with dedicated calibration procedures even though they are still under development and require additional research and development to increase their level of applicability and efficiency.

ACKNOWLEDGEMENTS

MOLAND Version 3 is a result of the EC/JRC contract nr. 21512-2003-12 F1SP ISP NL carried out by RIKS in 2004.

REFERENCES

Allen, P. and Sanglier, M. (1979a) A dynamical model of growth in a central place system, *Geographical Analysis*, 11: 256–272.

Allen, P. and Sanglier, M. (1979b) A dynamic model of a central place system-II, *Geographical Analysis*, 13: 149–164.

Barredo, J.I. and Demicheli, L. (2003) Urban sustainability in developing countries' megacities: modelling and predicting future urban growth in Lagos, *Cities*, 20(5): 297–310.

Barredo, J.I., Demicheli, L., Lavalle, C., Kasanko, M. and McCormick, N. (2004) Modelling future urban scenarios in developing countries: an application case study in Lagos, Nigeria. *Environment and Planning B*, 32: 65–84.

Barredo, J.I., Lavalle, C., Sagris, V. and Engelen, G. (2007) Representing future urban and regional scenarios for flood hazard mitigation, *Annals of Regional Science*, (submitted).

Batty, M. and Xie, Y. (1994) From cells to cities, *Environment and Planning B*, 21: 31–48.

Couclelis, H. (1985) Cellular worlds: a framework for modelling micro-macro dynamics, *Environment and Planning A*, 17: 585–596.

De Nijs, A.C.M., Engelen, G., White, R., van Delden, H. and Uljee, I. (2001) *De LeefOmgevingsVerkenner, Technische Documentatie*, RIVM-Report 408505007/2001, National Institute of Public Health and the Environment, Bilthoven, The Netherlands.

Engelen, G., White, R., Uljee, I. and Drazan, P. (1995) Using cellular automata for integrated modelling of socio-environmental systems, *Environmental Monitoring and Assessment*, 30: 203–214.

Engelen, G., Uljee, I. and van de Ven, K. (2003), WadBOS: Integrating knowledge to support policy-making in the Dutch Wadden Sea, in Geertman, S. and Stillwell, J. (eds) *Planning Support Systems in Practice*, Springer, Berlin, pp. 513–537.

Engelen, G., White, R., Uljee, I., Hagen, A., van Loon, J., van der Meulen, M. and Hurkens, J. (2004) *The Moland Model for Urban and Regional Growth*, Research Institute for Knowledge Systems, Maastricht, The Netherlands.

Engelen, G., Hagen-Zanker, A., de Nijs, A.C.M., Maas, A., van Loon, J., Straatman, B., White, R., van der Meulen, M., and Hurkens, J. (2005) *Kalibratie en Validatie van de LeefOmgevingsVerkenner, Report 2550016006/2005*, Environmental Planning Agency, Bilthoven, The Netherlands.

Frankhauser, P. (1994) *La Fractilite des Structures Urbaines*, Economica, Paris.

Hagen, A. (2003) Fuzzy set approach to assessing similarity of categorical maps, *International Journal of Geographical Information Science*, 17(3): 235–249.

Lavalle, C., Demicelli, L., Kasanko, M., McCormich, N., Barredo, J.I., Turchini, M., da Graça Saraiva, M., Nunes da Silva, F., Loupa Ramos, I. and Pinto Monteiro, F. (2002) *Towards an Urban Atlas*, Environmental issue report n° 30, European Environmental Agency, Copenhagen, Denmark.

McGarical, K., Cusham, S.A., Neel, M.C. and Ene, E. (2002) FRAGSTATS: Spatial Pattern Analysis Program for Categorical Maps. Computer software program produced by the authors at the University of Massachusetts, Amherst. Available at: http://www.umass.edu/landeco/research/fragstats/fragstats.html/

Monserud, R.A. and Leemans, R. (1992) Comparing global vegetation maps with the Kappa statistic, *Ecological Modelling*, 62: 275–293.

Power, C., Simms, A. and White, R. (2001) Hierarchical fuzzy pattern matching for the regional comparison of land use maps, *International Journal of Geographical Information Science*, 15(1): 77–100.

Van Loon, J. (2004) *Towards model-specific automatic calibration routines for the LeefOmgevingsVerkenner model*, Thesis report, RIKS, Maastricht, The Netherlands.

Van Delden, H., Engelen, G., Uljee, I., Hagen, A., van der Meulen, M. and Vanhout, R. (2005) *PRELUDE Quantification and Spatial Modelling of Land Use/Land Cover Changes*, RIKS report, Maastricht, The Netherlands.

Van den Steen, I. (2005) *Cartographie, évolution et modélisation de l'utilisation du sol en milieu urbain. Le cas de Bruxelles*, PhD Thesis, Université Libre de Bruxelles, Laboratoire de Géographie Humaine, Brussels, Belgium.

Visser, H. (2004) *The MAP COMPARISON KIT: Methods, Software and Applications*, RIVM-report 550002005, National Institute of Public Health and the Environment, Bilthoven, The Netherlands.

White, R. (1977) Dynamic central place theory: results of a simulation approach, *Geographical Analysis*, 9: 227–243.

White, R. (1978) The simulation of central place dynamics: two sector systems and the rank-size distribution, *Geographical Analysis*, 10: 201–208.

White, R. and Engelen, G. (1993) Cellular automata and fractal urban form: a cellular modelling approach to the evolution of urban land use patterns, *Environment and Planning A*, 25: 1175–1199.

White, R. and Engelen, G. (1997) Cellular automata as the basis of integrated dynamic regional modelling, *Environment and Planning B*, 24: 235–246.

Wilson, A.G. (1974) *Urban and Regional Models in Geography and Planning*, John Wiley and Sons, London.

Chapter 18

DYNAMIC SIMULATION OF LAND-USE CHANGE TRAJECTORIES WITH THE *CLUE*-s MODEL

P.H. Verburg and K.P. Overmars
Department of Environmental Sciences, Wageningen University, The Netherlands

Abstract: The *CLUE* (Conversion of Land Use and its Effects) model is one of the most widely applied models with approximately 30 applications in different regions of the globe focusing on a wide range of land-use change trajectories including agricultural intensification, deforestation, land abandonment and urbanisation. The model is a tool to better understand the processes that determine changes in the spatial pattern of land use and to explore possible future changes in land use at the regional scale. This chapter describes the functioning of the model and illustrates the potential of the model for scenario-based simulation of land-use change trajectories with two case studies, one which is a rural landscape in the eastern part of the Netherlands and one which is a strongly urbanized watershed surrounding Kuala Lumpur in Malaysia.

Key words: Land use; modelling; competition; Achterhoek; allocation; cellular automata; Kuala Lumpur.

1.   INTRODUCTION

The *CLUE* (Conversion of Land Use and its Effects) modelling framework was developed at Wageningen University and first published in 1996 (Veldkamp and Fresco, 1996). The model is now one of the most widely applied models with approximately 30 applications spread over the different regions of the globe addressing a wide range of land-use change trajectories including agricultural intensification, deforestation, land abandonment and urbanisation. The model is a tool to better understand the

E. Koomen et al. (eds.), Modelling Land-Use Change, 321–335.
© 2007 *Springer.*

processes that determine changes in the spatial pattern of land use and explore possible future changes in land use at the regional scale. The methodology links the spatially explicit analysis of relations between land use and its driving factors to a dynamic simulation technique to explore changes in land use under scenario conditions.

Two different versions of the *CLUE* model have been developed to account for differences in the structure of land-use data. Traditionally, land-use data for study areas with a relatively small spatial extent are based on land-use maps or remote sensing images that denote land-use types respectively by homogeneous polygons or classified pixels. This results in a land-use map in which only the dominant land-use type occupies each unit of analysis. The validity of this data representation depends on the spatial and categorical resolution. In a landscape with high spatial variability, e.g. a small scale agricultural landscape with patches of natural vegetation, such a representation is only valid if the data have a high spatial resolution (Moody and Woodcock, 1994; Schmit *et al.,* 2006). If high resolution data are not available or a high spatial resolution is not feasible due to, e.g. a very large extent of the study area, it is better to represent land use in a so-called 'soft-classified' system (Pontius and Cheuk, 2006). In such a system, each spatial unit, either a polygon or pixel, denotes the fraction of the different land-use or land-cover types. The data can be generated through aggregation of fine resolution data or based on a sample of high-resolution information such as agricultural census data.

The *CLUE* model was originally developed to use soft-classified data and the earlier applications were all based on census data for land use (Veldkamp and Fresco, 1996; de Koning *et al.,* 1999; Verburg *et al.,* 1999) while the extent was national (Ecuador, Costa-Rica, China) to supra-national (Central-America) with spatial resolutions varying between 7x7 kilometres to 32x32 kilometres. Another land-use model that uses soft-classified data is the Land Use Scanner, which was specifically developed for the Netherlands (Hilferink and Rietveld, 1999). For the identification of critical areas of land-use change, this data representation is sufficient. However, when the impacts of land-use change on landscape pattern need to be assessed, a high-resolution representation of land use with homogeneous spatial units is more appropriate. This data representation also provides the possibility to visualize the spatial pattern of various land-use types in a single map. Furthermore, for impact assessment, it is often needed to exactly know the characteristics of the location of change, e.g. the erosion potential after a change in land use depends on the location of the land-use change. Therefore, a modified version of the *CLUE* model, the *CLUE-s* model, was developed that is based on high-resolution data in which each pixel only contains one land-use type (Verburg *et al.,* 2002; Verburg and Veldkamp,

2004a; Overmars *et al.,* 2007). This version has mainly been used in sub-national case studies with a local to regional extent and a resolution ranging from 20 to 1,000 metres. However, since computer capacity and data availability have improved considerably during the last decade, it is now also possible to simulate high-resolution land-use changes for larger areas. Examples of applications of the model at the continental scale include the studies for the European Union (Verburg *et al.,* 2006) and Latin America (Wassenaar *et al.,* 2006).

This chapter describes the functioning of the *CLUE-s* model and provides insights into the range of possible applications. The functioning of the model is illustrated based on two newly developed applications of the model in respectively the eastern part of the Netherlands and the Klang-Langat watershed in Malaysia.

2.   MODEL DESCRIPTION

The modelling framework is based on the dynamic simulation of spatial patterns of land-use change in reaction to pre-defined changes in demand for land by different sectors (e.g. agriculture, wood harvesting). Figure 18-1 provides a schematic representation of the functioning of the modelling framework. During each time step, the model determines for each location (grid cell) the most preferred land use based on a combination of the suitability of the location itself and the competitive advantage of the different land-use types, which is a function of the land requirements (or demand). If the most preferred land-use type requires a land-use conversion which is not realistic or not allowed due to spatial policies and restrictions, the next most preferred land-use type is selected. Spatial policies, restrictions and a matrix listing which conversions are possible need to be specified by the user in advance (Verburg and Veldkamp, 2004a). After allocating the preferred land use to all locations in the study area, the aggregate demand for land use is compared with the allocated areas. If the demand is not correctly allocated, the competitive advantage of the different land-use types is modified in such a way that the land uses for which the demand was not met obtain a higher preference. Overrepresented land uses get a lower preference. This procedure is repeated iteratively until the demand equals the allocated area. The demand for the different land-use types needs to be specified by the user in advance. The calculation of the changes in demand is exogenous to the *CLUE-s* modelling framework and can be based on different techniques ranging from simple trend extrapolation to advanced multi-sectoral modelling.

The algorithm of the *CLUE-s* model is applicable to different scales and capable of simulating different types of land-use change trajectories. The actual configuration of the model is dependent on the case study, the dominant land-use change processes and the available information and data. Especially the specification of the location specific preference is different for each case study. The location specific preference is determined by a selection or combination of four different methods (Figure 18-1).

(1) In case information on the (proximate) determinants of land-use change is lacking or if the scale of application is much coarser than the scale of the understanding of the land-use change processes, empirical methods are used. The empirical analysis is used to estimate the contribution of different location characteristics (explanatory variables), such as soil conditions and accessibility, to the suitability of a location for a specific land-use type. The possible determinants are selected based on either theory or knowledge of the study area and the relations are quantified with logit models (Verburg *et al.,* 2004d). Upon a change in the value of one of the determinants (e.g. improved accessibility), the new location preferences can be calculated. The empirical specification of location preference does not necessarily lead to causal relations and assumes that the relations based on the current land-use pattern remain valid during the simulation period.

(2) In cases with more information on the determinants of the location preference of land-use types, the empirical analysis can be replaced by decision rules that reflect the knowledge of the processes. These decision rules should specify the relative preference of the different land-use types for each location as a function of the land-use history, biophysical and/or socio-economic conditions of the location (Overmars *et al.,* 2005).

(3) Land-use data often show spatial autocorrelation (Anselin, 2002; Munroe *et al.,* 2002). This can be a result of trends in the underlying factors e.g. the land tenure structure or spatial processes like agglomeration effects in residential land use and imitation among farmers. Spatial interactions can be represented in the model by the incorporation of a neighbourhood function as a determinant of the location preference. The relation between the location preference and the neighbourhood composition is determined either by (calibrated) decision rules, as in most cellular automata models, or based on statistical analysis (Verburg *et al.,* 2004c). At each time step of the analysis, the location preferences need to be updated to account for changes in land-use structure.

(4) Finally, current land use is in many cases an important determinant of the location preference. Land conversions are often costly, e.g. after establishing a plantation of fruit trees, the owner will not consider to 'move' his plantation to a nearby location due to the high investment costs involved. Other conversions are almost irreversible, e.g. residential area is not likely to

be converted back into agricultural area. Therefore, conversion elasticity is assigned to each land-use type, which increases the preference for locations where this land-use type is found in the current situation. During the simulation, these preferences are updated to account for the simulated changes in land use.

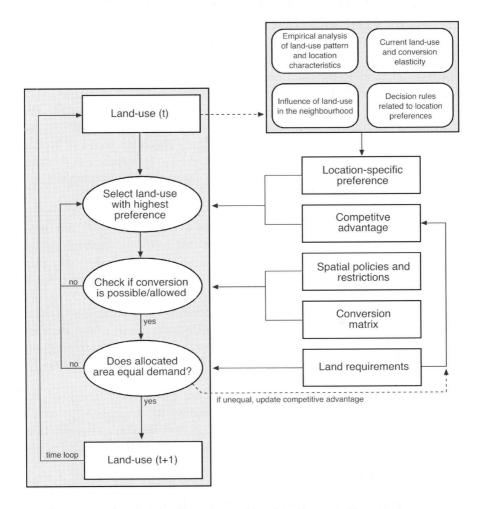

Figure 18-1. Overview of the *CLUE-s* modelling procedure.

The final location specific preference used in the simulations is determined by a mix of these four components based on the preferences of the model user and the available knowledge and data. This mix can be different for each land-use type and each application. This makes the modelling framework very flexible and enables configurations that classify the model as either a (constrained) cellular automata model, an empirical-statistical model or a dynamic simulation model (Lambin *et al.*, 2000).

3.   CASE STUDY 1: THE 'ACHTERHOEK' REGION IN THE NETHERLANDS

This case is a study in the framework of a research project "Sustainable multiple functions in the rural countryside", which is jointly undertaken by WUR (Wageningen University and Research Centre, Netherlands) and INRA (l'Institute National de la Recherche Agronomique, France). The study area comprises four municipalities in 'de Achterhoek' in the Netherlands and covers approximately 420 km^2. The main land-use type in this rural region is dairy farming. Currently, the Government is purchasing areas for nature conservation and subsidizing farmers to use farming regimes favourable for nature conservation.

In this case study, the *CLUE-s* model is used to compare two scenarios. Both scenarios are based on the same quantity of land-use change, but differ with respect to the assumed spatial policies that influence land allocation. The timeframe for the simulations is from 2000 to 2018 with yearly time steps and a spatial resolution of 50x50 metres. Initial land use is presented in Figure 18-2. Several land-use types in the area are considered to have high values for biodiversity and landscape conservation. The areas classified as nature consist of forests and other natural vegetations; management areas are agricultural lands that are managed to increase the natural value; reserves are newly purchased areas which are actively managed to create conditions for the development of natural vegetations.

The land requirements (i.e. demands) are determined based on the objectives of the provincial authorities regarding the improvement of the ecological quality of the area and include an overall decrease in agricultural area and increases in the land-use types nature, management, reserves and residential.

For the land-use types management, reserve, nature and residential, the location characteristics are determined by calculating the location specific preferences for each land-use type based upon a logistic regression with the current land use as dependent and location factors as independent variables. The factors included in the analysis to explain the location preference are the distance to highway, distance to provincial road, distance to cities, distance to creeks and streams, slope, elevation, soil characteristics and ground water level. For the land-use type agriculture, no logistic regression analysis was made, instead the preference of each location for agriculture (mainly grassland in this region) was assumed to be dependent on the production potential for grass derived from a quantitative land evaluation system.

In addition to the derived location preferences, the following neighbourhood functions were included: land-use changes into management and reserve are considered to be more preferred in the neighbourhood of other management, reserve and nature areas. New nature areas are considered to be preferred close to other nature areas while new residential areas will most likely be positively influenced by the quantity of other residential areas in the neighbourhood. The neighbourhood values make up 10% of the total location preference used in the simulations.

The difference between the two scenarios in this study is the incorporation of spatial policies. In scenario 1, no spatial policies were included. The spatial policy used in scenario 2 is based on the EHS (Ecological Main Structure) that aims to develop nature preferably at locations that enable connections between existing natural areas and thus reduce fragmentation of nature. This spatial policy cannot be implemented as a strict rule since it is often implemented as higher efforts to purchase land or higher subsidies for changes in farm management. Therefore, in scenario 2, the location preference for the land-use types management, reserve and nature are increased in the designated EHS areas.

The results of this analysis are presented in Figure 18-2 by maps that only show the areas that changed between 2000 and 2018. The designated EHS areas of scenario 2 are indicated as well. The changes to reserve in scenario 2 are all located within the EHS areas. In area A, conversions to reserve area are dominant in scenario 1, whilst in scenario 2, there is hardly any change in this area. On the other hand, in area B, changes into nature are found in both scenarios. Area C does not show changes in both scenarios. This may indicate that it is relatively difficult and costly to achieve changes in this region. Area D shows changes to more natural land use types in both scenarios. Establishing a corridor in this region is probably much easier as compared to area C.

The results visualize an important issue for policy making concerning the selection of locations to target subsidies and spatial policies. Given the trends in land use, some areas within the designated EHS area will change into reserve or management even without explicit spatial policies or support. Other areas only change when the policies are implemented and some areas will not change anyway. For these areas higher subsidies can be considered. Another option is to target resources to other locations where change is achieved more easily. Herewith, the model can be used to assess, discuss and fine-tune policies.

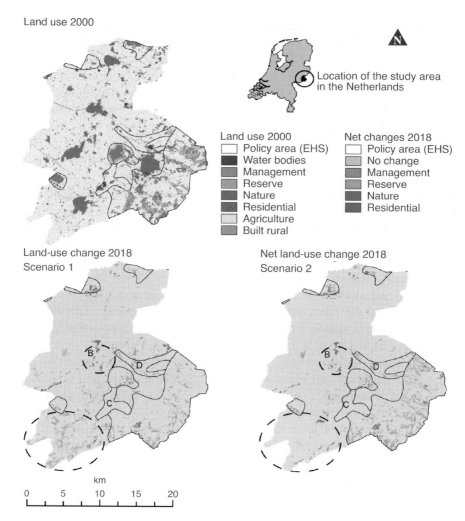

Figure 18-2. Observed land use in the study area in 2000 and land-use change as result of scenario 1 and scenario 2. (See also Plate 22 in the Colour Plate Section)

4.   CASE STUDY 2: THE KLANG-LANGAT WATERSHED IN MALAYSIA

The Klang-Langat Watershed is located in the mid-western part of Peninsular Malaysia and covers eight administrative districts of Selangor, Negeri Sembilan and the Federal Territory of Kuala Lumpur. The Klang-Langat Watershed represents the most highly urbanized region of the country. Kuala Lumpur, the capital city of about 1.5 million people is located at the confluence of Klang and Gombak rivers. The suburbs of Kuala Lumpur extend southward to the coast of Selangor and westward into the Langat river basin. The spillover development of Kuala Lumpur has spread into the surrounding districts of Selangor which have become highly urbanised. The whole region has about 4.2 million inhabitants. Some of the urban developments have been located in abandoned mining areas. The area was once a centre for tin mining activities. The study area has several patches of upland forest, lowland forest and mangrove forest. Besides the built-up areas and forests, the other major land use type in the area is agriculture. The main agricultural crops are oil palm, natural rubber, coffee, cocoa and coconut. Oil palm and rubber estates have been established through forest conversion and currently, many of these estates are in their turn converted into urban, residential, recreational and industrial areas. The Klang-Langat Watershed is also vital in terms of domestic water supply to the most densely populated area in the country.

The *CLUE-s* model application for this region was made to test different frequently used land-use allocation mechanisms in a region with opposing agricultural and urban developments. Eight different land-use types were considered and two different allocation mechanisms for urban land were tested. Both applications related to the period 1990-1999, a period during which the area of urban land expanded very rapidly (Figure 18-3).

Simulation 1 determines the location of change as a function of the neighbouring land uses. Based on observed land-use patterns, a set of neighbourhood rules was derived that define the location specific preferences for urban land. These preferences are updated each year, which makes the model function similar to a constrained cellular automata model.

In simulation 2, another approach is chosen to determine the location specific preferences for urban area. Here a logistic regression analysis based on the location of urban land in 1990 is used to quantify relations between the location of urban land and location characteristics such as accessibility and slope.

The results were quantitatively compared with the observed land-use change during the 1990-1999 period using the multi-resolution method for map comparison (Costanza, 1989; Pontius *et al.*, 2004). The results were also compared with two different reference models: the first assumes a complete persistence of land use at the 1990 situation and the second assumes random allocation of the observed changes. Both simulations with *CLUE-s* performed better than both reference models. Simulation 2 was much better able to predict the spatial pattern of urban development as compared to simulation 1. Obviously, the neighbourhood relations only capture part of the processes that govern urban expansion, while the empirical approach based on location conditions was better able to capture the full complexity of the involved processes. The estimated preferences for urban areas, related to the land-use pattern of 1990, apparently offered a reasonable prediction for the processes over the dynamic period 1990-1999.

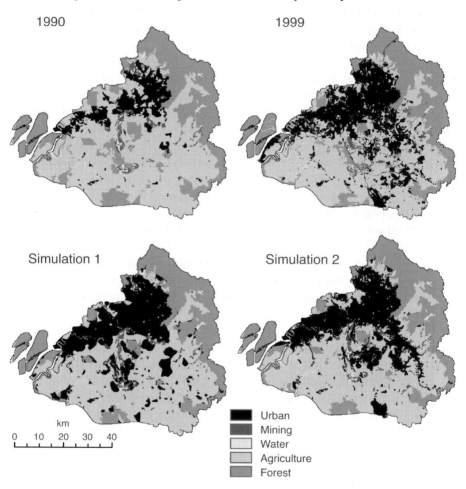

Figure 18-3. Observed land use in the Klang-Langat watershed in 1990 and 1999 and simulated land use for 1999 using two different algorithms.

The case study is an illustration of the use of the model to test different representations of the land-use change processes in a modelling environment and evaluate the outcomes. Such analysis may inform further research on the processes of land-use change and on the techniques needed to represent these processes in models.

5. DISCUSSION

This chapter described one particular tool for land-use change modelling. However, this land-use model is illustrative for a much larger group of land-use change models. Approximately one decade ago, the choice of land-use models was fairly small and most models were specifically developed for a particular case study. Currently, a much larger selection of land-use models is available (Briassoulis, 2000; Agarwal *et al.,* 2001; Verburg *et al.,* 2004b). In spite of conceptual differences, there is a tendency to develop models as frameworks in which the user can specify the representation of the land-use processes considered. The recent version of the *CLUE-s* model presented in this chapter is representative of this trend: the user can decide on the relevance of either empirically derived relations between location factors and land use, neighbourhood rules or the use of location preference maps derived from theoretical or observed processes of land-use change. There has always been a relatively strict subdivision between urban and rural models of land-use change. The modelling procedures used in urban and rural models were distinctly different: urban models were dominantly based on neighbourhood functions such as cellular automata and infrastructure access (White *et al.,* 1997; Clarke and Gaydos, 1998; Agarwal *et al.,* 2001; Dietzel *et al.,* 2005) while rural land-use models primarily used land-quality assessments as the basis of simulation (Veldkamp and Fresco, 1997; Alcamo *et al.,* 1998; Pontius *et al.,* 2001). The possibility to combine different procedures for land allocation within one modelling framework allows for a more integrated assessment of land-use dynamics that can explicitly address the competition between land uses in the urban fringe (Dietzel and Clarke, 2006). The development of such integrated approaches also facilitates the collaboration of urban and rural modellers with different expertise and complementary knowledge on the land-use system.

In many spatially explicit models, such as the *CLUE-s* model, the unit of analysis is an area of land, either a polygon representing a field, plot or census track, or a pixel as part of a raster-based representation. The disadvantage of this 'land-based' approach is the poor match with the agents of land-use change. Individual farmers or plot owners are usually not represented explicitly and the simulations usually do not match with the units of decision making. A rapidly expanding group of models use individual agents as units of simulation (Parker *et al.,* 2003; Bousquet and Le Page, 2004). These so-called multi-agent systems emphasize the decision-making process of the agents and the social organization and landscape in which these individuals are embedded. An agent can represent any level of organization (a herd, a village, an institution), and does not necessarily have to be an individual. A disadvantage of the agent as the basic unit of simulation is the difficulty to adequately represent agent behaviour and to link it to the actual land areas. Well established multi-agent models to predict changes for real landscapes have become available only recently (Huigen, 2004; Castella *et al.,* 2005). Such models are very data demanding and the specification of realistic agent behaviour and diversity is very challenging, especially when it comes to the behaviour of agents at higher levels of organizations such as institutions. The choice of either a pixel-based or an agent-based model is therefore dependent on the objectives of the study and the available information and resources. In many cases, it may even be most appropriate to use different model approaches to study the same region. Comparing the outcomes of such models may lead to a better and more complete understanding of the system dynamics and inform each other (Castella *et al.,* 2007). Agent-based models can explore mechanisms that can, later on, be included in spatial simulation models.

It is important to acknowledge that no single technique or approach can sufficiently describe the different processes at all spatial and temporal scales. Recent validation studies have indicated that most spatial models still contain a high level of uncertainty while multi-agent models sometimes have not been validated at all (Pontius *et al.,* 2004; Pontius *et al.,* 2007). The wide selection of models and modelling approaches that has become available provides the researcher with the opportunity to select the modelling approach that best fits the research questions and characteristics of the study area. The *CLUE-s* model enables the user to test different mechanisms of simulation in one single modelling framework. The model can easily be implemented in new case study areas and is used in policy support studies (Klijn *et al.,* 2005). However, the model is first and foremost a tool to support the analysis of land-use processes and interactions and evaluate different mechanisms of land allocation. Therefore, it is not possible to classify this new generation of 'hybrid' models into a single model category as proposed by Lambin (1997; 2000), Agarwal *et al.* (2001) or Briassoulis (2000). It is certain that *CLUE* and other modelling approaches will see more development and the implementation of new modelling concepts in the future. However, it is important to keep in mind that applications of such modelling frameworks will only be successful if a good balance between transparency of the model and representation of the complexity of the land-use system is achieved.

ACKNOWLEDGEMENTS

The authors would like to thank all that have contributed to the development of the *CLUE-s* model and specifically Alias Mohd Sood for his contribution to the Kuala Lumpur case study.

REFERENCES

Agarwal, C., Green, G.M., Grove, J.M., Evans, T.P. and Schweik, C.M. (2001) *A Review and Assessment of Land Use Change Models. Dynamics of Space, Time, and Human Choice.* Center for the Study of Institutions, Population, and Environmental Change, Indiana University and USDA Forest Service, Bloomington and South Burlington.

Alcamo, J., Leemans, R. and Kreileman, E. (1998) Global Change Scenarios of the 21st Century, Results from the IMAGE 2.1 Model, Elsevier, London.

Anselin, L. (2002) Under the hood: issues in the specification and interpretation of spatial regression models, *Agricultural Economics*, 27: 247–267.

Bousquet, F. and Le Page, C. (2004) Multi-agent simulations and ecosystem management: a review, *Ecological Modelling*, 176: 313–332.

Briassoulis, H. (2000) Analysis of land use change: theoretical and modelling approaches, in Loveridge, S. (ed) *The Web Book of Regional Science*, at http://www.rri.wvu.edu/ regscweb.htm. West Virginia University, Morgantown.

Castella, J.-C., Boissau, S., Trung, T.N. and Quang, D.D. (2005) Agrarian transition and lowland-upland interactions in mountain areas in northern Vietnam: application of a multi-agent simulation model, *Agricultural Systems*, 86: 312–332.

Castella, J.-C., Pheng Kam, S., Dinh Quang, D., Verburg, P.H. and Thai Hoanh, C. (2007) Combining top-down and bottom-up modelling approaches of land use/cover change to support public policies: Application to sustainable management of natural resources in northern Vietnam, *Land Use Policy*, (in press). doi:10.1016/j.landusepol.2005.09.009

Clarke, K.C. and Gaydos, L.J. (1998) Loose-coupling a cellular automaton model and GIS: long-term urban growth prediction for San Francisco and Washington/Baltimore, *International Journal of Geographical Information Science*, 12: 699–714.

Costanza, R. (1989) Model goodness of fit: a multiple resolution procedure, *Ecological Modelling*, 47: 199–215.

de Koning, G.H.J., Verburg, P.H., Veldkamp, A. and Fresco, L.O. (1999) Multi-scale modelling of land use change dynamics for Ecuador, *Agricultural Systems*, 61: 77–93.

Dietzel, C., Herold, M., Hemphill, J.J. and Clarke, K.C. (2005) Spatio-temporal dynamics in California's Central Valley: empirical links to urban theory, *International Journal of Geographical Information Science*, 19: 175–195.

Dietzel, C. and Clarke, K. (2006) The effect of disaggregating land use categories in cellular automata during model calibration and forecasting, *Computers, Environment and Urban Systems*, 30: 78–101.

Hilferink, M. and Rietveld, P. (1999) LAND USE SCANNER: An integrated GIS based model for long term projections of land use in urban and rural areas, *Journal of Geographical Systems*, 1: 155–177.

Huigen, M.G.A. (2004) First principles of the MameLuke multi-actor modelling framework for land use change, illustrated with a Philippine case study, *Journal of Environmental Management*, 72: 5–21.

Klijn, J.A., Vullings, L.A.E., van de Berg, M., van Meijl, H., van Lammeren, R., van Rheenen, T., Eickhout, B., Veldkamp, A., Verburg, P.H. and Westhoek, H. (2005) EURURALIS 1.0: A scenario study on Europe's Rural Areas to support policy discussion. Background document. Alterra report 1196, Wageningen University and Research Centre/Environmental Assessment Agency (RIVM).

Lambin, E.F. (1997) Modelling and monitoring land-cover change processes in tropical regions, *Progress in Physical Geography*, 21: 375–393.

Lambin, E.F., Rounsevell, M.D.A. and Geist, H.J. (2000) Are agricultural land-use models able to predict changes in land-use intensity? *Agriculture, Ecosystems and Environment*, 82: 321–331.

Moody, A. and Woodcock, C.E. (1994) Scale-dependent errors in the estimation of land-cover proportions: implications for global land-cover data sets, *Photogrammetric Engineering & Remote Sensing*, 60: 585–594.

Munroe, D.K., Southworth, J. and Tucker, C.M. (2002) The dynamics of land-cover change in western Honduras: exploring spatial and temporal complexity, *Agricultural Economics*, 27: 355–369.

Overmars, K.P., Verburg, P.H. and Veldkamp, A. (2007) Comparison of a deductive and an inductive approach to specify land suitability in a spatially explicit land use model, *Land Use Policy*, (in press). doi:10.1016/j.landusepol.2005.09.008

Parker, D.C., Manson, S.M., Janssen, M.A., Hoffman, M. and Deadman, P. (2003) Multi-agent systems for the simulation of land-use and land-cover change: a review, *Annals of the Association of American Geographers*, 93: 314–337.

Pontius, R.G., Cornell, J.D. and Hall, C.A.S. (2001) Modelling the spatial pattern of land-use change with GEOMOD2: application and validation for Costa Rica, *Agriculture, Ecosystems and Environment*, 85: 191–203.

Pontius, R.G., Huffaker, D. and Denman, K. (2004) Useful techniques of validation for spatially explicit land-change models, *Ecological Modelling*, 179: 445–461.

Pontius, R.G., Boersma, W., Castella, J.-C., Clarke, K., de Nijs, T., Dietzel, C., Duan, Z., Fotsing, E., Goldstein, N., Kok, K., Koomen, E., Lippitt, C.D., McConnell, W., Pijanowski, B., Pithadia, S., Sood, A.M., Sweeney, S., Trung, T.N. and Verburg, P.H. (2007) Comparing the input, output, and validation maps for several models of land change, *Annals of Regional Science*, (in press).

Pontius, R.G. and Cheuk, M.L. (2006) A generalized cross-tabulation matrix to compare soft-classified maps at multiple resolutions, *International Journal of Geographical Information Science*, 20: 1–30.

Schmit, C., Rounsevell, M.D.A. and La Jeunesse, I., (2006). The limitations of spatial land use data in environmental analysis, *Environmental Science and Policy*, 9(2): 174–188.

Veldkamp, A. and Fresco, L.O. (1996) CLUE-CR: an integrated multi-scale model to simulate land use change scenarios in Costa Rica, *Ecological Modelling*, 91: 231–248.

Veldkamp, A. and Fresco, L.O. (1997) Exploring land use scenarios, an alternative approach based on actual land use, *Agricultural Systems*, 55: 1–17.

Verburg, P.H., Veldkamp, A. and Fresco, L.O. (1999) Simulation of changes in the spatial pattern of land use in China, *Applied Geography*, 19: 211–233.

Verburg, P.H., Soepboer, W., Limpiada, R., Espaldon, M.V.O., Sharifa, M. and Veldkamp, A. (2002) Land use change modelling at the regional scale: the CLUE-S model, *Environmental Management*, 30: 391–405.

Verburg, P.H. and Veldkamp, A. (2004a) Projecting land use transitions at forest fringes in the Philippines at two spatial scales, *Landscape Ecology*, 19: 77–98.

Verburg, P.H., Schot, P., Dijst, M. and Veldkamp, A. (2004b) Land use change modelling: current practice and research priorities, *Geojournal*, 61: 309–324.

Verburg, P.H., de Nijs, T.C.M., Ritsema van Eck, J., Visser, H. and de Jong, K. (2004c) A method to analyse neighbourhood characteristics of land use patterns, *Computers, Environment and Urban Systems*, 28: 667–690.

Verburg, P.H., Ritsema van Eck, J., de Nijs, T., Dijst, M.J. and Schot, P. (2004d) Determinants of land use change patterns in the Netherlands, *Environment and Planning B*, 31: 125–150.

Verburg, P.H., Schulp, C.J.E., Witte, N. and Veldkamp, A. (2006) Downscaling of land use change scenarios to assess the dynamics of European landscapes, *Agriculture, Ecosystems & Environment*, 114(1): 39–56.

Wassenaar, T., Gerber, P., Verburg., P.H., Ibrahim, M. and Steinfeld, H. (2007) Projecting land-use changes in the neotropics: the geography of pasture expansion into forest, *Global Environmental Change*, (in press). doi:10.1016/j.gloenvcha.2006.03.007

White, R., Engelen, G. and Uijee, I. (1997) The use of constrained cellular automata for high-resolution modelling of urban land-use dynamics, *Environment and Planning B*, 24: 323–343.

PART VI: LAND-USE SIMULATION FOR POLICY ANALYSIS

Chapter 19

BEYOND GROWTH? DECLINE OF THE URBAN FABRIC IN EASTERN GERMANY

A spatially explicit modelling approach to predict residential vacancy and demolition priorities

D. Haase, A. Holzkämper and R. Seppelt
Department of Applied Landscape Ecology, Centre for Environmental Research (UFZ), Leipzig, Germany

Abstract: Urban growth has been replaced by decline of the urban fabric in many parts of Europe. Reasons for this shrinkage are to be found in processes of demographic change and migration in city regions. This chapter presents relevant indicators and a rule-based modelling approach to residential change and building demolition in Eastern Germany. The first part will focus on the research objectives and briefly discusses the urban decline phenomenon and the need for a new modelling approach in order to understand the current shifts in urban development. Besides this, relevant predictor variables for identifying spatial shrinkage and residential vacancy are discussed. Finally, their integration into a GIS-based spatially explicit model completes the chapter.

Key words: Urban shrinkage; demographic change; residential vacancy; demolition; rule-based cellular model.

1. INTRODUCTION

Within the last ten years, a shift in land-use development due to a rapid demographic change has become more and more obvious in many European countries (CEC, 1997; Cloet, 2003; Lutz, 2001). In recent studies of population development, an average population decrease of 11% within the EU-15 countries by 2050 is predicted, with extreme values of decline in Italy (25%), Spain, Switzerland and Austria (20%). For the enlarged EU-25, the average estimated decrease is 18% (Kröhnert *et al.*, 2004). This general

E. Koomen et al. (eds.), Modelling Land-Use Change, 339–353.
© 2007 *Springer.*

trend is highly correlated with continuously decreasing birth rates (currently 1.4 children/woman) and significant changes in the age distribution of populations. In city regions in particular, internal urban migration has taken place towards the urban periphery (residential suburbanisation) while simultaneously, inner city areas have become subject to processes of (partly extreme) abandonment (e.g. Heilig, 2002).

At the moment, we observe diverging processes of growth and decline in many Central and Eastern European city regions. Whereas in the 1980s and 1990s urban growth and suburban development occurred in these cities, partly accelerated by the demise of the socialist system, today they are faced with a general decline of population as well as an increase in life expectancy due to 'ageing' (Antrop, 2004; Cloet, 2003; Lutz, 2001). Thus, population re- and de-concentration processes and related pressures on the current land-use pattern determine the state of the environment and drive future land-use development strategies in particular in urban areas (Deutsch *et al.*, 2003; Ekins *et al.*, 2003; Haase and Nuissl, 2005). Figure 19-1 (Plate 23) exemplifies this with a projection of these diverging trends for Germany in 2020. The red areas will have further population growth; blue areas will have population decline.

Figure 19-1. Spatial re-concentration (blue) and de-concentration (red) trends up to 2020 in German regions. (See also Plate 23 in the Colour Plate Section)

Urban shrinkage, characterized by phenomena such as reconstruction, deconstruction or even demolition, is a process of marginalization of urban areas (cities and regions), where selective upgrading of the fabric and the infrastructure interact with de-industrialisation, massive out-migration, ageing of the population and decreasing birth rates (Cloet, 2003; Lutz, 2001). These processes are very relevant for spatial modelling because they may lead to high inner city discrepancies between growing and shrinking residential areas (with a strong gradient from peri-urban to inner city areas) with very different population densities.

In order to analyse land conversion and decline of the urban fabric in cities caused by the processes described above, models can be used as innovative tools to support spatial urban planning for sustainable development. Frequently used approaches in urban modelling are agent-based models (Miller *et al.*, 2004; Waddell, 2002), logit models of discrete choice (Landis and Zhang, 1998a; 1998b) and complex cellular automata models (Silva and Clarke, 2002; Clarke *et al.*, 1997; White *et al.*, 1997; Wu and Webster, 1998; Wu, 1998). Urban models that deal with interactions between urban land-use change and its socioeconomic driving forces are mainly implemented as agent-based models. These models often incorporate discrete choice theory (Ben-Akiva and Lerman, 1985; Horowitz *et al.*, 1986). Most model applications however deal with urban growth due to its present topicality, whereas the process of urban shrinkage is of growing importance in urban Europe today (Haase and Steinführer, 2005).

This chapter discusses a conceptual framework that analyses the spatial phenomenon of urban shrinkage and subsequently focuses on the related demolition of parts of the urban fabric. A set of predictor variables (social and spatial indices) is selected and incorporated into a rule-based modelling approach to model demolition priorities in a socialist prefabricated housing estate in Leipzig-Grünau, Germany. The findings of the model are then compared with an existing overall planning concept of the 'compact city' in the study area.

2. URBAN SHRINKAGE

To improve the knowledge of current urban shrinkage processes and to support urban spatial planning, the allocation of probable shrinkage and vacancy needs to be investigated. The phenomenon of shrinkage in residential areas in European cities is far from new: it has been noticed and already described (even if mostly theoretically) since the 1950s and 1960s (Couch *et al.*, 2005; van den Berg, 1982). In some European cities – former 'boomtowns' – that are now in decline, research on urban shrinkage has begun to take place. In most cases, this research focuses on population decline and a decrease of total population density in the central parts of the city based on socio-economic statistics. In most of the cases, references are made to urban structures (structural types) such as single family houses, prefabricated housing estates, historical districts, commercial sites as well as (old) industrial sites.

Couch *et al.* (2005) compare Liverpool in Great Britain, and Leipzig in Germany, as two cities undergoing a profound functional transformation from former industrial metropolitan agglomerations to service economy-based cities. Another study of Ivanovo in Russia examines a city in post-socialist economic transition in the aftermath of 'Perestroika' (Sitar and Sverdlov, 2004). Booza *et al.* (2004) focus on Detroit as an American metropolis now experiencing shrinkage. Dura-Guimera (2003) investigates the processes of urban deconcentration and, simultaneously, the dynamics of urban sprawl, including social processes in the 'Barcelona Metropolitan Area' in Catalonia. From recent statistics on population development, in- and out-migration with respect to the urban development of Mediterranean cities in Europe, Dura-Guimera describes population decline in the central urban area (city centre), population increase in the urban periphery and population expansion of the dispersed city. The process is described as the phenomenon of urban 'perforation' in Germany.

Until recently, shrinkage was tackled specifically as a problem of restructuring the old-industrialised areas, neighbourhoods and boroughs of a city in the sense of being part of regeneration processes. On the contrary, today, we find accelerating residential vacancies in housing estates in inner city areas and in (socialist) prefabricated housing estates. In spite of the urgency to tackle this problem and to improve spatial planning policies towards acknowledging non-growth and shrinkage as current urban phenomena in many parts of Europe, the paradigm of growth is still on the agenda (Müller and Siedentop, 2004). Urban modelling is also very restricted to growth and has not embraced the shift in core areas of urban dynamic development (Antrop, 2004; Haase and Magnucki, 2004). However, the effects of shrinkage will have an enormous influence on the development of cities in the near future. As mentioned above, population decline will lead to decreasing demand in the residential market. Related economic shrinkage processes due to declining purchasing power and low investment rates in urban regions will have enormous consequences on the attractiveness of places for new economic investments as well as for in-migration. A comparison of the regional distribution and the variance of the social variables such as population growth rate, net migration and fertility for all German cities with more than 200,000 inhabitants, learns that high residential vacancy rates are found in Eastern Germany (Halle, Chemnitz, Leipzig) and the old-industrialised Ruhr area (Gelsenkirchen, Krefeld, Dortmund). These cities are characterized by a population decline, caused by strong out-migration (5-25% per year). The fertility rates are approximately the same for all big German cities (around 1%).

For German cities, the current situation seems to be rather paradoxical: on the one hand, available data indicates dynamic suburban land-use growth (in single and semi-detached house settlements, new 'housing parks') with adjacent construction activities (trade and industry) in the urban fringe although population declines. On the other hand, there is an increasing process of shrinkage and perforation in the form of residential vacancies in the inner city areas although re-population seems to be a new trend (Figure 19-2). Here, mainly the old industrialised areas and the urban centres are affected. Previous study (Haase and Magnucki, 2004) has shown that the shrinkage phenomenon in the form of residential vacancies is strongly related to socio-demographic variables such as high unemployment rates or reduced (low) household income leading to a total net out-migration, especially in the inner city.

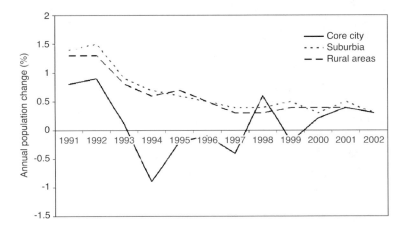

Figure 19-2. Average population change along the rural-urban gradient in German cities with more than 200,000 inhabitants, 1995-2000.
Source: Modified after Müller and Siedentop (2004); INKAR (2003)

3.   MODEL DEVELOPMENT

The model for predicting residential vacancies and related demolition priorities is based on a selected set of variables that have been incorporated into a raster-based modelling approach. It is specifically developed to model the urban development of the Leipzig-Grünau region. This section starts with a description of the demographic developments in this region. We then discuss the selection of relevant predictor variables and describe the model concept and implementation.

3.1   Study area

At present, there are over 55,000 empty flats in the administrative area of Leipzig. The area of Leipzig-Grünau, one of the largest prefabricated socialist residential areas of the former GDR, is heavily affected by urban shrinkage and population decline (Figure 19-3). Moreover, Grünau belongs to those areas in Leipzig where strategic demolition programmes have been carried out and others are planned. The district mainly consists of prefabricated housing estates, two single and semi-detached housing estates, recreational and trade areas (Haase and Nuissl, 2006). The prefabricated housing estates (or Wohn-Komplexen, WK, in German) can be divided into large polygonal blocks in the heart and at the periphery of Leipzig-Grünau (WK 4, 5.2, 7 and 8). Other housing estates (WK 1, 2, 5.1) are dominated by the so-called 'point' skyscrapers with 18 floors and a quadratic ground plan. The size of the housing estates ranges from 1,300 to 9,200 building units (apartments).

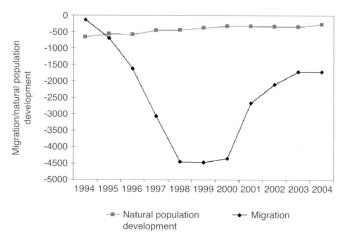

Figure 19-3. Demography of Leipzig-Grünau, 1994-2004.
Source: City of Leipzig (2004a; 2004b)

After German re-unification, Leipzig-Grünau suffered from an extreme population decline due to a negative natural population balance and most of all, an enormous out-migration in the late 1990s. This led to a high residential vacancy share and net out-migration concentrated in the peripheral and central parts (WK 4, 5.2, 7 and 8). This extreme decline process only stagnated in the last 2-3 years. It is still unclear in what direction the large residential area of Leipzig-Grünau will further develop. However, demolition and de-construction will characterise the face of this part of the city for the coming decades. Grünau serves as a study area for micro-level investigations of urban housing and restructuring by many institutions and spatially explicit social data are therefore expected to become available for calibrating the chosen variables utilised in the model.

3.2 Selection of predictor variables

Based on a broad literature review, social and spatial predictor variables for vacancy and urban shrinkage were selected and discussed with social scientists from the Centre for Environmental Research carrying out social empirical work in Leipzig-Grünau (Kabisch, 2005; Kabisch *et al.,* 2004). Strategic planning of demolition in Eastern German cities is mostly based on socioeconomic indices which result from the long-term monitoring systems of each city. In Dresden, the capital of Saxony, for example, the need for action in the form of demolition is assessed by using variables such as residential vacancy, structure type, share of unemployed people, share of welfare recipients and share of inhabitants older than 60 years (City of Dresden, 2003). Following this input from the discussions, the variables listed in Table 19-1 were considered as predictor variables.

Correlations between the selected variables and share of residential vacancy and correlations among variables were tested to select the set of statistically significant variables for the model. The correlation coefficients in Table 19-1 give an idea of how well vacancy and shrinkage can be explained by social features. Based on these outcomes, the variables 'out-migration' and 'share of population above 65 years' were selected, as they have the strongest correlations with vacancy and are not correlated to each other. Unemployment data as well as data on social welfare recipients are seen as relevant, but are not included as the basic idea of the model approach is to reduce the total number of predictor variables to about five.

Based on expert knowledge three additional variables were selected. These include the two spatial variables 'distance to urban sub-centre' and 'adjacent open areas' to represent the hypothesis that proximity to local facilities and green spaces is preferred. Furthermore the variable 'urban structure type' was included, following expert interviews that stated the importance of incorporating housing preferences and property rights. The complete set of selected variables thus consists of: out-migration (%), share of people above 65 years (%), distance to urban sub-centers (metres), adjacent open areas (%) and urban structural type.

Data for these variables are available on different spatial scales. Urban land-use data, including 30 categories, was derived from a land-use map 1:25,000 (Haase and Magnucki, 2004) with a resolution of five metres. This map was produced for the planning authority and provides detailed information about the urban structural pattern. Social data (e.g. migration, age groups) were derived from the social report (City of Leipzig, 2003) and were only available for each borough.

For the variable 'distance to urban sub-centre', the mean distance to the nearest sub-centre with an administrative function (municipal offices, post-office, shopping malls) is calculated for each grid cell five. The variable 'proportion of grid cell edges not adjacent to open area' is calculated by dividing the number of cell edges that are not adjacent to an open area by four, the total number of cell edges. Both variables are normalized to a range of values between 0 and 100. The 'structure type' variable is based on the urban land-use map where each residential land use was assigned a value between 0 and 100 according to its popularity and property status, where: '0' means no demolition possible due to private property; '50' means demolition not preferred because of good quality of houses and preferred flat properties; '100' means demolition possible in case of vacancy. Reconstructed old built-up areas and villas are supposed to have a better image than GDR-time prefabricated housing estates. It is assumed that residential types with higher values are less preferred than those with lower values.

Table 19-1. Correlations between vacancy and potential social predictor variables

Variable	R^2 to vacancy
Out-migration	0.7*
Foreigners	0.4
% married people	- 0.6
Unemployment	0.6*
% car owner	- 0.6*
Age group >65	0.8**
Age group <15	0.1
Social welfare recipients	0.6*

** 1% significance level % * 5% significance level (Pearson's r^2)

3.3   Model concept and implementation

The *Spatially Explicit Landscape Event Simulator (SELES)* modelling environment (Fall, 2000) was used to build our urban development model. The study area is defined in the model as a regular grid of 5x5 metre cells consisting of 1,500 rows and 3,000 columns. The model is based on the rasterised land-use map of the study area of which seven residential land-use types are selected for the simulation of demolition priority. Each residential grid cell is identified by a unique ID and holds information on its land use and on the local predictor variables. In the static expert-based and rule-based modelling approach, demolition priorities are given to these cells, based on the weighted sum of all the normalized predictor variables (see Eq. 1). Thus, we assume a linear relationship between predictor variables and the demolition probability. Figure 19-4 presents an overview of the model concept.

$$D(z) = \sum_{k=1}^{5} w_k \cdot I_k(z) \quad \text{with} \quad \sum_{k=1}^{5} w_k = 1 \tag{1}$$

where:

$D(z)$ is the demolition priority of gridcell z;

w_k is the weight for predictor variable k;

$I_k(z)$ is predictor variable k at location z;

$I_{[k=1]}$ is out-migration;

$I_{[k=2]}$ is the share of population above 65 years (%);

$I_{[k=3]}$ is the mean distance to urban sub-centre (m);

$I_{[k=4]}$ is the share of grid cell adjacent to open area (%); and

$I_{[k=5]}$ is the urban structural type.

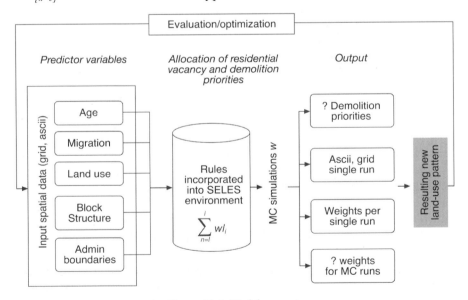

Figure 19-4. Model concept.

Demolition is allocated to the grid cells with the highest priorities for demolition until a given proportion of residential area (D_{max} %) is reached.

Here, we follow the overall concepts of the municipality of Leipzig (agencies for spatial urban planning and domestic construction) and the Federal State of Saxony, where a final demolition rate of $D_{max} = 10\%$ (until 2005) and $D_{max} = 30\%$ (until 2009) is planned based on a recent demographic and migration prognosis.

4.   RESULTS

The spatially explicit model was implemented first to test the applicability of verbally formulated reference scenarios stated by the Department of Urban Planning in Leipzig. Urban scenarios are often based on urban planning related storylines (or narrative descriptions) of a range of plausible, alternative future options within an exploratory framework (Alcamo, 2001). As one alternative pathway of urban development a reference scenario of urban shrinkage and demolition, called 'Maintenance of the urban city centre', was formulated. In this conceptual model of the compact city, the centre of the city will be preserved as an urban core to maintain urban life in compact structures. Demolition is supposed to be concentrated at the periphery. A scenario is built based on Eq. (2).

$$D = D_{CC} \qquad\qquad\qquad\qquad\qquad (2)$$

where:

D is the demolition priority; and
D_{CC} is the mean distance to city centre.

As part of a sensitivity analysis, 1,000 Monte Carlo (MC) simulations were run with randomly varying variable weightings. Integer values between 0 and 10 were selected as reasonable and differentiable weights, similar to the way an urban planner would value them. The effect of the variable weightings largely depends on the spatial distribution of the predictor variables. If all variables are equally weighted, demolition sites are allocated with respect to all variables in equal measure. The effects of all predictor variables are additive. The 1,000 MC runs were used to examine if and under which conditions (described through variable weightings) the overall concept 'maintenance of urban city centre' can be realized according to the local situation. Therefore, the mean distances to city centre from the demolished grid cells were listed for each MC run. The distributions of these values in each run were then compared to the distribution of values derived from the reference scenario of the 'compact city'.

The first results of the 1,000 Monte Carlo runs show that none of the runs produces a demolition pattern similar to the reference scenario of the 'compact city'. Compared to the reference scenario, the distances to the city centre from the demolished grid cells are considerably lower for all 1,000 runs (Figure 19-5). The median of all 1,000 mean distances from the demolished grid cells to the city centre is 7.7 kilometres as compared to 8.4 kilometres for the reference scenario. Each dot in Figure 19-5 on the right represents a single grid cell that is supposedly demolished, the dot on the left indicates the reference scenario of the alternative pathway described in Eq. (2). The main differences in variable weightings between those runs lie in the weights for out-migration, distance to sub-centres and urban structure type. In general, all runs are quite similar and demolition is mostly allocated in the central and the north-western part of Leipzig-Grünau (Figure 19-6 at right), where both 'share of inhabitants above 65 years' and the 'out-migration potential' are high.

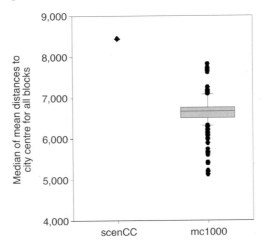

Figure 19-5. Median of mean distances to the city centre from all demolished grid cells of 1,000 MC runs (right) compared to the 'Maintenance of the city centre' scenario (left).

To indicate the validity of the simulation results we calculated a high demolition priority probability per housing estate (WK) for which we divided the number of locations where more than half of the 1000 MC runs allocated demolition by the total number of housing units per WK. This probability was then compared with the observed residential vacancy rate in 2003 and the share of individual building units that were actually demolished in 2004 or planned to be demolished in 2005 (recordings according to the Department of Urban Planning of Leipzig). Figure 19-7 shows that the WK's with a relatively high demolition priority coincide partly with high observed vacancy rates and share of demolished units. This correlation is also indicated by the correlation coefficients of the high demolition priority probability with existing residential vacancy in 2003 ($R^2 = 0.39$) as well as the share of actually demolished housing units ($R^2 = 0.54$).

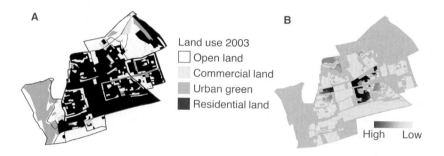

Figure 19-6 Land use 2003 (A) and spatial distribution of demolition priorities within 1,000 MC runs (B). (See also Plate 24 in the Colour Plate Section)

The analysis proves that residential vacancy and demolition in large cities cannot be seen as a single spatial problem of the urban prefabricated periphery (mainly consisting of housing estates of the socialist period). In Leipzig-Grünau, mainly central parts of the residential area are concerned. Thus it can be concluded, that the overall concept of the 'compact city' covers only some specific sides of the reality of vacancy in the investigation area of Leipzig. The low variations of the median distance values of the demolished grid cells with regard to the variable weightings could be explained by the low spatial resolution of the social predictor variables the model is based on. A higher variability of the social data due to an enlarged investigation area or more detailed information on out-migration and share of age-groups in Leipzig-Grünau would, most likely, lead to more differentiated simulation results.

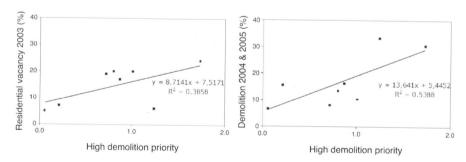

Figure 19-7. High demolition priority probability per housing estate (WK1 – WK8) compared to (A) residential vacancy rate in 2003 (%) and (B) share of actual (2004) and planned (2005) demolished building units (%).

5.   CONCLUSION

The first model results of simulated demolition priorities indicate that the overall concept of 'maintenance of city centre' is far from becoming reality in the chosen study area. None of the combinations of the identified predictor variables result in the envisaged compact city pattern. This conclusion is supported by observed and planned demolition in the study site. On the other hand, the apparent correlations of the MC analysis give an idea of how Grünau could develop in the next 5-10 years, as these correspond with the planned demolition within the housing complexes at the central and (north) western part of the city. Another general aspect which arose in the analysis is that deriving general conclusions for the whole city is not feasible, probably because the selected study area of Leipzig-Grünau was too small and the applied social statistical data were not detailed enough.

To improve the promising model results for Leipzig-Grünau, information from a housing estate specific social survey will be included to allow for a spatially more differentiated description of residential vacancy. The assumption of linear relationships between the predictor variables and demolition priority can be altered, as it is possible that there are important interactions between the variables. Furthermore, interactions and feedback need to be incorporated into the model. Dynamics can be incorporated by making demolition priorities depend spatial configuration of housing in the previous time step. Underlying population dynamics and the household pattern and behaviour (housing preferences, migration) can then be considered as dynamic input variables that change over time. The model can also be improved by incorporating more detailed social and demographic data at district level. For the 'city level' ($n_{district} = 63$), the annual statistical reports of Leipzig are supposed to be a sufficient database. Thus, it is proposed to apply the model to the whole city of Leipzig including the old, prefabricated and the new built-up housing estates. In addition, model transferability to other cities dealing with similar phenomena will also be tested.

ACKNOWLEDGEMENTS

We would like to thank Martha Doehler-Bezadi, Annegret Haase and Sigrun Kabisch for the stimulating discussions and kind cooperation.

REFERENCES

Alcamo, J. (2001) Scenarios as tools for international environmental assessment, European Environment Agency, Environmental Issue Report No. 24, European Environment Agency, Kongens Nytorv 6, DK-1050, Copenhagen, Denmark.

Antrop, M. (2004) Landscape change and the urbanisation process in Europe, *Landscape and Urban Planning*, 67: 9–26.

Ben-Akiva, M.E. and Lerman, S.R. (1985) *Discrete Choice Analysis: Theory and Application to Travel Demand*, MIT Press, Cambridge, MA.

Booza, J., Hagemann, A., Metzger, K. and Müller, N. (2004) Statistical data: Detroit, in *Shrinking Cities. A Project Initiated by the Federal Cultural Foundation, Germany in Cooperation with the Gallery for Contemporary Art Leipzig*, Bauhaus Foundation Dessau and the journal Archplus, Vol. 3 (Detroit), pp. 6–11.

CEC (1997) *Evolution Démographique Récente en Europe*, Strasbourg.

City of Dresden (2003) *Concept on Urban Development*, Report.

City of Leipzig (2003) *Social Report*, Agency for Statistics and Elections Leipzig.

City of Leipzig (2004a) *Statistical Report Leipzig* [Ortsteilkatalog der Stadt Leipzig 2004], Agency for Statistics and Elections Leipzig.

City of Leipzig (2004b) *Report of Grünau*.

Clarke, K.C., Hoppen, S. and Gaydos, L. (1997) A self-modifying cellular automaton model of historical urbanisation in the San Francisco Bay area, *Environment and Planning B: Planning and Design*, 24: 247–261.

Cloet, R. (2003) Population changes 1950–2050 in Europe and North America, Population Statstics.doc 3-03, 1–11.

Couch, C., Nuissl, H., Karecha, J. and Rink, D. (2005) Decline and sprawl; an evolving type of urban development, *European Planning Studies* (in print).

Deutsch, L., Folke, C. and Skånberg, K. (2003) The critical natural capital of ecosystem performance as insurance for human well-being, *Ecological Economics*, 44: 205–217.

Dura-Guimera, A. (2003) Population deconcentration and social restructuring in Barcelona, a European Mediterranean city, *Cities*, 20(6): 387–394.

Ekins, P., Folke, C. and De Groot, R. (2003) Identifying critical natural capital, *Ecological Economics*, 44: 159–163.

Fall, A. (2002) SELES Model Builder's Guide, Unpublished Report Gowland Technologies Ltd. (http://www.cs.sfu.ca/research/SEED/seles.htm)

Haase, D., and Nuissl, H. (2006) Does urban sprawl drive changes in the water balance and policy? The case of Leipzig (Germany) 1870–2003. *Landscape and Urban Planning*, (in press).

Haase, D. and Magnucki, K. (2004) Die Flächennutzungs- und Stadtentwicklung Leipzigs 1870 bis 2003. Statistischer Quartalsbericht 1/2004, Leipzig, pp. 29–31.

Haase, A. and Steinführer, A. (2005) *Cities in East Central Europe in the Aftermath of Post-socialist Transition. Some Conceptual Considerations about Future Challenges*, Series Institute of Geography and Spatial Planning, Polish Academy of Sciences, IgiPZ PAN (in print).

Heilig, G.K. (2002) *Stirbt der ländliche Raum?* IIASA Interim Report, Laxenburg.

Horowitz, J.L., Koppelman, F.S. and Lerman, S.R. (1986) Self-instructing course in disaggregate mode choise modeling. Technology Sharing Program, US Department of Transportation, Washington DC.

INKAR (2003) *Indicators and Maps for Spatial Development in Germany*, Statistical agencies of Germany and Ministry of Architecture and Regional Development.

Kabisch, S. (2005) Empirical analyses on housing vacancy and urban shrinkage, in Hurol, Y. Vestbro, R. and Wilkinson, N. (eds) *Methodologies in Housing Research*, The Urban International Press, Gateshead, pp. 188–205.

Kabisch, S., Bernt, M. and Peter, A. (2004) Stadtumbau unter Schrumpfungsbedingungen: Eine sozialwissenschaftliche Fallstudie. Wiesbaden, vs Verlag für Sozialwissenschaften, 194 S.

Kröhnert, S.N, van Olst, N. and Klingholz, R. (2004) *Deutschland 2020*, Berlin-Institut für Weltbevölkerung und Globale Entwicklung.

Landis, J. and Zhang, M. (1998a) The second generation of the California urban futures model. Part 2: Specification and calibration results of the land-use change submodel, *Environment and Planning B: Planning and Design*, 25: 795–824.

Landis, J. and Zhang, M. (1998b) The second generation of the California urban futures model. Part 1: Model logic and theory, *Environment and Planning A*, 30: 657–666.

Lutz, W. (2001) The end of world population growth, *Nature*, 412: 543–545.

Miller, E.J., Hunt, J.D., Abraham, J.E. and Salvini, P.A. (2004) Microsimulating urban systems, *Computers, Environment and Urban Systems*, 28: 9–44.

Müller, B. and Siedentop, S. (2004) Growth and shrinkage in Germany – trends, perspectives and challenges for spatial planning and environment, *German Journal of Urban Studies*, 43: 14–32.

Silva, E.A. and Clarke, K.C. (2002) Calibration of the SLEUTH urban growth model for Lisbon and Porto, Portugal, *Computers, Envonment and Urban Systems*, 26: 525–552.

Sitar, S. and Sverdlov, A. (2004) Shrinking cities: reinventing urbanism. A critical introduction to Ivanovo context from an urbanist perspective, in *Shrinking Cities. A project initiated by the Kulturstiftung des Bundes (Federal Cultural Foundation, Germany)*, in cooperation with the Gallery for Contemporary Art Leipzig, Bauhaus Foundation Dessau and the Journal Archplus, 1(Ivanovo), pp. 8–11.

Van der Berg, L. (1982) *Urban Europe*, Oxford, New York.

Waddell, P. (2002) Urbansim: modeling urban development for land use, transportation and environmental planning, *Journal of the American Planning Association*, 68(3): 297–314.

White, R., Engelen, G. and Uljee, I. (1997) The use of constrained cellular automata for high-resolution modelling of urban land-use dynamics, *Environment and Planning B: Planning and Design*, 24: 323–343.

Wu, F. and Webster, C.J. (1998) Simulation of land development through the integration of cellular automata and multicriteria evaluation, *Environment and Planning B: Planning and Design*, 25: 103–126.

Wu, F. (1998) Simulating urban encroachment on rural land with fuzzy-logic-controlled cellular automata in a geographical information system, *Journal of Environmental Management*, 53: 293–308.

Chapter 20

LAND-USE SIMULATION FOR WATER MANAGEMENT
Application of the Land Use Scanner in two large-scale scenario studies

J. Dekkers and E. Koomen
Department of Spatial Economics/SPINlab, Vrije Universiteit Amsterdam, The Netherlands

Abstract: Land use is one of the major components influencing local hydrological
            characteristics. Future land use is thus important in studies that focus on the
            upcoming challenges for water management. This chapter describes two
            applications of the *Land Use Scanner* model on a national or larger scale, in
            which the scenario method is used to simulate future land-use patterns.

Key words: Land use; spatial planning; spatial dynamics; water management.

1.   INTRODUCTION

Land use has a strong influence on the water balance of a given area; groundwater recharge varies per land-use type because of differences in infiltration and evaporation rates. In particular, changes in the urban surface, forest cover and agricultural use will influence the hydrological cycle. An increase in the built-up area will limit the infiltration of precipitation and therefore hamper the recharging of groundwater reservoirs. An increase in the sealed surface also leads to faster precipitation run-off, causing more pronounced peaks in river discharge which in turn might lead to flooding. An increase in forest cover, on the other hand, will limit discharge peaks. Agricultural practices (irrigation, crop choice, *et cetera*), in combination with soil type, strongly influence evaporation and infiltration rates. A combination of land-use change and hydrological models can provide information on possible future land-use patterns and related hydrological

E. Koomen et al. (eds.), Modelling Land-Use Change, 355–373.
© 2007 *Springer.*

impacts that is valuable for both planners and researchers dealing with water management (see de Roo *et al.*, 2003, for example).

We can typically distinguish three types of future land-use simulation for planning purposes: trend analysis, impact assessment and scenario studies. Trend analysis simulates the possible future state of a land-use system following current trends, developments and policies. This is a representation of autonomous developments that is, for example, helpful in enabling policymakers to make decisions on investments in the additional hydrologic infrastructure that is needed because of continuing urbanisation. Impact assessment typically analyses the possible developments that are caused by a specific, spatially explicit, plan or project. These studies might be combined with trend analysis to specifically assess the additional impact of the selected project. The scenario method is especially suited for long-term studies that deal with a wide array of possible developments and many related uncertainties. By systematically describing several opposing views of the future, we can simulate a broad range of spatial developments, thus offering a full overview of possible land-use alterations. Each individual outlook to the future will not necessarily contain the most likely prospects, but, as a whole, the simulations provide the bandwidth of possible changes as discussed by Dammers (2000). The individual scenarios should, in fact, not strive to be as probable as possible, but should stir the imagination and broaden the view of the future. Important elements are plausible unexpectedness and informational vividness (Xiang and Clarke, 2003). Policymakers can thus get an idea of the possible developments they face. Based on this knowledge, they can assess the need for action and select the most appropriate policy measures. Since the scenarios can also contain reference to actual or envisaged spatial policies, the simulations offer a depiction of their possible outcomes. Policymakers can thus be confronted with the likely outcomes of their decisions.

This chapter describes two water management related scenario studies of future land use in which the *Land Use Scanner* model is applied. We will first introduce the model and then discuss the case studies. The first study was recently finished and focused on the possibility of water shortage in the Netherlands. The second application describes work in progress on a new *Land Use Scanner* application for the Elbe river catchment area that covers large parts of Germany and the Czech Republic. This latter study focuses on the impact of climate change on the entire water system, including, for example, the risk of water shortages or the possibility of river flooding. Both cases show the potential of land-use modelling in large-scale scenario studies. In the concluding section, we will discuss some general issues related to this type of study and point out some recommendations for future research.

2. THE *LAND USE SCANNER*

The *Land Use Scanner* model used in the case studies was selected because of several practical advantages. Firstly, the model is *flexible* in the sense that new land-use types and their anticipated future demand can be incorporated easily. It is also relatively easy to include expert judgement reflecting the scenario assumptions. Secondly it is an *open* system that is available under a free, general public license. Full freedom is thus guaranteed to the users to share and change this software, provided that they do not pose any restrictions on future use. The model is furthermore *compatible* with standard GIS data formats making it easy to import and export data layers, and to include new geographical base data, for example. Its compatibility also ensures the possibility of transfering information to and from other models in the modelling chain of the case studies. Simulated land-use maps can be exported to other models to assess specific (hydrological) impacts, for example.

The *Land Use Scanner* is a GIS-based model that simulates future land use. It has been used for various policy-related research projects. Applications include, amongst others: the simulation of future land use following different scenarios (Schotten and Heunks, 2001), the evaluation of alternatives for a new national airport (Scholten *et al.*, 1999), the preparation of the Fifth National Physical Planning Report (Schotten *et al.*, 2001) and, more recently, an outlook for the prospects of agricultural land use in the Netherlands (Koomen *et al.*, 2005). A full account of the model is provided in Hilferink and Rietveld (1999).

The *Land Use Scanner* offers an integrated view of all types of land use. It deals with urban, natural and agricultural functions, normally distinguishing 15 different land-use categories. The model is grid based, covering the Netherlands in almost 200,000 cells of 500 by 500 metres each. Each cell describes the relative proportion of all present land-use types, thus presenting a highly disaggregated description of the whole country. Regional projections of land-use change are used as input for the model. These are land-use type specific and can be derived from sector-specific models of specialised institutes. The various land-use claims are allocated to individual grid cells based on their suitability. Unlike many other land-use models the objective of the *Land Use Scanner* is not to forecast the dimension of land-use change but rather to integrate and allocate future land-use claims from different sector-specific models or experts. The outcomes of the model should not be interpreted as fixed predictions for particular locations but rather as probable spatial patterns.

2.1   Mathematical formulation

The *Land Use Scanner* uses an allocation algorithm that is based on economic discrete choice theory to match the spatial claims of the different land-use types with the available land. The crucial variable for the allocation model is the suitability s_{cj} that represents the net benefits of land-use type j in cell c. The greater the suitability for land-use type j, the greater the probability that the cell will be used for this type. In the simplest version of the model, a logit type approach is used to determine this probability.

The model is constrained by two conditions: (1) the overall demand for each land-use function, and (2) the amount of land which is available. By imposing these conditions, a doubly constrained logit model is established which yields as a side-product the shadow prices of land in the cells. This is discussed in more detail in Koomen and Buurman (2002). In the doubly constrained model, the expected amount of land in cell c that will be used for land-use type j can be formulated as:

$$M_{cj} = a_j \cdot b_c \cdot \exp(\beta \cdot s_{cj}) \qquad (1)$$

in which:

M_{cj} is the amount of land in cell *c* expected to be used for land-use type *j;*

a_j is the demand balancing factor (condition 1) that ensures that the total amount of allocated land for land-use type *j* equals the sectoral claim;

b_c is the supply balancing factor (condition 2) that makes sure the total amount of allocated land in cell *c* does not exceed the amount of land that is available for that particular cell;

β is a parameter that allows for the tuning of the model. A high value for β makes the suitability more important in the allocation and will lead to a more mixed land use pattern; and

s_{cj} is the suitability of cell *c* for land-use type *j*, based on its physical properties, operative policies and neighbourhood relations.

2.2   Implementation in a geographical information system

The *Land Use Scanner* model is implemented in an information system using the Data and Model Server software made available by the Object Vision company (www.objectvision.nl). The resulting geographical information system (GIS) allows for storage, manipulation and presentation of the geographical data that are used in the model. Furthermore, it contains the necessary arithmetic functions to implement the logit functions of the allocation model. The actual simulation is done in an iterative process that follows ten steps. This process is graphically depicted in Figure 20-1 and described in detail in Koomen (2002).

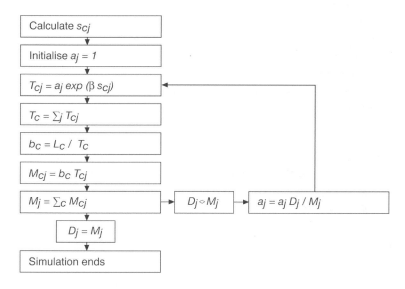

Figure 20-1. The *Land Use Scanner* simulation process.

The ten steps of the iterative simulation process are the following:
1. Calculate the suitability for every land-use type and cell (s_{cj}).
2. Initialise the demand balancing factors for every land-use type (a_j) at value 1.
3. Calculate the expected demand for every cell and land-use type (T_{cj}), β is a parameter with a chosen value (normally 1).
4. Summarise the total demand of all land-use types for land for every cell (T_c).
5. Calculate the supply balancing factor (b_c), L_c denotes the total amount of available land in a cell and is already known.
6. Calculate for every cell and land-use type the amount of allocated land (M_{cj}).
7. Summarise the total amount of allocated land for every land-use type (M_j).
8. Check for every land-use type whether the allocated amount of land is within a predefined range of the sectoral claim (D_j).
9. If the claim and allocated amount of land for a land-use type are not within the predefined range, a new value for the demand balancing factor is calculated and a new iteration starts again at step 3. This process leads

to a continuing increase in the a_j factor and can be considered as a bidding process.

10. The simulation is finished when the allocated amount of land is near enough to the sectoral claim.

The *Land Use Scanner* thus provides a powerful platform to simulate future land-use patterns through an integration of sector-specific land-use claims, large amounts of geographical data describing local characteristics and expert judgement on possible spatial developments.

3. NETHERLANDS DROUGHT STUDY

Water management in the Netherlands is normally concerned with the prevention of flooding, but the opposite problem, water shortage, is increasingly receiving attention. The idea of water shortage is not immediately combined with the wet appearance of the Netherlands, but indications arise for possible periods of water shortage when the overall demand for water is too high. Ground and surface water resources may at times be insufficient to allow for the combined use for such varied functions as transportation, irrigation, recreation and drinking water production. In order to assess the magnitude of this problem a water shortage study was carried out in the Netherlands (Klopstra *et al.,* 2005). Anticipated changes in land use were an important starting-point in this study. This section describes the drafting of the scenarios, their implementation in the model, the results of the simulation and their subsequent use in a hydrological model. A more extensive description of this case study is provided in Dekkers and Koomen (2005).

3.1   Designing scenarios

The starting-point for our part of the drought study was formed by three future scenarios that were developed by the International Centre for Integrative Studies at the University of Maastricht (ICIS, 2002): 'Environment matters', 'Government controls' and 'Market rules'. The proposed scenarios differ especially according to the factors that influence the problem of water shortage. They are related to the scenarios of prospected economic development that have previously been composed by the Netherlands Bureau for Economic Policy Analysis (CPB, 1996; 2001): 'Divided Europe', 'European Co-ordination' and 'Global Competition'. In this study we have added a reference scenario that strictly follows current trends and spatial policies. This provides us with a depiction of autonomous developments that helps assess the specific impact of the other scenarios.

The initial assumptions for this scenario are identical to the 'Government controls' scenario, but their implementation in the model is slightly different. Table 20-1 gives an overview of some of the basic assumptions attached to the scenarios.

To simulate future land use according to the scenarios, we need to specify both the magnitude (demand) and preferred locations (suitability) for the prospected spatial developments according to the storylines of the scenarios. This will be discussed in the following sub-sections.

Table 20-1. Basic assumptions of the four scenarios and related spatial implications

	Environment matters	Government controls and reference scenario	Market rules
Economic situation	Small-scale and clean industry ICT and eco-technology in services Biological farming and eco-recreation	Industry and services need more space	Large-scale industries Service sector grows rapidly Technological breakthroughs
Government intervention	Spatial policies determine land use	Spatial policies determine land use	Minimal spatial constraints
Climate change	Extreme changes Flood risk increased	Less extreme changes	No extreme changes
Spatial implications	Compact urban areas No residential land use allowed in green and wet areas. Nature development follows current policy More space for water	Residential land use around existing centres Commercial land use facilitated near major infrastructure Interweaving of urban and rural areas Nature development in rural areas, follows current policy	Urbanisation of rural areas is allowed Nature especially meant for recreation

Source: Adapted from ICIS (2002) and CPB (1996, 2001)

3.1.1   Obtaining land-use claims

The future demand for land is expressed as a set of additional regional claims for each of the land-use types that are to be simulated. As it is our intention to simulate the impact on the hydrological system, we have selected a number of land-use types that have specific infiltration and evaporation characteristics. These include two urban land-use types, seven types of agriculture and a combined class of nature and forest. The remaining types of land use (infrastructure and water) are fixed at their present location. The model allows for the possibility to include future developments at a predefined location, which can be used to add, for example, new motorways. In this study, however, this option was not used since new infrastructure developments were expected to have a limited and basically local impact.

The regional claims for our scenarios are based on a recent study that analysed the future prospects for nature in the Netherlands, *National Nature outlook 2030* (Natuurplanbureau, 2002; de Nijs *et al.*, 2002) as both their land-use typology and general scenario assumptions matched well with ours. This study, however, lacked a 'Reference scenario', so an adjusted version of the 'Government controls' scenario was used in which the anticipated amount of change for specific land-use types was lowered to bring it more in line with actual, recent developments. All additional land-use claims (Table 20-2) are added to the current land use in the *Land Use Scanner* to arrive at the expected total future area of the different land-use types.

Table 20-2. Overview of additional land-use claims, summarized at the national level

Land-use function	Environment matters	Government controls	Market rules	Reference scenario
Residential (incl. recreation)	86,281	90,573	151,258	90,573
Commercial	58,982	58,982	68,336	58,982
Meadow	-437,000	-370,000	-346,000	-196,000
Corn	14,000	-16,000	-26,000	-16,000
Arable farming	-252,189	-113,293	-283,702	-114,000
Greenhouses	0	0	0	0
Flower bulbs	9,133	199	5,956	199
Orchards	8,118	177	5,295	175
Other agriculture	30,749	-375	21,750	-374
Nature and Forest	499,997	344,994	399,996	172,496
Infrastructure	0	0	0	0
Water	0	0	0	0
Total of additional claims	18,071	-4,743	-3,111	-3,949

Source: Adapted from de Nijs *et al.* (2002)

The total of all additional land-use claims should equal zero, since an increase in the amount of land used for one function leads to a corresponding decrease in other functions. This is also a prerequisite for the allocation algorithm to find a feasible solution. As the expected land-use changes originate from different experts and several sector-specific models that are not fully tuned to each other, this condition is rarely exactly met. We solve this problem through the distinction of minimum claims, for land-use types where more land may be allocated and maximum claims where the allocation of less land is allowed. In our case, we have assigned maximum claims to the agricultural land-use functions as we expect agriculture to provide the extra space needed for the other, economically more powerful land-use functions if needed. This leads to the allocation of less agricultural land in the 'Environment matters' scenario and the allocation of more urban and natural land in the other scenarios.

Note furthermore that the land-use claims in table 20-2 are summarized at the national level. In the actual model simulation, the land-use demand is expressed at the regional level. Two different regional divisions are used in this application: claims for nature/forest, residential and commercial land are available for the 40 socioeconomic (NUTS-3) regions in the country, whereas the agricultural claims are present at 14 relatively homogenous, agricultural (LEI14) regions. The model simulates the land use of each individual cell and keeps track of the original regional claims through the regionally specific demand balancing factor (a_j). Therefore, the mismatch of the regional divisions does not pose a problem to simulation.

3.1.2   Defining local suitability

Besides the magnitude of the spatial developments, we also need to specify a notion of spatial preference for the different types of land use. In the *Land Use Scanner*, this is done in the suitability maps. These are created for all types of land use and are based on the scenario assumptions. A full description of this part of the model application is documented in Koomen and Dekkers (2003). Table 20-3 shows how the assumptions are translated into suitability maps for the 'Market rules' scenario. This example shows the type of geographical information and spatial relations that can be included in the suitability definition. An important concept here is the application of distance decay functions that express the diminishing impact of proximity. The proximity to current residential land use is deemed important in the suitability for the future residential land use that is to be simulated, for example. This influence is strongest in the cells directly neighbouring current residential land use and will slowly decrease to zero at a distance of 20 cells (10 kilometres) following a distance decay function.

Table 20-3. Translation of 'Market rules' scenario assumptions into suitability maps

Land-use type	Implementation suitability maps
Residential	Attracted to: area of 10 km around current residential locations; 5 km around forest and 1 km around water
	No constraints on developments in areas that are currently designated for nature development or in wet areas as in the other scenarios
Commercial	Attracted to: area of 10 km around current commercial locations; 1 km around highways and 2.5 km around highway entries and exits
	No constraints are implemented here as is the case in the other scenarios
Nature	Attracted to: ecological main structure, existing nature areas and proximity of urban areas
Agriculture	Based on physical suitability (soil type and water level) maps of the individual crops

3.2   Land-use simulations

The initial results of the land-use simulations are maps that for every land-use type and scenario depict the expected number of hectares per grid cell in 2030. These maps reflect a direct transformation of the probability that a land-use type will be allocated to a grid cell. This probability can also be used differently: for example, to allocate only the most probable land-use type to a cell. To provide an easier to interpret overview, we take a similar approach. A dominant land-use map was generated for each scenario, indicating per grid cell only, which land-use type takes up the most hectares. In total, five different land-use types are distinguished in these maps (Figure 20-2).

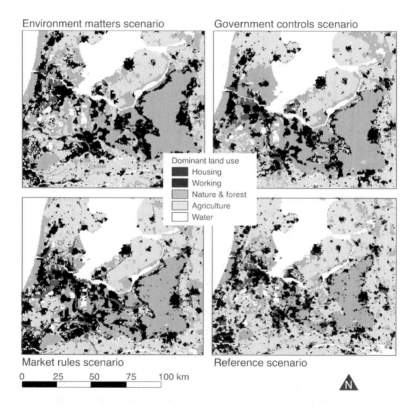

Figure 20-2. Dominant future land-use maps for each scenario. (See also Plate 25 in the Colour Plate Section)

The scenarios resulted in very diverse images of land use in the Netherlands in 2030. The 'Environment matters' scenario shows compact urban areas, small industrial growth near urban areas and ample space available for nature. The 'Government controls' scenario, in contrast, shows a large growth of commercial land-use functions near large urban areas in the west and south of the Netherlands, caused by the high density of train stations, highway entries and exits. The 'Market rules' scenario differs most from the current situation: residential land use has penetrated nature areas and commercial land use has spread alongside infrastructure corridors over large parts of the Randstad area and the province of Noord-Brabant. The 'Reference scenario' generally follows the assumptions of the 'Government controls' scenario, but relaxes the extent of government influence. This can be clearly seen from the reduced impact of the nature policies in the large nature areas in the central and east part of the country; current residential settlements and fragmentation of the nature areas persist.

3.3   Assessing the hydrological impact

The *Land Use Scanner* maps of future land are used as a starting point for further hydrological study. This consists of two subsequent phases.

Firstly, the resulting scenario maps from the *Land Use Scanner* simulation are used to start a discussion about possible future problems in water management in the Mid-West of the Netherlands with representatives of all relevant local stakeholders. These parties need information on how their region will develop within the next 30 years and, more specifically, what the influence of climate and socioeconomic changes will be on land use according to these different scenarios. The simulations give the involved parties insight into the type of developments they are to expect over the next years, the options they have for responding to these developments and the possible consequences of their responses.

Secondly, for each scenario, the resulting land-use maps are converted in order to be used as input for MOZART: a hydrological model that covers the upper, unsaturated soil zone (Peereboom, 2003). The model is used to provide information on the nature, severity and size of the water shortage problem in the whole country. The hydrological situation for the future scenarios is simulated using the current water management guidelines, meaning, for instance, that flood retention areas are included in the simulations of a scenario with increased flood risk. The hydrological model provides, amongst other information, future potential evaporation rates, based on the local crop and vegetation type provided by the land-use simulation and additional data on, for example, soil type. Evaporation is one of the crucial components of future water demand. Estimates of its future magnitude can, in combination with estimates of prospected water supply (precipitation), facilitate the assessment of the possibility of a soil moisture deficit. This deficit is essentially the difference between the potential (theoretical) evaporation of the crops, vegetation and soil and the actual evaporation that is possible, given the expected climate conditions. A shortage of soil moisture leads to sub-optimal conditions that hamper vegetation growth and in extreme cases may lead to the dying of plants.

Table 20-4 presents an overview of selected climate-indicator estimates and the related expected increase in soil moisture shortage for three scenarios at the national level. The climate indicators are selected from a moderate climate change scenario that is part of a set of standard Dutch future climate scenarios (Beersma *et al.*, 2004). The climate variables are kept constant in all scenarios to better understand the sensitivity to land-use change. The expected differences in moisture shortage for the scenarios are mostly related to a relocation of currently existing land-use types to areas with different hydrological conditions and to the introduction of new crop-types. Further analysis focused on obtaining the soil moisture shortage for six separate regions. Consequently, policy measures can now be developed in order to decrease the risk of damage caused by future moisture shortages. Also, estimations of water needed per region to avoid shortages can be computed, both in time and space (Klopstra *et al.*, 2005).

Table 20-4. General climate-indicator estimates for an average climatological year in 2050 and the expected change in soil moisture shortage according to the different land-use change scenarios

Variable	Environment matters	Reference scenario	Market rules
Temperature		$+1^0C$	
Precipitation (average per year)	year : +3%; winter: +6%; summer: +1.4%		
Evaporation (average per year)		+4%	
Sea level		+25 cm	
River discharge		-5%	
Change in soil moisture shortage	+3%	+10%	+7%

Source: Klopstra *et al.* (2005)

3.4   Conclusion

The *Land Use Scanner* is capable of generating diverse images of the future within a short time that are consistent with the scenario assumptions. The maps show the essence of the spatial developments that are associated to the scenarios. This case study can be typified as a quick-and-dirty application to inform water managers of the possible range of spatial developments they are facing. The validity of the exact amount of simulated change and projected locations is, of course, open to debate, but we achieved the main objective of this study: presenting possible spatial developments within a structured scenario framework. This helped initiate discussions about possible future water management problems and provided the necessary input to hydrological models to assess the magnitude of a possible future drought problem.

To facilitate a better integration of *Land Use Scanner* results with hydrological models, a number of improvements have been suggested by the experts of the Dutch Institute for Inland Water Management and Waste Water Treatment. These relate to the level of detail, the land-use typology and the heterogeneous character of the grid cells. The level of detail should preferably be changed from 500 by 500 metres to 50 by 50 metre grid cells. This, however, calls for more precise and better founded assumptions regarding future land-use demand and locational preferences that are very hard, if not impossible, to obtain for this type of long-term, large-scale studies.

The current land-use typology can be improved by, for example, including a distinction in various natural land-use types (open nature areas, deciduous and coniferous woods), since these have very different evaporation characteristics. An increase in the number of land-use types can be fairly easily achieved, as is demonstrated in another *Land Use Scanner* application (Borsboom-van Beurden *et al.*, Chapter 16 in this book) that simulates future land use for 28 types. But also in this respect we have to consider the contradiction between the need for thematic detail and the uncertainties that are associated with pinpointing spatial developments for all these categories.

The final suggested improvement is the output of homogenous grid cells relating to only one type of land use, instead of the current heterogeneous approach that assigns fractions of the cell to a number of land-use types. Homogenous cells can be created on the basis of the current output, for example, by assigning cells to the dominant land-use types. But this normally causes a bias towards a number of prevailing land-use types that often take up most, but not all, of a cell. The newest version of the *Land Use Scanner* (4.70), that has recently been finalized (Bouwman *et al.*, 2006), offers a new allocation algorithm that directly delivers homogeneous cells in a 100 by 100 metre grid.

4.   GLOWA-ELBE

The aim of the GLOWA project is to assess the impact of climate and social change on the water system of river basins within various climatic regions with large precipitation differences. GLOWA is the German abbreviation for the federal research program on 'Global Change in the Hydrological Cycle'. The GLOWA-Elbe project is a research initiative of the German Federal Ministry of Education and Research (BMBF) and is directed at the Elbe river basin.

The most important aspects that will be examined are changes in precipitation extremes in the region, the effect of social changes on water demand and on nutrient and pollution emissions, and the direct and indirect effects of global change on good quality surface water. The BMBF needs an answer to the following questions: which alternative policy strategies exist for the Elbe basin? How can these alternatives be evaluated and what strategy should be recommended to stakeholders?

The Elbe catchment area has relatively low precipitation and has the lowest mean water availability in Germany. The flooding of Hamburg in 2002, however, shows that short periods of water excess are also present and that these can have disastrous effects. The project will research plausible changes in this region and will provide policymakers with information on the effects of certain policy strategies on problems of drought and flooding.

The *Land Use Scanner* model is used to disaggregate projected agricultural and environmental developments and social changes to a more detailed geographical scale level and to assess, through land-use simulations, possible future water demand and possible changes in hydrology and water quality (Figure 20-3). The outlook for these possible future changes is structured through the use of scenarios that systematically take into account the possible changes in climate and society (e.g. population growth). Simultaneously, all kinds of relevant policy options can be deployed in the scenarios. Starting point in this approach is a single climate-change scenario. Different socio-economic developments are subsequently based on the anticipated climatic changes. Land use is then simulated according to these socio-economic scenarios in two variations. Simulations will be made applying current environmental policy as well as following more strict environmental policies.

The future demand for land for each sector is calculated using sector-specific models. The outcomes of these models are fed into the *Land Use Scanner*, which then simulates future land use using its economic mechanisms of demand and supply. The future water demand is extracted from the simulated land-use patterns and subsequently used in models that analyse the ecological and socioeconomic impact of the changes in water availability and the quality of the surface water. In the end, multi-criteria and cost-efficiency analyses are performed to *"...deliver recommendations for adapting water management to the challenges of global change in the Elbe basin"* while balancing *"...the conflicting interests of policymakers, stakeholders and society at large"* (AG GLOWA-Elbe, 2005: p.2).

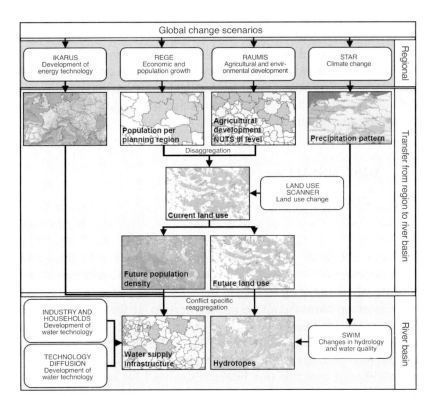

Figure 20-3. GLOWA-Elbe modelling framework and performed scale transformations. (See also Plate 26 in the Colour Plate Section)

Source: Hartje *et al.* (2005).

For this project, 16 land-use types are modelled in a 250 by 250 metre grid consisting of over 2.3 million grid cells. Current land use from the year 2000 is taken from the European CORINE land-cover database, which is based on satellite imagery. At this moment, various spatial policy maps relating to, for example, natural protection and urban development have been added to the model, as well as a map containing the agricultural yield potential for each grid cell and distance maps for roads, railroads and railway stations. Three levels of administrative regions (Bundesländer,

Raumordnungsregionen and Kreise) have now been included that will be used as the regional divisions for the different land-use claims. To get grip on the wide range of possible future changes, a trend scenario is defined that extrapolates current developments and policies. After analysing possible social global changes, alternative scenarios will be defined that deviate from the trend scenario.

In summary, the *Land Use Scanner* is applied in this extensive research project to integrate the future projections of different specialised models and institutes. This integration and subsequent disaggregation allow for a hydrological impact assessment at a more detailed level, making the model a crucial component in chain of instruments that try to unravel the intricacies of global change at the local level.

5.   DISCUSSION

From our two case studies, we conclude that land-use models, such as the *Land Use Scanner,* are adequate instruments to simulate future land-use patterns within a scenario-analysis framework. The *Land Use Scanner* proved to be a flexible and open system that allows for the integration of different information sources and the combination with other models for the additional assessment of hydrological impacts.

An important contradiction in the application of the *Land Use Scanner* and most other land-use models relates to the need for highly detailed simulations, in terms of both spatial (grid cell) and thematic (number of land-use types) resolution, in long-term studies where many uncertainties exist. Solving this contradiction between the need for local-level detail and large-scale uncertainties regarding future developments is a major task for further research. Interesting related attempts include the construction of multi-scale models (e.g. Veldkamp *et al.*, 2001 or Verburg *et al.*, Chapter 18 in this book) and the increased attention for agent-based modelling, as is discussed by Ettema *et al.,* in Chapter 14. The scenario method offers a structured approach to deal with the many uncertainties related to future developments, but has some limitations as discussed below.

Scenario simulation results are heavily reliant on expert judgement, so their validity is questionable. This is inherent to the scenario approach, as is also discussed by Klosterman (1999) in his description of the *What if?* scenario-based planning support system. The limited validity is not a serious problem as long as the simulation outcomes are treated for what they are worth: images depicting possible future developments following a large number of scenario-related assumptions. This guideline for interpreting the results is quite often ignored in subsequent presentation and use of the outcomes. Devising new ways of presenting uncertain information might help overcome this difficulty.

A way to provide information on the validity of the simulation outcomes is to calibrate and validate land-use models. This provides valuable information on the performance of the models, but only applies to the short periods for which they are tuned. So the use of such rigorous validation exercises is limited for a scenario study that is normally strongly dependent on imagination. We do, however, in accordance with Verburg *et al.* (2004), feel that serious testing is needed of the model's capacity to produce sensible outcomes and have carried out a preliminary validation exercise (documented in Pontius *et al.*, 2007). Similar work is in progress.

An additional worry for this type of study, which relies on the consecutive use of several models, is the propagation of errors. Initial model results are themselves used as input in a number of subsequent model steps, that each may add extra uncertainties, assumptions and related errors to the original outcomes. This is a well-known issue in GIS (e.g. Heuvelink, 1998) that, however, has received limited attention in the application of land-use models. An initial sensitivity analysis has been carried out for the *Land Use Scanner* (Dekkers, 2005), but more work, especially on the consequences of the repeated transformations of model results and the consistency of the scenario-related assumptions in the different models needs to be done.

Another issue in model improvement is increasing the usability of the simulation results. Options that can help presenting and interpreting modelling outcomes include: a more interactive user interface, 3D visualisations of results and the construction of indicators that summarize results in relation to policy issues. The implementation of these and other options is beyond the scope of this chapter.

ACKNOWLEDGEMENTS

The authors thank the Netherlands Environmental Assessment Agency (MNP) for allowing us to use a preliminary updated version of the *Land Use Scanner*. The authors are also grateful for the input and feedback of Niels Vlaanderen and Ivar Peereboom of the Institute for Inland Water Management and Waste Water Treatment (RIZA). Finally, the authors like to thank the German Federal Ministry of Education and Research (BMBF) and the Technischen Universität Berlin for funding and co-operating in the GLOWA-ELBE II research project. We especially appreciate the valuable comments of Jana Borgwardt on a preliminary version of this text.

REFERENCES

AG GLOWA-Elbe (2005) *Global change impacts on the water cycle in the Elbe river basin–risks and options*, Project summary, www.glowa-elbe.de

Beersma, J.J., Buishand, T.A. and Buiteveld, H. (2004) *Droog, droger, droogst; KNMI/RIZA-bijdrage aan de tweede fase Droogtestudie Nederland*, KNMI-publicatie 199-II, De Bilt

Bouwman, A.A., Kuiper, R. and Tijbosch H.W. (2006) *Ruimtelijke beelden voor Zuid-Holland*, Rapport 500074002/2006, Milieu- en Natuurplanbureau, Bilthoven.

CPB (1996) *Omgevingsscenario's Lange Termijn Verkenning 1995-2020*, werkdocument No 89, Centraal Planbureau, Den Haag.

CPB (2001) *De Ruimtevraag tot 2030 in twee scenario's*, Centraal Planbureau, Den Haag.

Dammers, E. (2000) *Leren van de toekomst: over de rol van scenario's bij strategische beleidsvorming*, Eburon, Delft.

Dekkers, J.E.C. (2005) *Grondprijzen, geschiktheidkaarten en parameterinstelling in de RuimteScanner. Technisch achtergrondrapport bij Ruimtelijke Beelden*, RIVM report 550016005, Bilthoven.

Dekkers, J.E.C. and Koomen, E. (2005) *Simulations of Future Land Use for Water Management: Assessing the suitability for scenario-based modelling*, Paper for the 45th congress of the European Regional Science Association, Amsterdam, August 23–27.

de Nijs, T., Crommentuijn, L., Farjon, H., Leneman, H., Ligtvoet, W., de Niet, R. and Schotten, K. (2002) *Vier scenario's van het Landgebruik in 2030, Achtergrondrapport bij de Nationale Natuurverkenning 2*, RIVM rapport 408764 003/ 2002, Bilthoven.

de Roo, A., Schmuck, G., Perdigao, V. and Thielen, J. (2003) The influence of historic land use changes and future planned land use scenarios on floods in the Oder catchment, *Physics and Chemistry of the Earth*, 28: 1291–1300.

Hartje, V., Klaphake, A., Grossmann, M., Mutafoglu, K., Borgwardt, J., Blazejczak, J., Gornig, M., Ansmann, T., Koomen, E. and Dekkers, J. (2005) *Regional Projection of Water Demand and Nutrient Emissions – The GLOWA-Elbe approach*, Poster presented at the Statuskonferenz Glowa-Elbe II, Köln May 18–19, 2005, www.glowa-elbe.de/pdf/status_mai2005/04poster-tub.pdf.

Heuvelink, G. (1998) *Error Propagation in Environmental Modelling with GIS*, Taylor & Francis, London.

Hilferink, M. and Rietveld, P. (1999) Land Use Scanner: an integrated GIS based model for long term projections of land use in Urban and rural areas, *Journal of Geographical Systems*, 1(2): 155–177.

ICIS (2002) *Droogtescenario's vertaald in modelinput*, concept-memo, Maastricht.

Klosterman, R.E. (1999) The What-If? collaborative planning support system, *Environment and Planning B*, 26: 393–408.

Koomen, E. (2002) *De Ruimtescanner verkend; kwaliteitsaspecten van het informatiesysteem Ruimtescanner*, Vrije Universiteit Amsterdam.

Koomen, E. and Buurman, J. (2002) Economic theory and land prices in land use modeling, in Ruiz, M., Gould, M., Ramon, J. (eds) *5th AGILE Conference on Geographic Information Science Proceedings*, Universitat de les Illes Balears, Palma (Illes Balears), pp. 265–270.

Koomen, E. and Dekkers, J.E.C. (2003) *Landgebruikssimulatie voor Droogtestudie*, RIZA werkdocument 2003.141x, Ministerie van Verkeer en Waterstaat/RIZA, Lelystad.

Koomen, E., Kuhlman, T., Groen, J. and Bouwman, A. (2005) Simulating the future of agricultural land use in the Netherlands, *Tijdschrift voor Economische en Sociale Geografie*, 96(2): 218–224.

Klopstra, D., Versteeg, R. and Kroon, T. (2005) *Aard, ernst en omvang van watertekorten in Nederland; eindrapport*, RIZA, HKV, Arcadis, KIWA, Korbee en Hovelynck,

Natuurplanbureau (2002) *Nationale Natuurverkenning 2, 2000-2030*, Kluwer, Alphen aan de Rijn.

Peereboom, I. (2003) *Van Ruimtescanner uitvoer naar MOZART invoer: De vertaling van landgebruikscenario's tot discrete landgebruiksklassen*, werkdocument 2003.142x, Ministerie van Verkeer en Waterstaat, Rijkswaterstaat/RIZA, Lelystad.

Pontius Jr., R.G., Boersma, W., Castella, J.C., Clarke, K., de Nijs, T., Dietzel, C., Duan, Z., Fotsing, E., Goldstein, N., Kok, K., Koomen, E., Lippitt, C.D., McConnell, W., Pijanowski, B., Pithadia, S., Mohd Sood, A., Sweeney, S., Ngoc Trung, T. and Verburg, P.H. (2007) Comparison of land change modeling applications with quantitative validation, *Annals of Regional Science* (submitted)

Scholten, H.J., van de Velde, R., Rietveld, P. and Hilferink, M. (1999) Spatial information infrastructure for scenario planning: the development of a land use planner for Holland, in Stillwell, J., Geertman, S. and Openshaw, S. (eds) *Geographical Information and Planning*, Springer-Verlag, Berlin, pp. 112–134.

Schotten, C.G.J. and Heunks, C. (2001) A national planning application of Euroscanner in the Netherlands, in Stillwell, J.C.H. and Scholten, H.J. (eds) *Land Use Simulation for Europe*, Kluwer Academic Publishers, Amsterdam, pp. 245–256.

Schotten, C.G.J., Goetgeluk, R., Hilferink, M. Rietveld, P. and Scholten, H.J. (2001) Residential construction, land use and the environment. Simulations for the Netherlands using a GIS-based land use model, *Environmental Modeling and Assessment*, 6: 133–143.

Verburg, P.H., Schot, P.P., Dijst, M.J. and Veldkamp, A. (2004) Land use change modelling: current practice and research priorities, *GeoJournal*, 61: 309–324.

Veldkamp, A., Verburg, P.H., Kok, K., de Koning, G.H.J., Priess, J. and Bergsma, A.R. (2001) The need for scale sensitive approaches in spatially explicit land use change modelling, *Environmental Modeling and Assessment*, 6: 111–121.

Xiang, W.N. and Clarke, K.C. (2003) The use of scenarios in land-use planning, *Environment and Planning B*, 30: 885–909.

Chapter 21

GIS-BASED MODELLING OF LAND-USE SYSTEMS

EU Common Agricultural Policy reform and its impact on agricultural land use and plant species richness

P. Sheridan[1], J.O. Schroers[1] and E. Rommelfanger[2]
[1]Institute of Agricultural and Food Systems Management and [2]Institute of Biometry and Population Genetics, Justus Liebig University, Gießen, Germany

Abstract: The chapter presents the land-use model *ProLand* and the fuzzy expert system *UPAL*. *ProLand* models the regional distribution of agricultural land-use systems whilst *UPAL* predicts the species richness of vascular plants. Linking land-use and ecological models allows us to assess socioeconomic and ecological effects of policy measures by identifying interactions and estimating potential trade-offs. The effects of the Common Agricultural Policy reform on land use, key economic figures, and plant species richness are modelled for a study area in Hesse, Germany. Results indicate that the reform positively influences ground rent and species richness.

Key words: Land-use modeling; fuzzy expert system; species richness.

1. INTRODUCTION

Landscapes generate multiple commodity and non-commodity outputs, e.g. agricultural produce or species habitats. As changes in landscapes arise from technological innovations, socioeconomic as well as political developments (Rounsevell *et al.*, 2003; Stoate *et al.*, 2001), land-use models and landscape evaluation frameworks need to estimate impacts of policy, socioeconomy and technology on both private and public outputs.

Agricultural policy affects the comparative advantage of land-use systems and thus land-use intensity and land-use patterns (Weinmann *et al.*,

E. Koomen et al. (eds.), Modelling Land-Use Change, 375–389.
© 2007 *Springer.*

2006), which in turn influence plant species richness (Waldhardt et al., 2003). The reform of the European Union's Common Agricultural Policy (CAP) replacing coupled with decoupled transfer payments (EC, 2004) forces land users to re-evaluate their production programs as payments are no longer granted for certain crop or animal species.

The model *ProLand* (Prognosis of Land use) simulates the distribution of agricultural land-use systems while *UPAL* (Unscharfe Prognose der Artenvielfalt Landwirtschaftlich genutzter Flächen) predicts associated plant species richness. The models allow the assessment of socioeconomic and ecological effects of policy measures by identifying interactions and estimating potential trade-offs. They are applied in two scenarios to estimate the effects of political conditions before (Agenda 2000) and after the CAP Reform in a less favoured area, the Lahn-Dill-Highlands in Hesse, Germany. The region of about 660 km² is characterized by unfavourable natural conditions in terms of water availability and temperature, and small agrarian structure.

2.   MODEL DESCRIPTIONS

The land-use model *ProLand* (Weinmann, 2002; Kuhlmann et al., 2002) and the phytodiversity model *UPAL* are developed at the collaborative research centre SFB 299 'Land-Use Options For Peripheral Regions' of the Justus Liebig University, Gießen. At the centre, researchers from multiple disciplines develop transferable models and land-use options for less favored areas. The resulting ITE²M (Integrated Tool for Economic and Ecological Modeling) is a model network covering economic, abiotic and biotic aspects. It allows the evaluation of multifunctional landscapes with actual and simulated land-use patterns based on scenarios addressing different political, socioeconomic, and natural conditions (Möller et al., 2002).

2.1   The land-use model *ProLand*

Land use and land-use changes are functions of natural, socioeconomic, political and technological variables. Land-use modelling qualitatively and quantitatively assesses the impact of changes in these variables on multiple landscape services. *ProLand* is a spatially explicit, deterministic, comparative-static model that simulates agricultural and silvicultural land-use patterns as end points of adaptation processes. Model predictions are based on small-scale data of physical, biological and socioeconomic characteristics of an area. The model divides regions into economic decision units, which can be grid cells or vector elements of discretionary size such as individual fields, without relying on specific farm structures. *ProLand* calculates key economic figures, data on socioeconomic and technological attributes, e.g. transfer payment volume or pesticide input, and determines the ground rent maximizing land-use system for each individual decision unit. Employed as an economic laboratory, it can analyze the effects of changes in political, technological and socioeconomic conditions. As all results are spatially explicit, they can be easily aggregated in common geographical information systems or database management systems. Also, they can be combined with ecological as well as hydrological indicators provided by respective models (Weber *et al.*, 2001; Möller *et al.*, 2002).

2.1.1 Agricultural and silvicultural land-use systems

Applying the entity-relationship data model (Chen, 1976), agricultural and silvicultural land-use systems consist, at the primary level, of the entity sets 'crops', 'field operations', 'animal husbandry', and their relations. These entities are described using biological and technological attributes, specific to each entity. Land-use systems are determined by political, socioeconomic, natural and technological conditions as well as their relations. A system at the secondary level is thus extended by these entity sets and applicable relations between all sets. Consequently, a comprehensive description of land-use systems requires data on all entity sets and relations (Schroers *et al.*, 2004).

The following example of dairy cow keeping illustrates this approach. A description of a corresponding land-use system requires data on the fodder crops grown (entity set 'crops'), how they are produced (entity set 'field operations'), and how the animals are kept (entity set 'animal husbandry'). To comprehensively describe the system, additional data are required, e.g. factor and product prices, transfer payments, interest and wage rates, and production quotas. This structure is used for all agricultural and forest production processes.

ProLand's land-use systems database reflects the biological, socioeconomic and political attributes of agricultural production. Spatially explicit modelling requires additional site-specific data on natural, structural and political attributes that influence the costs and benefits of land-use systems, e.g. plant available water and temperature as non-controllable growth factors, site specific transfer payments, slope and field size.

The relational databases and the direct link to a GIS enable *ProLand* to simulate spatially variant interventions in land structure, market policy, available land-use systems and land-use restrictions. Variable options of market price support are incorporated as scenario-based price structures for marketable cash crops and processed products such as milk and beef.

Coupled and decoupled payments are stored for every crop or animal species, respectively every decision unit, allowing for simulations with spatially referenced direct payments and animal premiums. This allows the integration of virtually all land-use systems including renewable energies, as well as conservation measures and the estimation of, for example, opportunity costs of conservation programs in selected sub-regions.

2.1.2   Objective function

The model assumes ground rent maximizing behaviour of land users. Ground rent is defined as revenues minus costs including opportunity costs for capital and labour in monetary units per area unit (Brinkmann, 1922). It represents the remuneration for land employed in agricultural or silvicultural production as follows:

$$GR = \frac{R - LC - IC - MC}{A} \tag{1}$$

where GR is ground rent; R is revenues; LC is the opportunity costs of labour; IC is the opportunity costs of capital; MC is material costs; and A is the area farmed by the land user. Ground rent is affected by spatially and temporally variant natural, political, technological and macroeconomic conditions. *ProLand* was developed to model the effects of variations in these conditions on land-use sytems and land-use patterns.

Revenues are determined by given prices and endogenous estimates of maximum realizable yields, calculated using linear-limitational yield functions. The functions describe the influence of the non-controllable growth factors, annual precipitation, usable field capacity and annual temperature sum on crop yield which is either limited by plant available water or temperature sum. Thus, maximum realizable yields are endogenous variables and a function of site-specific values of non-controllable growth factors.

The model calculates the ground rent maximizing land-use system for every individual decision unit in a region. Obviously, these decision units frequently have heterogeneous site conditions, e.g. soil composition, slope *et cetera,* which influence productivity and, consequently, ground rent. Assigning sub-polygons containing site-specific information derived from small-scale raster elements to the actual decision units retains high resolution information while modelling larger polygons.

The model estimates the ground rent for each of these sub-polygons and each land-use system and selects the ground rent maximizing alternative. As sub-polygons can be of different size, it then calculates the area weighted average of the ground rent for the entire decision unit. Equation (2) exemplifies this approach for the three land-use types that are used in the presented application: arable land, grassland and forest.

$$GR_{max,k} = Max\left[GR_{max,k}^{arable}, GR_{max,k}^{grassland}, GR_{max,k}^{forest}\right]$$

$$= Max\left[\sum_{a=1}^{w}\left(\sum_{i=1}^{n}A_i\left(B_{a,i} - C_{a,i}\right)\right), \sum_{i=1}^{n}A_i\left(B_{g,i} - C_{g,i}\right), \sum_{i=1}^{n}A_i\left(B_{f,i} - C_{f,i}\right)\right] \quad (2)$$

where:

$GR_{max,k}$ is the maximum ground rent on subpolygon k [€/ha];

$GR_{max,k}^{arable}$ is the ground rent of arable farming on subpolygon k [€/ha];

$GR_{max,k}^{grassland}$ is the ground rent of grassland on subpolygon k [€/ha];

$GR_{max,k}^{forest}$ is the ground rent of forest on subpolygon k [€/ha];

A_i is the area share of subpolygon i $(i=1,..,n)$ in polygon k;

$B_{a,i}$ is the revenues of cropping system a on subpolygon i [€/ha];

$C_{a,i}$ is the costs of cropping system a on subpolygon i [€/ha];

$B_{g,i}$ is the revenues of grassland system g on subpolygon i [€/ha];

$C_{g,i}$ is the costs of grassland system g on subpolygon i [€/ha];

$B_{f,i}$ is the revenues of forest system f on subpolygon i [€/ha]; and

$C_{f,i}$ is the costs of forest system f on subpolygon i [€/ha];

2.2 *UPAL – a fuzzy expert system for species richness*

The fuzzy expert system, *UPAL*, assesses the impact of land-use changes on the species richness of vascular plants. It derives the values of ecologically relevant parameters from several site-specific attributes and land-use operations. Land-use dependent site characteristics influencing plant species richness are derived from predictions generated by *ProLand*. Detailed information on crop rotation, fertilization and pesticide strategy, and field operations are considered. The expert system then classifies natural and land-use dependent site characteristics into aggregate factors. Based on a set of rules, it assigns the number of species to these classes and thus to the decision units.

2.2.1 Requirements

ProLand not only forecasts arable farming, grassland, or forest land uses but also detailed strategies including crop rotation and life stock management. Thus, *UPAL* has to differentiate between a large number of land-use systems. However, little or even no site-specific detailed empirical data on current land-use systems are available, only the land-use type, i.e. arable farming, grassland or forest. Evidently, it is virtually impossible to collect data on both the land-use system and natural attributes of a site. Thus, *UPAL* has to be able to assess species richness even if no specific data for a combination of the land-use system and natural parameters exists.

Also, the uncertainty of the assessment has to be calculated. Ecological systems are extremely complex, meaning that a large number of input parameters will be omitted in the modelling approach, obviously resulting in an uncertain estimate of the number of occurring species. This uncertainty has to be documented with the output and passed on to other models.

Additionally, the output should be understandable by non-ecologists. It has to include an explanation of what a low, medium, or high number of species means in a specific context.

An approach complying with all these requirements is a fuzzy expert system. A model based on rules predicting how land uses impact species richness in a given region even if no explicit data is available is capable of processing detailed land-use information.

2.2.2   Methods and design

Fuzzy expert systems are rule-based systems using expert knowledge and fuzzy logic. While classical sets can only allocate the membership values 0 and 1, fuzzy sets use membership values from 0 to 1 (Klir and Folger, 1988). The transformation from metric data to fuzzy sets is called fuzzification. These fuzzificated values enter a set of if-then rules called inference. The last step is the defuzzification of the fuzzy sets to metric data. All these steps are created with expert knowledge (Figure 21-1).

Every plant species has a certain ecological optimum. If natural parameters are at this optimum, the species can compete much better than under sub-optimal conditions. This optimum is defined by several ecological parameters derived from natural parameters. The most important parameters are moisture, nutrient availability, soil acidity, temperature and light impact. Soil salinity, soil heavy metal content and the climatic zone are also important. Some of these parameters are directly influenced by land use such as nutrient availability altered by fertilization and soil acidity altered by lime application. In agricultural areas, only a few ecological parameters depend on natural conditions: temperature, mainly influenced by altitude and solar insolation, light impact, influenced by solar insolation and the current vegetation on a site, and moisture.

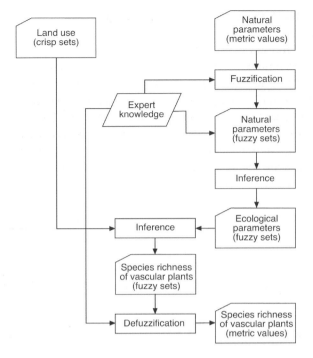

Figure 21-1. UPAL model structure.

UPAL consists of two modules: The first module assigns natural parameters to ecological parameters. The second module assesses what impact ecological parameters and land use have on species richness of vascular plants.

At present, *UPAL* considers the parameters of moisture, temperature and light impact in its calculations. Some parameters are disregarded like soil salinity and soil heavy metal content. The climatic zone parameter is considered constant in the research region. Others are derived only from land use based on *ProLand*'s specification that all farmers employ good agricultural practices. Because of this specification, *UPAL* assumes the nutrient availability and soil acidity parameters depend on land use.

The impact of land use on species richness is very complex. Different land uses influence the natural parameters and change the ecological environment for plant species in agricultural areas. Land use influences plant species occurrence with physical stress factors such as grazing and mowing on grassland or application of herbicides on crop fields. All influences of the forecasted land use have to be considered and their impacts have to be integrated in the rule base. While some impacts on plant species richness are obvious, such as an extremely negative influence of herbicide application, other impacts are more difficult to assess.

2.3   **Connecting *ProLand* and *UPAL***

Evaluating multiple, interdependent landscape functions requires information exchange between models. Such multi-criteria assessments allow to identify trade-offs and hot-spots but require spatially explicit information (Bockstael, 1996). *UPAL* and *ProLand* exchange data based on identical decision units through a common GIS (Figure 21-2). This configuration enables both models to share results and data from the underlying databases, e.g. land-use systems, among themselves and with other GIS-based models. Uncertainty and error propagation from *ProLand* to *UPAL* are still under investigation.

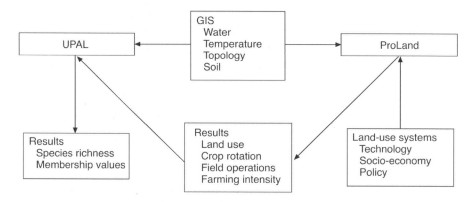

Figure 21-2. Data flow between *ProLand* and *UPAL.*

3.   **SCENARIO DESCRIPTION**

The CAP reform became effective in Germany in 2005. Key elements are the decoupling of transfer payments from the production program, requirements regarding 'Cross Compliance' and the redirection of funds from the first to the second pillar (Modulation) (BMVEL, 2005). By the year 2013, area payments for grass and arable land will be aligned at regional level as laid out by EC regulation 1782/2003 (EC, 2003). With the new CAP land users may choose among five options: (1) maintain the existing land-use program; (2) maintain the existing land-use program, but change its intensity; (3) switch to a different land-use system; (4) cease agricultural production but keep fields in 'good agricultural and ecological condition' in accordance with the Cross Compliance requirements; and (5) abandon the fields to natural succession and waive the area payments.

General political and economic conditions in the state of Hesse are reflected equally in both scenarios (Agenda 2000 and CAP reform). The most important legislative constraint concerns forests which must not be converted to other land uses. Quantity and input price structures reflect the situation in 2004 and are derived from data provided by the German agricultural market and price recording agency (ZMP, 2004a; ZMP, 2004b) and the consortium for agricultural technology (KTBL, 2003). The CAP reform scenario is defined with transfer payments set to the values projected for 2013. Transfer payments from the second pillar not affected by the reform, such as conservation programs, are not altered. Prices were adjusted based on projections by the model *AGRISIM II* (Borresch *et al.*, 2005).

4. RESULTS

4.1 Land use

ProLand generates detailed predictions of land use, land-use intensity, ground rent and other economic key figures for each decision unit. However, the description here focuses on land-use categories and aggregated figures.

Figure 21-3 illustrates the predicted land use for both scenarios. About 55% of the region is used as forest in both scenarios, mainly due to legislative constraints. Under the Agenda 2000 about a quarter of the area is grassland. Arable farming has a share of about 8%. The share of grassland is about 5% higher under the CAP reform at the cost of arable farming. Cutting (mulching) of permanent grassland is negligible. Overall, land use shows little change.

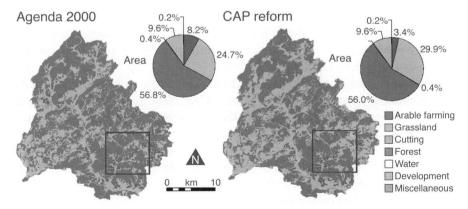

Figure 21-3. Predicted land uses in the Lahn-Dill region. (See also Plate 27 in the Colour Plate Section)

Source: Weinmann *et al.* (2006)

Predictions for sub-regions may differ considerably, e.g. in the area marked by the black rectangle and magnified in Figure 21-4 which shows significant reactions to agricultural policy changes.

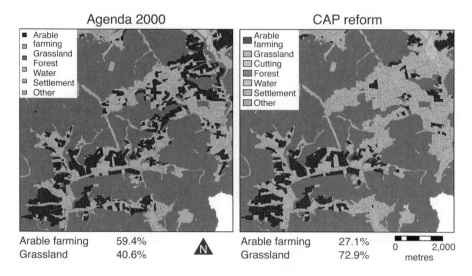

Figure 21-4. Predicted land uses for Agenda 2000 and CAP reform scenarios for the sub-region. (See also Plate 28 in the Colour Plate Section)

Source: Weinmann *et al.* (2006)

The sub-region's south-western corner shows little difference in land use contrary to the north-east. Here, a large share of land used for arable farming in the Agenda 2000 scenario is used as grassland in the CAP reform scenario. Inspecting the natural conditions reveals higher annual precipitation and lower annual temperature sums than in the south-west. The decoupling induces land users to switch to extensive land-use systems, predominantly extensive grassland. Coupled transfer payments may have distorted the land users' factor allocation decisions. Figure 21-5 supports this conclusion as is discussed below.

The cultural landscape in the Agenda 2000 scenario is a mixture of arable and grassland farming systems. In the CAP reform scenario, grassland becomes more dominant, affecting the landscapes aesthetic appearance. Other areas remain mostly unchanged as is the case in the south-west in Figure 21-3, where grassland dominates in both scenarios. Generalizing, fields are larger, precipitation is higher and temperatures are cooler here than in the region shown in Figure 21-4. Apparently, natural and structural conditions can offset policy effects.

The spatial distribution of transfer payments is heterogeneous. Overall, about 75% of agricultural land receive more payments after the CAP reform,

25% get less. Figure 21-5 shows the difference of transfer payments between Agenda 2000 and the reformed CAP for the same area shown in Figure 21-4.

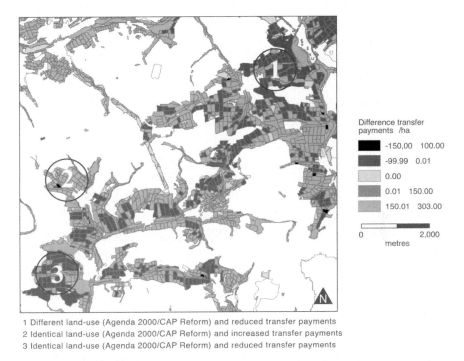

1 Different land-use (Agenda 2000/CAP Reform) and reduced transfer payments
2 Identical land-use (Agenda 2000/CAP Reform) and increased transfer payments
3 Identical land-use (Agenda 2000/CAP Reform) and reduced transfer payments

Figure 21-5. Transfer payment differences between Agenda 2000 and CAP reform.

Fields located in area 1 show differing land use in the two scenarios and receive less payments. Apparently, Agenda 2000 transfers set an incentive to adopt arable farming systems by improving their comparative advantage. Fields in area 2 receive more transfer payments after the reform but show no land-use changes. Monetary incentives are higher than necessary if the objective is to avoid land-use changes or land abandonment. Area 3 includes sites with identical land-use category in both scenarios and decreased transfer payments, indicating that incentives may have been unnecessarily high under Agenda 2000 conditions given the above targets.

The site-specific differences in ground rents between land-use systems vary over space. Therefore, land use or intensity may change in certain areas as a reaction to small variations in e.g. political conditions while others remain relatively stable. Consequently, the uncertainty and sensitivity of the model results depend on the level of detail (e.g. crop rotation/intensity compared to land-use category) and site under investigation. Sensitivity and uncertainty analyses performed with *ProLand* confirmed the results' resilience against parameter value changes at the level of land-use categories for most of the region (see Sheridan, 2006, for a detailed discussion).

Table 21-1 lists selected key figures aggregated over the entire region. Ground rent as a representation of a landscape's economic performance is significantly higher in the CAP reform than in the Agenda 2000 scenario. Transfer payments account for most of the difference. Labour input remains relatively constant.

Table 21-1. Economic key figures for Agenda 2000 and CAP reform scenarios

Land use	Agenda 2000			CAP reform		
	Ground rent [€]	Cpld. pay-ments [€]	Labour input [h]	Ground rent [€]	Decpld. pay-ments [€]	Labour input [h]
Arable	2,913,787	1,740,208	181,509	975,566	624,702	35,953
Grassland	9,121,083	1,062,489	663,32	14,819,067	5,193,159	821,591
Cutting	0	0	0	77,830	88,805	216
Forest	2,033,786	0	27,987	1,655,507	0	22,312
Sum	14,068,656	2,802,697	872,822	17,527,970	5,906,666	880,072

Source: Weinmann et al., 2006.

4.2   Species richness

The medial species richness estimated for the CAP reform scenario is 15% higher than for the Agenda 2000 scenario. Grassland replaces intensive crop rotations at numerous sites. These land-use changes mainly occur in areas used for arable farming in the Agenda 2000 scenario but with natural conditions favouring grassland. Intensive crop rotations include a routine application of herbicides obviously lowering the expected number of species. This may be the main cause for the forecasted increase of medial plant species richness. Figure 21-6 presents the difference in species richness between Agenda 2000 and CAP reform for the region in Figure 21-5. Most fields show no difference in species richness. The north-east quadrant clearly shows a positive influence, however. This is in line with the overall results as most fields in that area are predicted as grassland in the CAP scenario and arable land in the Agenda 2000 scenario.

The overall increase is analyzed further with regard to land-use differences on sites with a combination of ecological parameters favorable for species richness. The medial species richness increases 10% on sites with a membership higher than 50% for moisture class 'dry'. On sites with a membership value higher than 50% for moisture class 'wet', plant species richness increases 5%.

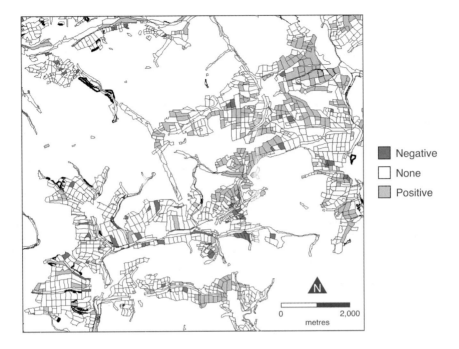

Figure 21-6. Difference in species richness between Agenda 2000 and CAP reform scenarios.

5. CONCLUSION

Landscapes have to fulfill a multifunctional role. The CAP reform's objectives include: *"helping agriculture produce safe and healthy food, contribute to sustainable development of rural areas, and protect and enhance the status of the farmed environment and its biodiversity"* (EC, 2004). The simulation runs indicate that the CAP reform will assist in achieving these goals. Incentives to intensify production are removed as payments are no longer a function of product output. Instead, they are linked to multiple objectives such as food safety, animal and plant health and animal welfare, as well as the requirement to keep all farmland in good agricultural and environmental condition (cross-compliance).

The CAP reform has positive effects on ground rent, labour input and plant species richness in this less favoured area. As output prices changed only by fractions of a percent, these effects are largely attributable to the decoupling of transfer payments. The extensive land-use systems' comparative advantage increases in this region. Especially grassland systems profit. Contrary to Agenda 2000, they receive decoupled area payments after the CAP reform. Marginal sites such as dry sites with high insolation are not abandoned or afforested but kept in use as extensive grassland. As shown, specific sites used intensively under Agenda 2000 conditions are extensified. This extensification has a positive influence on plant species richness.

The described effects vary throughout the region. Some areas profit both economically in terms of ground rent and ecologically in terms of plant species richness. Others show no change or are worse off in economic terms. Aggregated over the region results are positive. However, results do not necessarily apply to other regions. Especially regions with intensive arable farming may show different reactions in land use and species richness.

The scenarios illustrate the relevance of political factors in land-use and biodiversity modelling. The approach allows to identify sensitive sites that show reactions in land use, farming intensity and plant species richness. The research presented here is exemplary for the overall collaboration in the ITE²M model framework developed at the SFB 299 (compare Bach and Frede, 2004). Results are available for other ecological and hydrological models as well as evaluation concepts. The model structures are such that new research results can be incorporated. Model transfer to other regions is a subject of further research.

REFERENCES

Bach, M. and Frede, H.-G. (2004) Agricultural economy, ecology, and hydrology – modelling regional land use and trade-offs, in Simmering, D. (ed) *Proceedings of the GfÖ 2004*, Berlin, p. 310.

BMVEL (2005) *Meilensteine der Agrarpolitik - Umsetzung der europäischen Agrarreform in Deutschland*, Bundesministerium für Verbraucherschutz, Ernährung und Landwirtschaft, Berlin.

Bockstael, N.E. (1996) Modelling economics and ecology: the importance of a spatial perspective, *American Journal of Agricultural Economics*, 78: 1168–1180.

Borresch, R., Schmitz, K., Schmitz, P. M. and Wronka, T. C. (2005) CHOICE – ein integriert ökonomisch-ökologisches Konzept zur Bewertung von Multifunktionalität, in Umwelt- und Produktqualität im Agrarbereich, Gesellschaft für Wirtschafts- und Sozialwissenschaften des Landbaus, ed. (in press).

Brinkmann, T. (1922) *Grundriß der Sozialökonomik*, VII. Abteilung. J. C. B. Mohr, Tübingen.

Chen, P. (1976) The entity-relationship model – toward a unified view of data, *Transactions on Database Systems*, 1(1): 9–36.

EC (2003) Council Regulation (EC) No 1782/2003, European Commission, Brussels.

EC (2004) *The Common Agricultural Policy Explained*, European Commission, Brussels.

Klir, G.J. and Folger, T.A. (1988) *Fuzzy Sets, Uncertaint and Information*, Prentice-Hall, London.

KTBL (2003) *Datensammlung Betriebsplanung Landwirtschaft 2003/2003*, Kuratorium für Technik und Bauwesen in der Landwirtschaft e.V. Darmstadt.

Kuhlmann, F., Möller, D. and Weinmann, B. (2002) Modellierung der Landnutzung: Regionshöfe oder Raster-Landschaft? *Berichte über Landwirtschaft*, 80(3): 351-392.

Möller, D., Fohrer, N. and Steiner, N. (2002) Quantifizierung regionaler Multifunktionalität land- und forstwirtschaftlicher Nutzungssysteme, *Berichte über Landwirtschaft*, 80(3): 393–418.

Rounsevell, M.D.A., Annetts, J.E., Audsley, E., Mayr, T. and Reginster, I. (2003) Modelling the spatial distribution of agricultural land use at the regional scale, *Agriculture, Ecosystems and Environment*, 95: 465–479.

Schroers, J. O. and Sheridan, P. (2004) GIS-basierte Landnutzungsmodellierung mit ProLand. in Schiefer, G. (ed) *Lecture Notes in Informatics – Proceedings*, 49: 347–350.

Sheridan, P. (2006) Experimentelle Standortwirkungsforschung: Raumbezogene Sensitivitätsanalysen im Landnutzungsmodell ProLand. Dissertation, in prep., Gießen.

Stoate, C., Boatman, N.D., Borralho, R.J., Rio Carvalho, C., de Snoo, G.R. and Eden, P. (2001) Ecological impacts of arable intensification in Europe, *Journal of Environmental Management*, 63: 337–365.

Waldhardt, R., Simmering, D. and Albrecht, H. (2003) Floristic diversity at the habitat scale in agricultural landscapes of Central Europe - summary, conclusions and perspectives, *Agriculture, Ecosystems and Environment*, 98: 79–85.

Weber, A., Fohrer, N., and Möller, D. (2001) Long-term land use changes in a mesoscale watershed due to socio-economic factors – effects on landscape structures and functions. *Ecological Modelling*, 140: 125–140.

Weinmann, B. (2002) Mathematische Konzeption und Implementierung eines Modells zur Simulation regionaler Landnutzungsprogramme, Agrarwirtschaft Sonderheft 174, Agrimedia, Frankfurt.

Weinmann, B., Schroers, J.O. and Sheridan, P. (2006) Simulating the effects of decoupled transfer payments using the land use model ProLand, *Agrarwirtschaft*, 55(5/6).

ZMP (2004a) ZMP-Marktbilanz Getreide, Ölsaaten, Futtermittel, 2002-2004, Zentrale Markt- und Preisberichtsstelle GmbH Bonn.

ZMP (2004b) ZMP-Marktbilanz Vieh und Fleisch, 2002–2004, Zentrale Markt- und Preisberichtsstelle GmbH Bonn.

Colour Plate Section

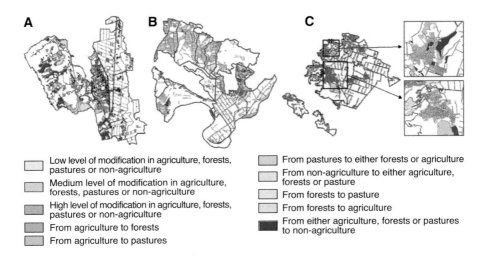

A   **B**   **C**

☐ Low level of modification in agriculture, forests, pastures or non-agriculture

☐ Medium level of modification in agriculture, forests, pastures or non-agriculture

☐ High level of modification in agriculture, forests, pastures or non-agriculture

☐ From agriculture to forests

☐ From agriculture to pastures

☐ From pastures to either forests or agriculture

☐ From non-agriculture to either agriculture, forests or pasture

☐ From forests to pasture

☐ From forests to agriculture

☐ From either agriculture, forests or pastures to non-agriculture

Plate 1. See also Figure 2-3 on page 33

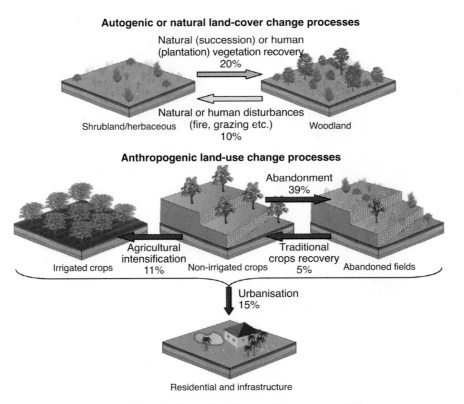

Autogenic or natural land-cover change processes

Natural (succession) or human (plantation) vegetation recovery
20%

Natural or human disturbances (fire, grazing etc.)
10%

Shrubland/herbaceous   Woodland

Anthropogenic land-use change processes

Abandonment
39%

Agricultural intensification
11%

Traditional crops recovery
5%

Irrigated crops   Non-irrigated crops   Abandoned fields

Urbanisation
15%

Residential and infrastructure

Plate 2. See also Figure 6-4 on page 106

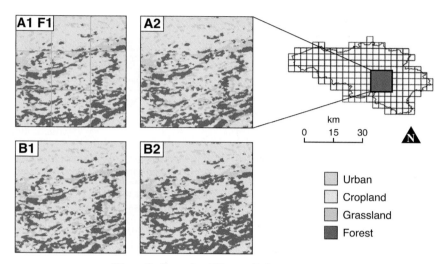

Plate 3. See also Figure 7-2 on page 127

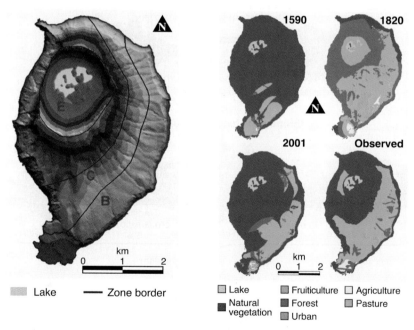

Plate 4. See also Figure 8-1 on page 134   Plate 5. See also Figure 8-5 on page 142

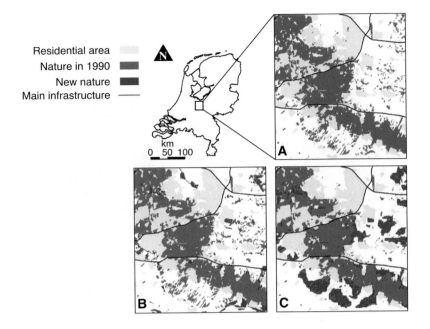

Residential area
Nature in 1990
New nature
Main infrastructure ———

km
0 50 100

A
B
C

Plate 6. See also Figure 9-2 on page 160

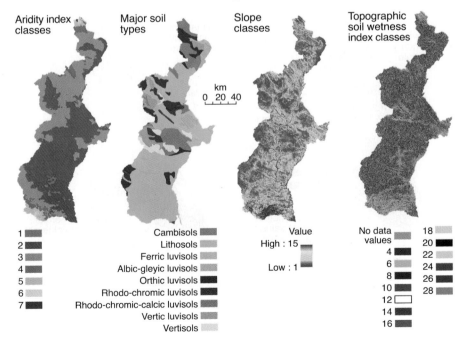

Aridity index classes

Major soil types

Slope classes

Topographic soil wetness index classes

km
0 20 40

1
2
3
4
5
6
7

Cambisols
Lithosols
Ferric luvisols
Albic-gleyic luvisols
Orthic luvisols
Rhodo-chromic luvisols
Rhodo-chromic-calcic luvisols
Vertic luvisols
Vertisols

Value
High : 15

Low : 1

No data values
4
6
8
10
12
14
16

18
20
22
24
26
28

Plate 7. See also Figure 11-3 on page 190

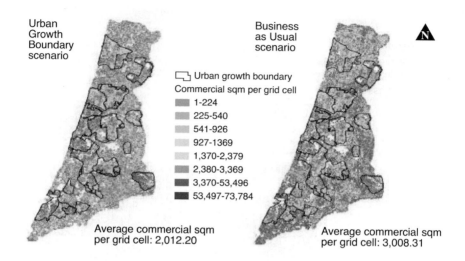

Plate 8. See also Figure 12-4 on page 210

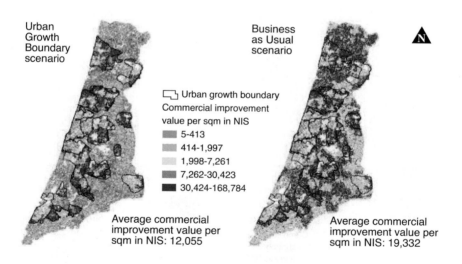

Plate 9. See also Figure 12-5 on page 210

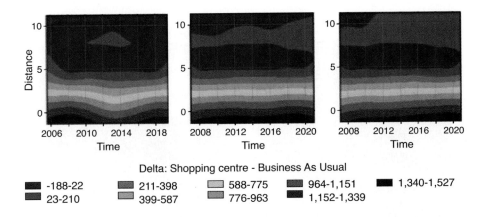

Delta: Shopping centre - Business As Usual

- -188-22
- 23-210
- 211-398
- 399-587
- 588-775
- 776-963
- 964-1,151
- 1,152-1,339
- 1,340-1,527

Plate 10. See also Figure 12-7 on page 214

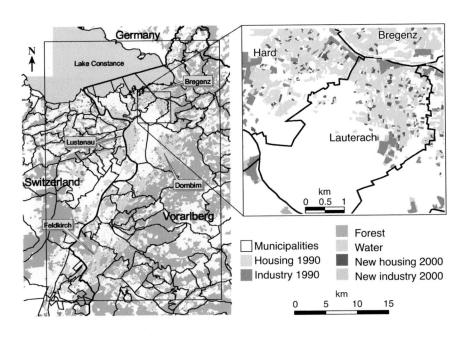

Plate 11. See also Figure 13-2 on page 224

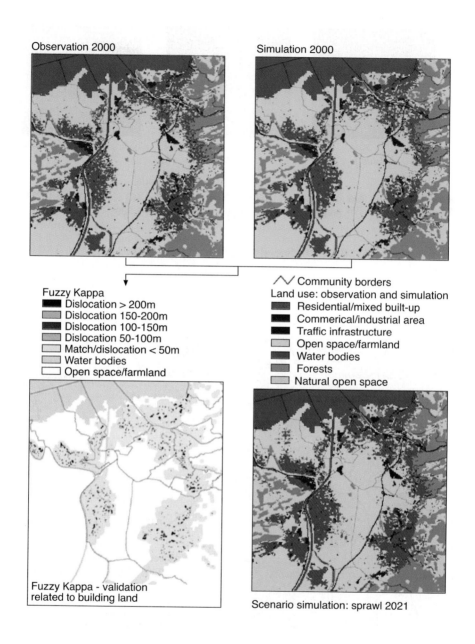

Plate 12. See also Figure 13-5 on page 233

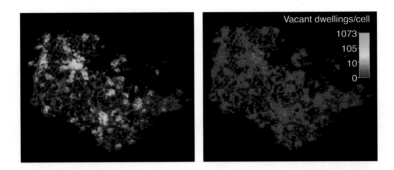

Plate 13. See also Figure 14-3 on page 256

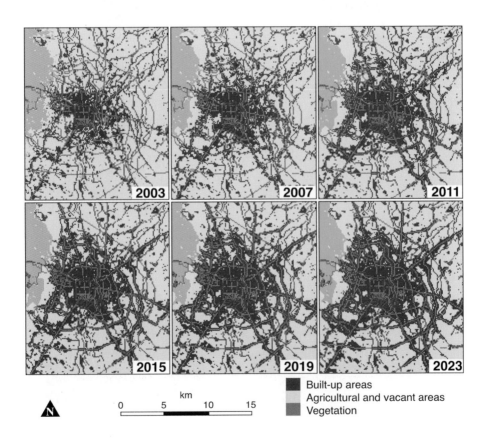

Plate 14. See also Figure 15-5 on page 274

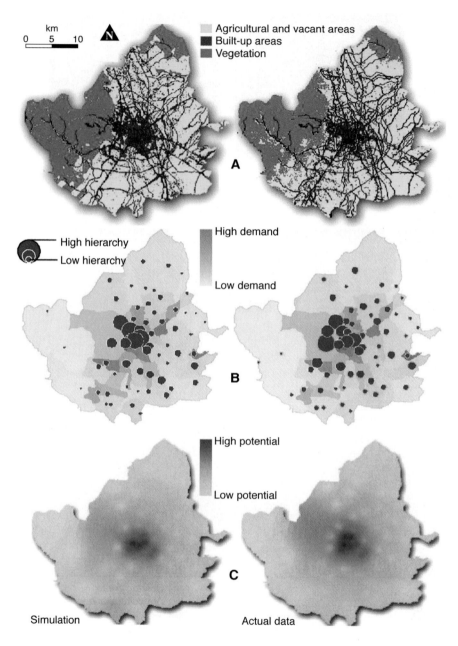

Plate 15. See also Figure 15-4 on page 272

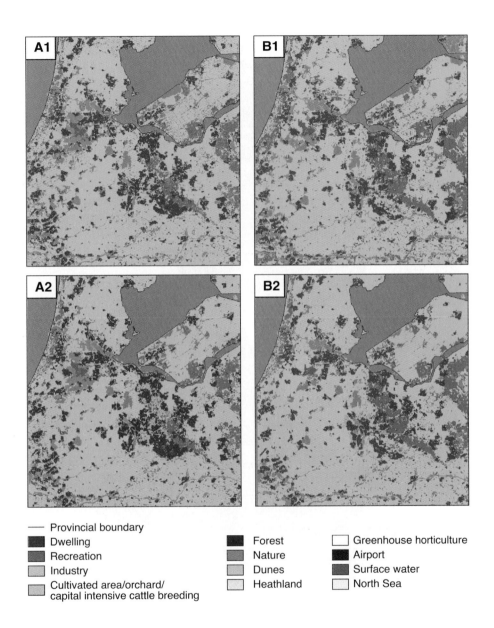

Provincial boundary
Dwelling
Recreation
Industry
Cultivated area/orchard/
capital intensive cattle breeding

Forest
Nature
Dunes
Heathland

Greenhouse horticulture
Airport
Surface water
North Sea

Plate 16. See also Figure 16-3 on page 288

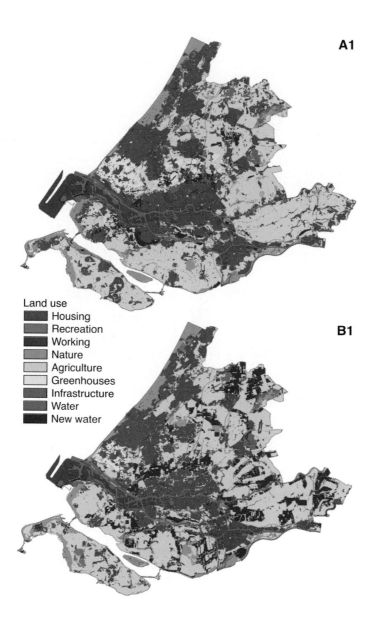

Plate 17. See also Figure 16-4 on page 291

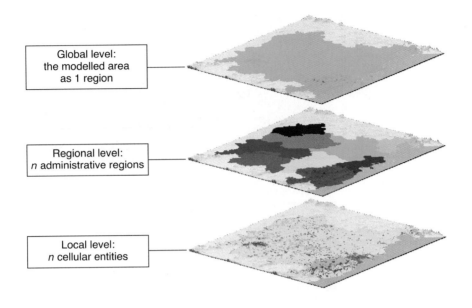

Plate 18. See also Figure 17-2 on page 304

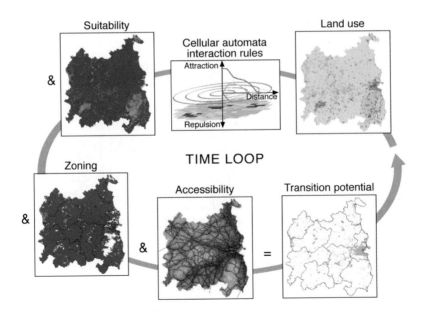

Plate 19. See also Figure 17-3 on page 308

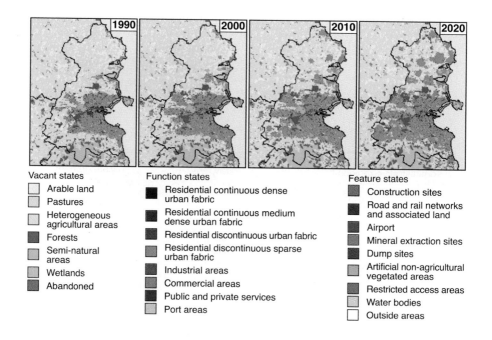

Vacant states
- Arable land
- Pastures
- Heterogeneous agricultural areas
- Forests
- Semi-natural areas
- Wetlands
- Abandoned

Function states
- Residential continuous dense urban fabric
- Residential continuous medium dense urban fabric
- Residential discontinuous urban fabric
- Residential discontinuous sparse urban fabric
- Industrial areas
- Commercial areas
- Public and private services
- Port areas

Feature states
- Construction sites
- Road and rail networks and associated land
- Airport
- Mineral extraction sites
- Dump sites
- Artificial non-agricultural vegetated areas
- Restricted access areas
- Water bodies
- Outside areas

Plate 20. See also Figure 17-6 on page 314

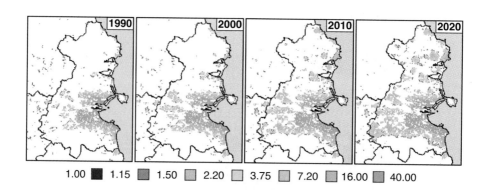

1.00 ■ 1.15 ■ 1.50 ☐ 2.20 ☐ 3.75 ☐ 7.20 ☐ 16.00 ■ 40.00

Plate 21. See also Figure 17-7 on page 315

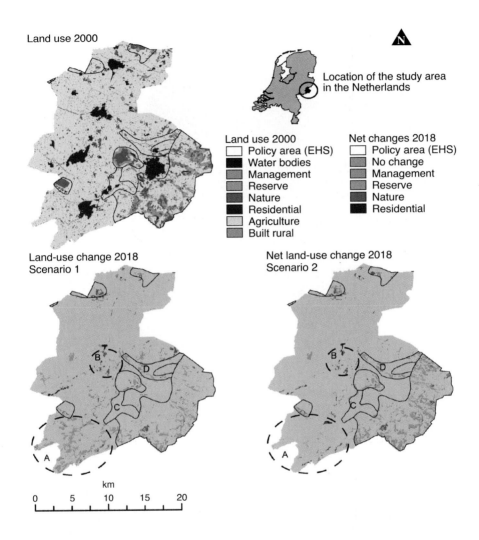

Plate 22. See also Figure 18-2 on page 328

Plate 23. See also Figure 19-1 on page 340

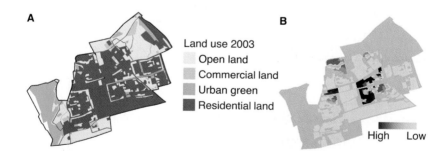

Plate 24. See also Figure 19-6 on page 350

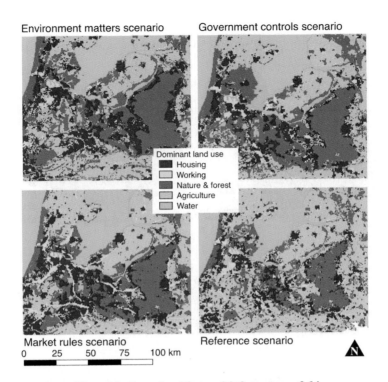

Plate 25. See also Figure 20-2 on page 364

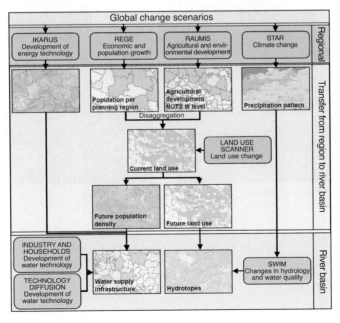

Plate 26. See also Figure 20-3 on page 369

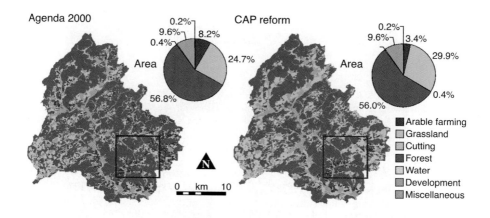

Plate 27. See also Figure 21-3 on page 383

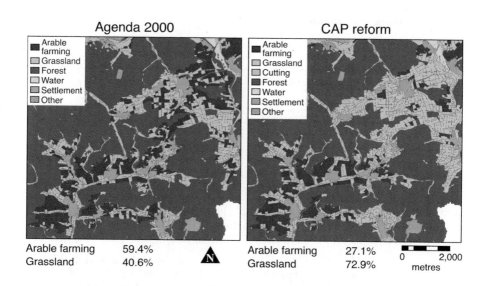

Plate 28. See also Figure 21-4 on page 384

Index

The GeoJournal Library

21. V.I. Ilyichev and V.V. Anikiev (eds.): *Oceanic and Anthropogenic Controls of Life in the Pacific Ocean.* 1992   ISBN 0-7923-1854-4
22. A.K. Dutt and F.J. Costa (eds.): *Perspectives on Planning and Urban Development in Belgium.* 1992   ISBN 0-7923-1885-4
23. J. Portugali: *Implicate Relations.* Society and Space in the Israeli-Palestinian Conflict. 1993   ISBN 0-7923-1886-2
24. M.J.C. de Lepper, H.J. Scholten and R.M. Stern (eds.): *The Added Value of Geographical Information Systems in Public and Environmental Health.* 1995
ISBN 0-7923-1887-0
25. J.P. Dorian, P.A. Minakir and V.T. Borisovich (eds.): *CIS Energy and Minerals Development.* Prospects, Problems and Opportunities for International Cooperation. 1993
ISBN 0-7923-2323-8
26. P.P. Wong (ed.): *Tourism vs Environment: The Case for Coastal Areas.* 1993
ISBN 0-7923-2404-8
27. G.B. Benko and U. Strohmayer (eds.): *Geography, History and Social Sciences.* 1995   ISBN 0-7923-2543-5
28. A. Faludi and A. der Valk: *Rule and Order. Dutch Planning Doctrine in the Twentieth Century.* 1994   ISBN 0-7923-2619-9
29. B.C. Hewitson and R.G. Crane (eds.): *Neural Nets: Applications in Geography.* 1994
ISBN 0-7923-2746-2
30. A.K. Dutt, F.J. Costa, S. Aggarwal and A.G. Noble (eds.): *The Asian City: Processes of Development, Characteristics and Planning.* 1994   ISBN 0-7923-3135-4
31. R. Laulajainen and H.A. Stafford: *Corporate Geography. Business Location Principles and Cases.* 1995   ISBN 0-7923-3326-8
32. J. Portugali (ed.): *The Construction of Cognitive Maps.* 1996   ISBN 0-7923-3949-5
33. E. Biagini: *Northern Ireland and Beyond.* Social and Geographical Issues. 1996
ISBN 0-7923-4046-9
34. A.K. Dutt (ed.): *Southeast Asia: A Ten Nation Region.* 1996   ISBN 0-7923-4171-6
35. J. Settele, C. Margules, P. Poschlod and K. Henle (eds.): *Species Survival in Fragmented Landscapes.* 1996   ISBN 0-7923-4239-9
36. M. Yoshino, M. Domrös, A. Douguédroit, J. Paszynski and L.D. Nkemdirim (eds.): *Climates and Societies – A Climatological Perspective.* A Contribution on Global Change and Related Problems Prepared by the Commission on Climatology of the International Geographical Union. 1997   ISBN 0-7923-4324-7
37. D. Borri, A. Khakee and C. Lacirignola (eds.): *Evaluating Theory-Practice and Urban-Rural Interplay in Planning.* 1997   ISBN 0-7923-4326-3
38. J.A.A. Jones, C. Liu, M-K.Woo and H-T. Kung (eds.): *Regional Hydrological Response to Climate Change.* 1996   ISBN 0-7923-4329-8
39. R. Lloyd: *Spatial Cognition.* Geographic Environments. 1997   ISBN 0-7923-4375-1
40. I. Lyons Murphy: *The Danube: A River Basin in Transition.* 1997
ISBN 0-7923-4558-4
41. H.J. Bruins and H. Lithwick (eds.): *The Arid Frontier.* Interactive Management of Environment and Development. 1998   ISBN 0-7923-4227-5
42. G. Lipshitz: *Country on the Move: Migration to and within Israel, 1948–1995.* 1998
ISBN 0-7923-4850-8

The GeoJournal Library

The GeoJournal Library

65. C.M. Hall and A.M. Williams (eds.): *Tourism and Migration.* NewRelationships between Production and Consumption. 2002   ISBN 1-4020-0454-0
66. I.R. Bowler, C.R. Bryant and C. Cocklin (eds.): *The Sustainability of Rural Systems.* Geographical Interpretations. 2002   ISBN 1-4020-0513-X
67. O. Yiftachel, J. Little, D. Hedgcock and I. Alexander (eds.): *The Power of Planning.* Spaces of Control and Transformation. 2001   ISBN Hb; 1-4020-0533-4
     ISBN Pb; 1-4020-0534-2
68. K. Hewitt, M.-L. Byrne, M. English and G. Young (eds.): *Landscapes of Transition.* Landform Assemblages and Transformations in Cold Regions. 2002
     ISBN 1-4020-0663-2
69. M. Romanos and C. Auffrey (eds.): *Managing Intermediate Size Cities.* Sustainable Development in a Growth Region of Thailand. 2002   ISBN 1-4020-0818-X
70. B. Boots, A. Okabe and R. Thomas (eds.): *Modelling Geographical Systems.* Statistical and Computational Applications. 2003   ISBN 1-4020-0821-X
71. R. Gerber and M. Williams (eds.): *Geography, Culture and Education.* 2002
     ISBN 1-4020-0878-3
72. D. Felsenstein, E.W. Schamp and A. Shachar (eds.): *Emerging Nodes in the Global Economy: Frankfurt and Tel Aviv Compared.* 2002   ISBN 1-4020-0924-0
73. R. Gerber (ed.): *International Handbook on Geographical Education.* 2003
     ISBN 1-4020-1019-2
74. M. de Jong, K. Lalenis and V. Mamadouh (eds.): *The Theory and Practice of Institutional Transplantation.* Experiences with the Transfer of Policy Institutions. 2002
     ISBN 1-4020-1049-4
75. A.K. Dutt, A.G. Noble, G. Venugopal and S. Subbiah (eds.): *Challenges to Asian Urbanization in the 21st Century.* 2003   ISBN 1-4020-1576-3
76. I. Baud, J. Post and C. Furedy (eds.): *Solid Waste Management and Recycling.* Actors, Partnerships and Policies in Hyderabad, India and Nairobi, Kenya. 2004
     ISBN 1-4020-1975-0
77. A. Bailly and L.J. Gibson (eds.): *Applied Geography.* A World Perspective. 2004
     ISBN 1-4020-2441-X
78. H.D. Smith (ed.): *The Oceans: Key Issues in Marine Affairs.* 2004
     ISBN 1-4020-2746-X
79. M. Ramutsindela: *Parks and People in Postcolonial Societies.* Experiences in Southern Africa. 2004   ISBN 1-4020-2542-4
80. R.A. Boschma and R.C. Kloosterman (eds.): *Learning from Clusters.* A Critical Assessment from an Economic-Geographical Perspective. 2005
     ISBN 1-4020-3671-X
81. G. Humphrys and M. Williams (eds.): *Presenting and Representing Environments.* 2005   ISBN 1-4020-3813-5
82. D. Rumley, V.L. Forbes and C. Griffin (eds.): *Australia's Arc of Instability.* The Political and Cultural Dynamics of Regional Security. 2006   ISBN 1-4020-3825-9
83. R. Schneider-Sliwa (ed.): *Cities in Transition.* Globalization, Political Change and Urban Development. 2006   ISBN 1-4020-3866-6
84. B.G.V. Robert (ed.): *Dynamic Trip Modelling.* From Shopping Centres to the Internet Series. 2006   ISBN: 1-4020-4345-7

The GeoJournal Library

85. L. John and W. Michael (eds.): *Geographical Education in a Changing World*. Past Experience, Current Trends and Future Challenges Series. 2006
ISBN: 1-4020-4806-8
86. G.D. Jay and R. Neil (eds.): *Enterprising Worlds*. A Geographic Perspective on Economics, Environments & Ethics Series. 2007 ISBN: 1-4020-5225-1
87. Y.K.W. Albert and H.G. Brent (eds.): *Spatial Database Systems*. Design, Implementation and Project Management Series. 2006 ISBN: 1-4020-5391-6
88. H.J. Miller (ed.): *Societies and Cities in the Age of Instant Access*. 2007.
ISBN: 1-4020-5426-2
89. J.L. Wescoat, Jr and D.M. Johnston (eds.): *Political Economies of Landscape Change*. 2007 ISBN: 1-4020-5848-6
90. E. Koomen, J. Stillwell, A. Bakema and H.J. Scholten (eds.): *Modelling Land-Use Change*. Progress and Applications. 2007 ISBN: 1-4020-5647-8
91. E.Razin, M.Dijst and C.Vázquez (eds.): *Employment Deconcentration in European Metropolitan Areas*. Market Forces versus Planning Regulations. 2007
ISBN: 1-4020-5761-X